Molecular Diagnostics

HISTOPATHOLOGY

EDITED BY Guy Orchard & Brian Nation

SECOND EDITION

CLINICAL IMMUNOLOGY

EDITED BY Angela Hall, Chris Scott & Matthew Buckland

SECOND EDITION

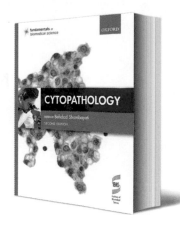

CYTOPATHOLOGY

EDITED BY Behdad Shambayati

SECOND EDITION

BIOMEDICAL SCIENCE PRACTICE

EXPERIMENTAL & PROFESSIONAL SKILLS

EDITED BY Nessar Ahmed, Hedley Glencross, and Qiuyu Wang

SECOND EDITION

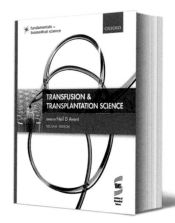

TRANSFUSION & TRANSPLANTATION SCIENCE

EDITED BY Neil D Avent

SECOND EDITION

HAEMATOLOGY

Gary Moore, Gavin Knight & Andrew Blann

SECOND EDITION

MEDICAL MICROBIOLOGY

EDITED BY Michael Ford

SECOND EDITION

CELL STRUCTURE & FUNCTION

EDITED BY Guy Orchard & Brian Nation

CLINICAL BIOCHEMISTRY

EDITED BY Nessar Ahmed

SECOND EDITION

fundamentals OF
biomedical science

Fundamentals of Biomedical Science

Molecular Diagnostics

Edited by
Anthony Warford and Nadège Presneau
University of Westminster, London, UK

OXFORD
UNIVERSITY PRESS

Introduction

This book is a new addition to the Oxford University Press Fundamentals of Biomedical Science series. The majority of the titles can be easily identified as 'discipline specific', but *Molecular Diagnostics* is purposely different. Why is this? The answer is in the title.

Molecular diagnostics concerns the application of technologies to reveal molecules linked to disease that assist in the diagnosis, stratification, and targeted treatment of the patient. In current terminology, this is known as precision medicine. This deliberately broad definition ensures that all relevant molecules are included. Accordingly, this book covers DNA, RNA, and proteins, and seeks to open to the reader the wider vista of OMICS. Accordingly, areas such as genomics (DNA), transcriptomics (RNA), and proteomics (proteins), together with the presently more esoteric metabolomics that can involve the interrogation of peptides are covered.

At a technical level it is important to note that the methodologies can be divided into those that are applied to intact cells and tissues, and others that use homogenates and fractions of these as the starting point. In the first part of this book, Chapters 1–7, these technologies are described. In Chapter 1 the requirement for optimal sample preparation is emphasized and the special requirements of a molecular laboratory are covered in Chapter 2. These represent the initial and essential requirements that need to be in place before successful molecular analysis can be undertaken. Some of the content in these, and Chapters 3 and 4 may be familiar to readers who have already dipped into other titles in this series, but here they are brought together to emphasize the need to apply many diverse technologies for the purpose of molecular diagnosis. In addition, other technologies that are either already contributing to molecular diagnosis or are on the cusp of doing so are also included. Thus, mass spectrometry (Chapter 4), sequencing, including high-throughput next-generation sequencing (Chapter 5), and the application of bioinformatics to its analysis (Chapter 6) are described. The last chapter in this section of the book is largely forward-looking and covers the new concept of liquid biopsy and application of digital PCR.

In the second part of the book, Chapters 8–12, the application of these technologies to cancer is presented. It is here that the 'headlines' are being made, with an increasing number of molecular techniques being used to guide targeted therapy using small molecule drugs and monoclonal antibodies to interact with the driving molecular pathways causing cancers. Chapters 8 and 9 cover haematological disorders, while epithelial cancers and melanoma are discussed in Chapters 10 and 11. In these chapters, and as especially highlighted in Chapter 11, treatment based on the inhibition of immune checkpoint inhibitors now offers targeted therapy hitting the interaction of programmed cell death receptor-1 (PD-1) and programmed cell death ligand-1 (PD-L1) across several tumour types. Chapter 12 covers mesenchymal cancers and demonstrates the importance of the application of molecular diagnostics to a relatively diverse and rare set of tumours.

This book will be relevant to undergraduate students on a wide variety of biomedical pathways. It should open a vista to molecular diagnostics that cuts across traditional 'discipline' boundaries and that is already making significant inroads into the practise of diagnostic pathology. This is only going to increase.

In the early part of this century the draft of the human genome, which took nearly 10 years to complete using Sanger sequencing, was published (International Human Genome Sequencing Consortium, 2004). At that time, the term 'junk DNA' was used to cover the sequence outside of the gene boundaries. Since then, we have learnt that a good proportion of this DNA is read and forms non-coding RNA. Included in this are numerous micro-RNA sequences that control gene to protein translation. We are only now beginning to unravel how this impacts on disease. In the future, the 100,000 genomes project (Samuel and Farsides, 2017) should further clarify the precise role of DNA, RNA, and by implication, the role of proteins, in health and disease. The Human Genome Project has also been a great platform to enhance the sequencing methods, such as what we now call 'next generation sequencing' and the bioinformatics tools to analyse the sequencing results with their 'big data sets'. So, molecular diagnostics will only increase in its scope and importance and, accordingly, this book provides a window on this exciting area.

Given the fast-moving nature of the field, can this book be 'up to date?' In terms of technologies, the answer is a qualified 'yes'. However, merging of methodologies is occurring with the divide between intact sample analysis and that of homogenate preparations being anticipated by bringing together mass spectrometry applied to frozen sections (see Chapter 4, Section 4.4) and sequencing of gene expression in intact single cells now described (see Chapter 5, Section 5.5). These are today's experimental methods, but they, or other new technologies, may become the routine of tomorrow.

In respect of diagnostic applications, the second section of this book is selective. As already highlighted, in Chapters 8–12 the impact of molecular diagnostic methods on cancer is explored, while, in the last chapter of this book, its application to pre- and postnatal testing is included to illustrate that molecular diagnostics is certainly not only cancer focused. Indeed, the reader is encouraged to look at Chapters 13 and 16 in *Medical Microbiology* in the Fundamentals of Biomedical Science series (Ford, 2014) to note the further application of molecular diagnostics to virology. Accordingly, this book does not cover all possible applications of molecular diagnostics, but we hope that the reader will find much of interest and be stimulated not only to explore the subject more deeply, but also to contribute, in due course, directly to its development.

Finally, we would like to thank all the authors who have contributed to this book. It has been a challenge as it breaks new ground, but we trust it has been worthwhile.

Bon voyage!

Anthony Warford and Nadège Presneau

July 2018

 # References

- Ford M (ed.). *Medical Microbiology*, Fundamentals of Biomedical Science series, 2nd edn. Oxford, Oxford University Press, 2014.

- International Human Genome Sequencing Consortium. Finishing the euchromatic sequence of the human genome. Nature 2004; 431: 931–945.

- Samuel GN, Farsides B. The UK's 100,000 Genomes Project: manifesting policymakers' expectations. New Genet Soc 2017; 36: 336–353.

1

Sample Preparation

Anthony Warford and Sandra Hing

After studying this chapter, you should be able to:

- Understand the processes that degrade cells and tissues following removal of nutrient and oxygen supply.
- Identify the different types of sample suitable for both genomic and proteomic analysis.
- Discuss the different methods by which samples may be optimally prepared and stored for subsequent analysis.
- Describe the variety of methods that can be used to establish the quality and quantity of DNA, RNA, and protein in samples.

1.1 Introduction

With very few exceptions, the analysis of cells and tissues is undertaken after the removal of the nutrient and oxygen supply. In analysing these non-living samples, care must be taken to preserve them in as life-like a state as possible. The more care that is taken, the more extensive the range of options there will be for analysis and for that to accurately reflect their former status. Accordingly, to avoid the situation where subsequent investigation and analysis cannot be 'properly performed', sample preparation must be accorded very high importance as the first step in any molecular analysis method.

This chapter considers the ways in which samples may be degraded, and provides practically based advice for optimal sample preparation and storage, together with options for how samples can be quality assessed for suitability for molecular analysis.

> ### Key Point
> Sample preparation determines the quality and, therefore, the scope of subsequent molecular analysis.

1.2 **Degradation of cells and tissues**

Several terms are used to describe the processes that cause the degradation of samples.

Ischaemia is used as the term to define the interval phase between the removal of nutrients and oxygen supply, and the cessation of the degradation of the sample after it is placed in a preservative or **fixative**. During surgery, a period of *warm ischaemia* occurs once the blood supply is severed from the tissue that is to be removed. This will be followed by the cold *ischaemic* phase after the tissue has been removed.

Autolysis is a process that describes the pH-induced destruction of cells. Following the removal of nutrients and oxygen, cells will become progressively more acidic and, around pH 5, this will trigger the rupture of lysosomes with the release of a cascade of enzymes that will result in the self-destruction of the cells.

Putrefaction is used to describe the external bacterial attack on cells where the homeostatic processes that maintain their integrity have ceased to function. The epithelial lining of the alimentary tract is particularly susceptible to this form of degradation and extensive loss of cells can occur very quickly (Figure 1.1).

In some tissues other factors may hasten degradation. For example, in the pancreas the caseation of normal metabolism will quickly result in the self-destruction of exocrine acinar cells, as the digestive enzymes they normally produce are liberated. The consequence can be an organ that appears to be intact, but on subsequent microscopic examination, retains only the supporting connective tissue framework and islets of Langerhans. Another organ that is quickly affected is the bladder, where the presence of urine will quickly remove the delicate internal transitional epithelial lining.

Practically, the longer that these degradative processes occur, the more prone the sample is to morphological and/or molecular degradation. From the latter perspective, different

Fixative

A chemical that is used to bring about the fixation of a sample. This may be cross-linking, joining proteins to proteins or proteins to nucleic acids. The most frequently used fixative of this type is formaldehyde, which is used in aqueous solution to fix tissues. Alternatively, the fixative may be non-additive. In this instance, the fixation process involves the removal of water. Alcohol is a commonly used precipitating fixative and is suited to the fixation of individual cells.

Autolysis

is a process that describes the pH-induced destruction of cells.

(A)

(B)

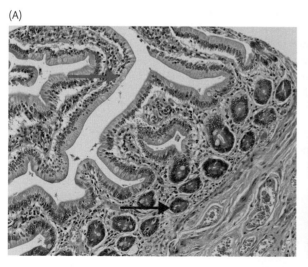
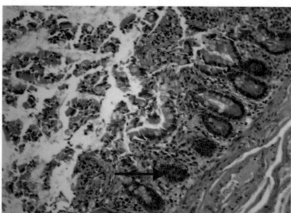

FIGURE 1.1

Alimentary tract: effects of putrefaction. Examples of haematoxylin and eosin-stained FFPE samples of small intestine (×200 magnification). (A) Surgical resection sample; note preservation of crypts (red arrow) and villus epithelium (green arrow). (B) Cadaveric sample; note that only the crypt epithelium is well preserved .

components are more susceptible to these degradative processes. Phosphorylated proteins are amongst the fastest components to be degraded (Baker et al., 2005), protein and mRNA intermediate (Atkin et al., 2006), and DNA often the most resilient, as judged by its suitability for forensic analysis. Interestingly, microRNA appears to be more stable than mRNA to degradation (Hall et al., 2012).

The take-home message is clear; keep the interval between severance of nutrient and oxygen supply to halting the degradative processes to a minimum. This is, of course, more easily done when the scientist is in control of events, such as in the case of blood samples, cell culture, or the harvesting of tissue from animal model studies. With proper procedures in place it can, and sometimes must, be controlled (see Box 1.1) for surgically removed tissue, but for cadaveric samples the preservation may be extensively compromised.

BOX 1.1 Guidelines for controlling ischaemia

Assessment of cancer tissue for the expression of molecular markers that indicate their suitability for **targeted therapy** is now being regularly undertaken using histopathological samples. For invasive breast cancer, assessment of hormone receptor status linked to Tamoxifen™ or aromatase inhibition therapy and HER2 status linked to HER2 inhibition therapies such as Herceptin™ are covered by international guidelines that have been issued to ensure the standardization of analysis. These guidelines (Hammond et al., 2010; Wolff et al., 2014; Rakha et al., 2015) include limitation of the cold ischaemic interval to less than 1 hour. With other therapeutic regimes being introduced for the targeted treatment of cancers in other organs, it is highly likely that guidance will also recommend control of the ischaemic time interval.

Targeted therapy
Specific treatment for a cancer according to its molecular profile. Also known as precision medicine. Treatment using small molecule drugs or monoclonal antibodies involves specific interaction with molecules to inhibit cancer growth.

Key Point

Once nutrient and oxygen supplies are removed, cells begin to degrade. Accordingly, to maintain sample quality, steps must be taken to halt this process as soon as possible.

1.1 SELF-CHECK QUESTION

What is meant by the term ischaemia and what are the two ways in which cells suffer degradation?

1.3 Sample preparation and storage

A considerable range of preparative procedures are available, being mainly driven by sample type (Figure 1.2). In this section we highlight processes for each of these that should result in optimal preparation together with consideration of appropriate storage methods.

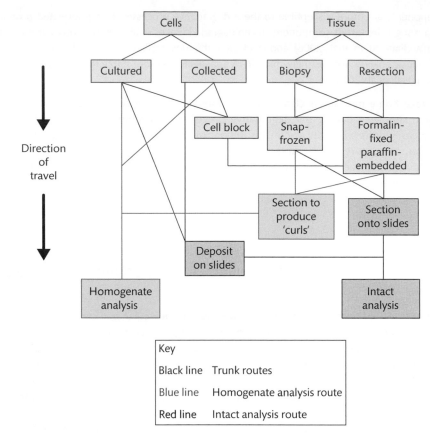

FIGURE 1.2
Sample type route map.

1.3.1 Cell-based preparations

In the majority of instances, the scientist should be in control of this type of sample preparation. When this is not the case, the provision of **standard operating procedures** to others directly handling or handing over cell samples should ensure their optimal conservation.

Cell culture preparations perhaps represent the easiest type of sample to handle as, with the proviso that they are harvested when in peak condition, they should provide optimal material for analysis. When non-adherent cells have been cultured to an optimal concentration and condition, as determined, respectively, by haemocytometer and Trypan blue exclusion (Strober, 2015), then the first step in sample preparation will usually be centrifugation to form a concentrate of intact cells. In the case of adherent cells, these should be harvested when they are confluent mono-layers, but before the cells heap up on one another. They must be carefully dislodged either by mechanical means or by light enzymatic digestion from the media they have been grown on, then centrifuged as for non-adherent cells. Subsequent handling steps may then follow common pathways as outlined in Figure 1.3.

Cells are routinely cultured in Roswell Park Memorial Institute medium (RPMI) and 10% foetal calf serum (FCS), and must be cultured under sterile conditions in a Class 2 laminar flow

Standard operating procedure

A controlled document that provides a method that must be followed without deviation.

Cell culture

A method of growing cells outside the body in an environment that enables cell division to continue. It is also known as *in vitro* culture.

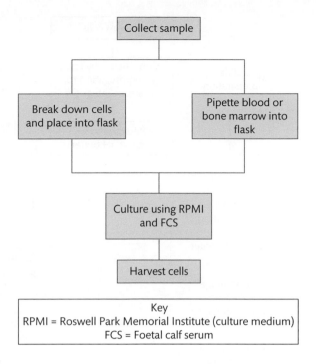

FIGURE 1.3
Cell culture.

cabinet. It may be useful to include an antibiotic in the growth medium to inhibit contamination introduced from the host. Cell culture can take from 2 days to 2 weeks, dependent upon the type of sample, cell type, and rate of cell division. The number of times a primary culture can undergo cell division is limited due to nutrient requirements and tissue culture conditions. During this period, cells require regular 'feeding' and examination under a microscope to monitor growth and the condition of the cells.

Once adherent cells are confluent they can be harvested (Figure 1.4). Confluence is described as the percentage of cells attached to the tissue culture flask. For example, 100% confluence means that the surface is completely covered by cells, whereas 50% confluence is where roughly half of the surface is covered.

A blood sample, generally referred to as venous blood or whole blood, is usually obtained from a vein inside the elbow joint. It is an excellent sample type for use in **genomic** diagnostic purposes as it is easy to collect and can be used in a wide variety of downstream applications. It is abundant in nucleic acid-based **biomarkers** and therefore can be used to detect a wide variety of **genetic disorders**, which are germline or sporadic/somatic.

The blood is collected into various vacutainers with different coloured tops that indicate the type of anticoagulants present within them (Table 1.1). Anticoagulants are used to inhibit the blood from clotting and the tubes must be inverted thoroughly to allow the blood to mix with the anticoagulant. This then allows a variety of tests to be performed using the whole blood, including DNA extraction. Investigations often require specific nucleic acid extraction protocols or downstream manipulation, and therefore special anticoagulants are sometimes required. Providing the correct anticoagulant has been selected, high quality nucleic acid isolation should be possible.

Genomic
Pertaining to genes or the genome.

Biomarkers
Constituents of cells that are used to measure change occurring in them. They are frequently used as specific indicators of a disease or to assess its course.

Genetic disorders
Conditions that arise from changes in the DNA. These can involve major structural changes, such as the gain or loss of chromosomes, or they can be minute, such as single nucleotide polymorphism. When germline, they are inheritable, when sporadic/somatic, they begin in a cell that is capable of division during life.

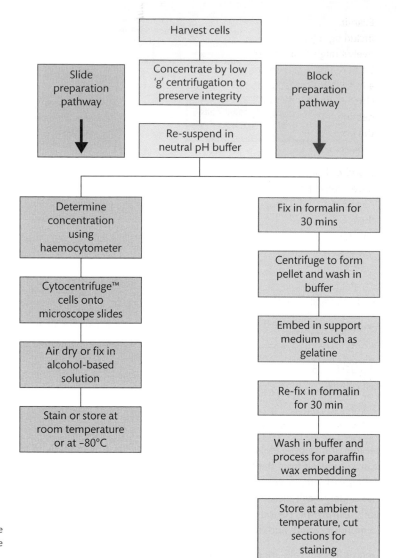

FIGURE 1.4

Cultured cell preparation pathways.

Note: The Cytoblock™ system can be used to prepare cell blocks. The Cytospin™ system and Cytoblock™ are ThermoFisher products

TABLE 1.1 Types of anticoagulants suitable for use in molecular analysis

Colour	Type	Additive
Purple	EDTA	Ethylene diamine tetra-acetic acid or EDTA, as it is commonly referred to, is commonly used for nucleic acid extraction. It prevents clotting by binding to calcium. Most types of downstream molecular analysis can be performed using DNA or RNA extracted from EDTA samples.
Green	Heparin	Allows analysis of white blood cells and is therefore suitable for DNA extraction, cytogenetic analysis, and immunophenotyping.
Blue	Sodium citrate	An anticoagulant that allows measurement of coagulation proteins. It acts by forming a loose and reversible ionic complex with calcium ions. Sodium citrate blood samples are required for genetic tests for Factor V and prothrombin.

Examination and genomic analysis of bone marrow is required for a number of disorders, including myeloma, leukaemia, and aplastic anaemia. While venous blood samples are relatively straightforward to collect, a bone marrow aspirate can be difficult and requires a clinically trained specialist. The process is described as bone marrow aspiration and involves insertion of a heavy gauge needle into a particular bone, such as the hip or breast to aspirate the marrow. Once aspirated, the sample is collected into either ethylenediaminetetraacetic acid (EDTA) or heparin. As with blood samples, the collection tube must be inverted several times to enable the bone marrow to mix with the anticoagulant. Once collected, bone marrow samples can be processed in the same way as blood.

Blood or bone marrow samples are received to the laboratory at room temperature and may have been in transit for 2–3 days although a maximum of 24 hours is the ideal scenario. If samples need to be stored prior to transport to the laboratory, refrigeration at 4°C is a requirement with the exception of samples to be used for RNA extraction or any **cell-free DNA (cfDNA)** analysis, which must be stored at room temperature. Samples received from other laboratories may also have been stored at room temperature for an unspecified time. The type of analysis requested and, therefore, the requirement for DNA or RNA extraction will determine if these conditions are satisfactory. Samples for RNA extraction, **Lymphoprep™** (Stem Cell Technologies, Cambridge), or T cell separation need to be processed within 72 hours of being collected.

It should be noted that blood or bone marrow samples contain very few dividing cells and, therefore, need to be cultured in the laboratory in order to generate sufficient cell count for particular genomic applications. Requirements for tissue culture are minimal and as little as a 0.25-mL sample is required. Preferably, these need to be processed within 24 hours of the sample being taken in order to maximize the number of viable cells. However, up to 5–7 days after collection there may be enough viability for a sufficient cell count to be achieved. Tissue viability can be improved if the tissue sample is transported in a transport medium, such as RPMI with 10% FCS supplied by the laboratory.

Mononuclear cells (MNCs) i.e., monocytes and lymphocytes can be isolated from blood and bone marrow using a density gradient medium (Figure 1.5). This process enriches the cells for genomic analysis which may be required for a number of genetic disorders such as those of **myeloproliferative** origin. The principle of isolation is based on the fact that mononuclear cells have a lower buoyant density than the erythrocytes and granulocytes, and therefore they can be separated by centrifugation using a suitable density-gradient medium. The method is simple, quick, and effective, and provides cells suitable for subsequent nucleic acid extraction.

Another cell separation method uses immunomagnetic T cell beads to isolate highly purified T cells from whole blood, mononuclear cells or buffy coat layers. This can be performed either manually or using an automated platform. Cells are bound to magnetic particles using antibody/antigen then separated from the unwanted cells using a magnet. Finally, they are released from the complex and are now ready for downstream processing. The separation of CD3+ cells, which are a component of the T cell population, is required to monitor bone marrow transplant patients.

For **cytology** samples, consideration must be given to the need to undertake microscopy, as well as to conserve molecular constituents. The screening of cervical epithelial cells for cancer has been adopted by many countries (https://healthcaredelivery.cancer.gov/icsn/cervical/screening.html). Originally, cells were collected by use of a spatula and these were smeared onto a microscope slide before **fixation** and staining. This process has now been replaced by the adoption of the liquid-based thin preparation method (Strander et al., 2007). In contrast to the smear procedure the cells are collected using a brush that is then immersed in a

Cross reference
See Chapter 8 for more details.

Cell-free DNA (cfDNA)
DNA that is liberated from a cell into a body fluid. Its identity and concentration can be used to identify and track the course of a disease.

Lymphoprep™
A medium that can be used to separate mononuclear cells from other cell types, such as granulocytes and erythrocytes, which have a higher density.

Mononuclear cells (MNCs)
i.e. monocytes and lymphocytes can be isolated from blood and bone marrow using a density gradient medium.

Myeloproliferative
Proliferation of blood cell types derived from the bone marrow.

Cross reference
See Chapters 8 and 9 for more details.

Cytology
The microscopic observation of individual intact cells.

Fixation
A process by which the degradation of cells and tissues is irreversibly halted. Optimal fixation will result in the conservation of cytological and morphological features, as well as the retention of chemical constituents for analysis.

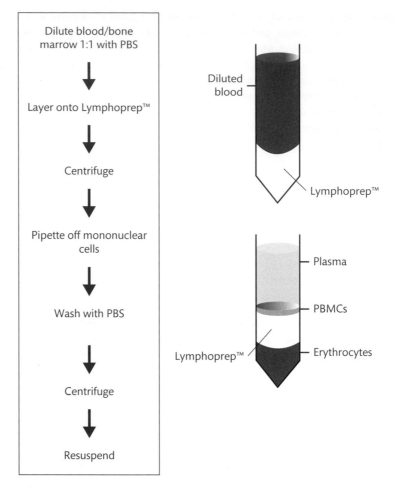

Dilute blood/bone marrow 1:1 with PBS

↓

Layer onto Lymphoprep™

↓

Centrifuge

↓

Pipette off mononuclear cells

↓

Wash with PBS

↓

Centrifuge

↓

Resuspend

Diluted blood

Lymphoprep™

Plasma

PBMCs

Lymphoprep™

Erythrocytes

FIGURE 1.5

Mononuclear cell preparation using the Lymphoprep™ method.

PBS, phosphate buffered saline.

TABLE 1.2 Cytology-based preparation options for non-gynaecological samples

Site	Collection	Preparation
Body cavities	Aspiration; concentration	Cytospins or cell blocks
Breast	Fine needle aspiration*	Smears
Cerebrospinal fluid	Fine needle aspiration	Cytospin
Lung	Sputum	Smears
	Endobronchial ultrasound-guided transbronchial needle aspiration (EBUS-TBNA)	Smears or cell blocks
Lymphoid	Dab or imprint of bisected sample onto microscope slides	
Thyroid	Fine needle aspiration	Smears

*Fine needle aspiration: a method involving the guided insertion of a narrow-gauge needle into a suspected lesion to remove cells for cytological examination.

preservative fluid. Subsequently, the cells are centrifuged and automatically deposited as a monolayer on a microscope slide. Instruments are available to automate the process and a distinct advantage is that any residual cells held in the preservative can be processed to provide additional slides for diagnostic or research purposes. For non-gynaecological sites, a variety of cell collection methods are available. These are summarized in Table 1.2.

For **karyotyping** by **G banding**, and the demonstration of gene and chromosome territories by *in situ* hybridization cells from a variety of sources are cultured, then arrested in mitosis and chromosome spreads made. The principles of this sample preparation method are outlined in Box 1.2.

Karyotyping

A method that enables the complement of chromosomes from a cell to be arranged in order and analysed visually to determine gross abnormalities.

G banding

A technique used to stain chromosomes with Giemsa to enable visualization of regions of chromosomes to aid karyotyping.

Cross reference

See Chapter 3 for more details.

BOX 1.2 Chromosome preparations for cytogenetic analysis

Chromosomes preparations for cytogenetic analysis—either conventional chromosome banding or fluorescence *in situ* hybridization (FISH)—can be obtained by arresting dividing cells in metaphase. While chromosomes exist as separate entities during the entire cell cycle, it is during metaphase that they are most condensed and most clearly visible as individual elements. Chromosome banding is the collective name used to indicate a number of techniques for the differential staining of metaphase chromosomes upon which, from a methodological perspective, clinical cytogenetics has very heavily relied until recently.

Chromosome banding is a sort of 'barcoding' for homologous chromosome pairs, which allows visualization at once of the entire chromosomal complement of a cell (or karyotype) and organizes it in an orderly layout or karyogram. The study of karyotypes or karyotyping is still applied in clinical diagnostics as a reference genetic test which, while inherently limited in resolution, permits genome-wide assessment of chromosomes in terms of numerical and structural integrity.

For postnatal karyotypic analysis, for instance, in cases where there is clinical suspicion of a chromosomal disorder or in cases of couples with reduced fertility, which might be harbouring undisclosed chromosomal aberrations, chromosomes are normally 'harvested' from short-term cultures of peripheral blood.

Venous blood samples (2–5 mL) are collected in tubes coated with an anticoagulant, typically sodium heparin, and subsequently inoculated in small amounts in tissue culture flasks containing nutrient-enriched cell culture medium. The cell cultures are then incubated in 5% CO_2 at 37°C. The medium of choice is normally RPMI 1640, nutritionally enriched by the addition of foetal calf FCS.

Phyto-haemagglutinin (PHA) is also added to the cultures to stimulate mitotic division of lymphocytes. After 72 hours incubation and before proceeding with cell fixation, the cultures are treated with Colcemid, a 'spindle poison', which by inhibiting tubulin polymerization during mitosis, halts the cells in metaphase. This is a key step in the procedure, which ensures accumulation and relative numerical increase of metaphase chromosome 'plates', albeit interphase nuclei remain the prevalent component of the cell population.

Following the incubation with Colcemid, the blood cultures are transferred from flasks to tubes and centrifuged. After removal of the supernatants, the cell pellets are subject to hypotonic shock by undergoing treatment in KCl. The ensuing osmotic swelling of the cellular volumes produces metaphase spreads of enhanced quality by reducing chromosome 'crowding' and overlaps. The hypotonic treatment is followed by repeated cycles of centrifugation and washes in Carnoy fixative (3:1 methanol/acetic acid).

Once fixed, the cells can be either stored at −20°C or used at once for 'dropping' on microscope slides. When air-dried on microscope slides and aged for a few days at room temperature, the chromosome preparations are ready for staining or *in situ* hybridization and microscope analysis. A simple staining of a test slide with **Giemsa** prior to processing allows quick and accurate assessment of the quality of the chromosome preparation, with regard to cell concentration, relative frequency of metaphasic versus interphasic cells, and extent and configuration of chromosome spreading in the metaphase plates. If the quality of the chromosome preparation is considered satisfactory, the slides can undergo further processing.

For postnatal karyotypic analysis, alternative samples to blood from which short-term cell cultures can be established may be used. These cultures may include:

- skin fibroblasts grown as monolayers;
- *bone marrow cells*: the tissue of choice for karyotypic analysis of patients with blood disorders whose lymphocytes fail to grow in culture;
- tumour specimens for the investigation of acquired, rather than constitutional, chromosomal abnormalities.

For prenatal karyotypic analysis, chromosomes are normally 'harvested' from either amniocytes, retrieved from amniotic fluid samples collected by amniocentesis, or chorionic villi, which requires disaggregation and harvesting of placental tissue *in vitro*. With minor modifications, mainly in relation to the choice of culture media, to ensure maximum cell proliferation with different cell types, and differences related to the way in which cells grow (suspension or monolayer) and physical handling, the principles of chromosome harvest procedures from these samples are consistent.

The most widely used banding technique for chromosomal analysis in clinical diagnostic settings is G banding. It revolves around the proteolytic ability of the enzyme trypsin to differentially 'digest' chromosomes and, following staining with a Giemsa solution in Sorensen buffer, to reveal specific longitudinal patterns of dark and pale alternating chromosome sections known as bands. Around 500 bands per entire karyotype can be demonstrated by standard G banding. As mentioned earlier, chromosome preparations can also be used for *in situ* hybridization to reveal the location of genes or chromosomal territories.

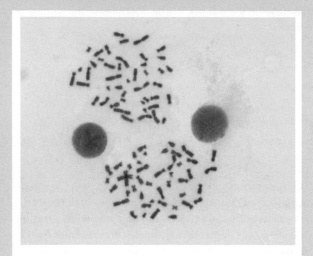

FIGURE 1.6

A Giemsa-stained metaphase spread showing a normal male complement of human chromosomes.

Cross reference

The principles and practice of in situ hybridization is discussed in Chapter 3.

Giemsa

A dye stain technique using methylene blue to stain nuclei and eosin the cytoplasm of cells.

Frozen tissue

For these samples, degradation is halted by snap freezing the tissue, usually by immersion in liquid nitrogen. For biopsies this can usually be done for the whole sample, but for resection samples representative blocks or cubes of tissue need to be cut out of the larger sample so that the freezing process is efficient.

1.3.2 Tissue preparations

In contrast to cell preparations tissues can be more challenging as ischaemic intervals for diagnostic samples are not always known. When explorative biopsy samples are collected the ischaemic interval should be short so they can be processed according to investigation without delay. However, when a resection sample is removed—for instance, during the removal of a segment of colon—then the much larger sample may have to wait while the surgical team attends to the patient undergoing surgery before it can be immersed in a fixative. Tissue collected post-mortem will inevitably be compromised, but if the cadaver has been kept in refrigerated storage then some degree of molecular analysis can still be performed. Accordingly, attention to the application of standardized sample preparation is still important. Two major preparation pathways are used for tissue, these being the frozen and the fixed pathways.

For **frozen tissue** samples, degradation is halted by snap freezing the tissue, usually by immersion in liquid nitrogen. For biopsies this can usually be done for the whole sample, but for resection samples representative blocks or cubes of tissue need to be cut out of the larger sample so that the freezing process is efficient. This is particularly important when subsequent morphological analysis is required to avoid the formation of vacuoles of ice (ice-crystal artefact) during the freezing process, which distorts the cytology and/or morphology of cells and tissues. When only molecular analysis is required the freezing of 1-cm^3 tissue cubes is acceptable. However, when microscopy is also required, the maximum depth of tissue should not exceed 3 mm and the area about 1 cm^2 if the morphological features are not to be disturbed by the presence of open spaces formed of ice during freezing.

The examination of frozen preparations of voluntary (skeletal) muscle by microscopy for cytological and enzyme perturbations in myopathies requires special care, as snap freezing of the biopsy samples in liquid nitrogen alone is not sufficient to prevent the formation of ice-crystal artefacts. In this situation, the biopsies are immersed in isopentane that has been precooled by surrounding its container with liquid nitrogen or solid carbon dioxide.

After tissue samples have been snap-frozen they should not be allowed to thaw before storage. Should this happen, then the sample should be allowed to thaw completely and the snap-freezing process then repeated. Before storage, either in a –80°C freezer or in vapour-phase liquid nitrogen, samples requiring microscopic analysis should be foil wrapped to exclude the surface action of air that can, over a period of prolonged storage, desiccate the outermost areas to a dust-like consistency.

For diagnostic histopathology formalin fixation (see Box 1.3) is followed by embedding in paraffin wax. The former irreversibly halts degradation of the sample; the latter provides an internal and external support to the tissue so that sections, typically between 2 and 4 µm thick, can be cut for staining and microscopic examination. The whole process is usually referred to as **formalin-fixed paraffin embedding (FFPE)**.

Formalin-fixed paraffin embedding (FFPE)

For diagnostic histopathology, formalin fixation is followed by embedding in paraffin wax. The former irreversibly halts degradation of the sample; the latter provides an internal and external support to the tissue so that sections, typically between 2- and 4-µm thick, can be cut for staining and microscopic examination.

BOX 1.3 Formalin fixation

Formaldehyde (methanal, HCHO) is the simplest mono-aldehyde and, as a gas, it is 37% soluble in aqueous solution. Unless freshly prepared from powdered paraformaldehyde, it will form polymers in an aqueous solution. Furthermore, on storage it will gradually transform into formic acid (methanoic acid, HCOOH). Typically, it is used for fixation at a 1/10 dilution of the concentrate when it is referred to as 4% formalin. In this diluted form it is either used buffered to neutrality (neutral-buffered formalin) or with sodium chloride added to 0.9% (formol saline).

Formaldehyde initially fixes by hydroxymethylation of basic amino acid groups of proteins and also reacts in an additive fashion with nucleic acid bases when they are not involved in base pairing. These are rapid and potentially reversible reactions. The subsequent formation of methylene bridges between hydroxymethylated proteins and formaldehyde-modified proteins and nucleic acids results in a cross-linked meshwork that stabilizes the morphology of the sample by either directly fixing tissue components or entrapping them in this framework. The formation of methylene bridges is a comparatively slow and irreversible reaction. While formalin remains the

most suitable fixative for the conservation of morphology and retention of the majority of molecular constituents, it does have deficiencies.

Epitopes are masked from interaction with antibodies by formalin fixation. This process is time-dependent and influenced by sample type. Cells in suspension, for example, will be fixed very quickly, while the interaction with tissue will be slower, as the fixative needs to first penetrate the sample. In both instances, antigen retrieval using proteolytic enzymes or superheating in a weak buffer will be required before incubation with an antibody. The practicalities of this procedure are discussed in Chapter 3, and with careful control the epitope-masking effects of formalin fixation can be successfully reversed. In a modified form, these antigen-retrieval methods have also been successfully employed for extracting proteins for Western blot and liquid chromatography/mass spectroscopy analysis (Shi et al., 2006; Vincenti and Murray, 2013).

The deleterious effect of formalin fixation on nucleic acids is permanent. It involves both fragmentation and modification/loss of nucleotide bases. As with proteins the effect of formalin is time- and sample-dependent. Fragmentation can be minimized by restricting the duration of fixation or using it at 4°C. Unfortunately, fragmentation is exacerbated by paraffin wax processing (Klopfleisch et al., 2011). The modification and loss of bases as a result of formalin fixation must be considered when undertaking sequence analysis (Do and Dobrovic, 2015). These effects do not preclude the use of formalin-fixed preparations for the demonstration of nucleic acids, either *in situ* or in extracts, but methodological changes are often required and care must be exercised in analysis of results.

Cross reference
See chapters 3 and 4 for more details.

When a fixative is not able to complete its reaction with a sample, then there will be poor conservation of morphology and loss of soluble molecular components. This condition of '*under-fixation*' is commonly caused by insufficient time of immersion in a fixative. A variation of under-fixation, called '*zonal fixation*' is caused by attempting to fix a block of tissue that is too thick for fixative to adequately penetrate, resulting in its central regions being incompletely fixed. Prolonging fixation using cross-linking fixatives will degrade nucleic acids and may render demonstration of proteins by analytical methods problematic. This condition, in which morphology is usually unaffected, is known as '*over-fixation*'.

With these considerations in mind, care should always be taken to match the duration of formalin fixation to the sample. Biopsy tissue will often be adequately fixed after immersion in formalin for a few hours at ambient temperature. In the case of large resection samples initial access of formalin to the tissue should be carefully considered. For most, slicing or 'bread boarding' (Orchard et al., 2015) will be sufficient to assist the penetration of formalin so that fixation can begin using a ratio of at least 10/1 v/v fixative to sample. Subsequently, for the adequate fixation of blocks of tissue dissected from a resection sample of $1 \times 1 \times 0.3$ cm dimensions, then 24 hours' additional fixation may be required.

Processing involves the replacement of the aqueous formalin fixative with paraffin wax. As water and wax are immiscible, processing involves the removal of water using ascending grades of alcohol, which is then replaced by a solvent such as xylene, which is finally replaced by infiltration with paraffin wax. The whole procedure is automated. For biopsy samples, rapid processing schedules that employ heat and vacuum replacement of the reagents are often used. However, with the larger resection blocks of tissue these are typically processed using an overnight schedule to ensure that all solutions are adequately exchanged. Failure to attend to the differing time requirements during paraffin processing and attending to the quality control of the solvents used can result in compromised morphology and inconsistencies in molecular results.

The final step in FFPE process is to surround the processed tissue with paraffin wax contained within a mould in the embedding step, in order to provide external support for sectioning. Two important criteria should be observed when embedding tissues. To avoid sample loss during sectioning, care must be taken during this step to ensure that the samples are embedded on the same plane. This is particularly important when biopsy samples are being embedded as, due to their small size, precious tissue could otherwise be lost during sectioning. It is also important to orientate some types of sample to ensure that relevant morphological features are present in the cut sections. For example, skin should be orientated so that epidermis and dermis are present in sections and alimentary tract embedded so that the mucosa and underlying structures can be viewed simultaneously in the stained section. Embedded FFPE blocks can be stored at ambient temperature, but in line with local conditions, air conditioning may be required to keep the wax from softening. Under optimal storage conditions FFPE tissue can remain stable for many years.

For subsequent analysis of intact tissue thin sections, usually between 2 and 4 μm thick, are cut and mounted on microscope slides. This process, known as microtomy, involves the use of a microtome that allows precision cutting of the tissue. For FFPE samples, microtomy is undertaken at ambient temperature, while for the production of frozen sections microtomy is undertaken within a cryostat, where all components are kept frozen. It is standard procedure to mount sections on adhesive-coated slides to minimize the possibility of their detachment during analysis. Unstained sections of FFPE and frozen tissues can, respectively, be stored at ambient temperature or in a freezer. In the case of some protein and nucleic acid targets reactivity may be diminished within days of cutting of FFPE samples. Accordingly, in these situations, standardization of the interval between cutting and immunocytochemical staining or *in situ* hybridization is important. For a more detailed explanation of the technical nuances that surround microtomy please consult texts as provided in the reading list.

When extraction of nucleic acids for analysis is required, FFPE sectioning may be undertaken at 5–10 μm with collection of the resultant 'curls' being placed in sterile microfuge tubes. These can then be stored indefinitely at ambient temperature as they await extraction. It is very important to avoid cross cross-contamination of samples during this process. As a minimum, gloves must be worn and a new microtome blade used for each new sample. The surrounding area, together with handling forceps, must also be scrupulously cleaned between each sample that is cut (Box 1.4).

BOX 1.4 Steps to prevent cross-contamination between samples when cutting sections for PCR and sequencing

1. Prior to cutting, the microtome should be thoroughly cleaned of all debris. This may include the use of xylene or alcohol.

2. If trimming of blocks is required, then step 1 will need to be repeated.

3. The blade to be used for cutting sections must be thoroughly cleaned using xylene or alcohol between each sample, and the blade moved along to a new area. If this cleaning cannot be assured, then a new blade should be used for each sample.

4. Transfer cut sections using disposable forceps for each sample or stringently cleaned metal forceps.

5. Place the sections for each sample into a sterile Eppendorf tube ready for extraction of nucleic acid and store appropriately.

Key Point

The process of sample preparation varies according to sample type and its analytical destination.

1.2 SELF-CHECK QUESTION

What are the main routes for the preparation of tissue for morphological analysis and blood cells for nucleic acid analysis following extraction?

1.4 **Extraction methods**

This section considers the principles for optimal preparation of homogenates for molecular analysis. For the optimal conservation and ease of extraction, fresh or frozen samples should be used as fixation, according to the type employed, will often degrade molecular constituents. As already noted formalin fixation is used for diagnostic histopathology and, while the mechanisms of molecular degradation are not fully understood, their effects are, and these must be considered before the use of samples exposed to this fixative (Box 1.3). In contrast, when non-additive denaturing fixatives, such as alcohol or acetone are employed, then proteins and nucleic acids are often not affected.

1.4.1 Proteomic methods

Cross reference
See Chapter 4 for more details.

Western blotting for proteins and mass spectroscopy (MS) for peptides represent major endpoints for protein extracts. With appropriate extraction protocol modification, proteins can be extracted from FFPE samples, and good equivalence in protein and peptide conservation has been demonstrated between these and fresh/frozen extracts (Shi et al., 2006; Sprung et al., 2009; Vincenti and Murray, 2013). A flow diagram for sample preparation for Western blotting and MS is provided in Figure 1.7. More detailed information is available in Tanca et al. (2015) and http://www.abcam.com/protocols/sample-preparation-for-western-blot.

1.4.1.1 Principles

There are some general principles that apply to all protein extraction methods.

- To avoid sample loss, keep sample preparation to a minimal number of steps.
- Tissue and cell disruption should be performed in such a way as to minimize proteolysis, thereby reducing degradation.
- Inclusion of an inhibitor to minimize degradation should be considered.
- Do not expose the samples to repeat freeze–thaw cycles; keep these to a minimum.
- Remove particulates prior to analysis, either by centrifugation or other forms of clean-up (e.g. salt removal).
- Never heat a sample containing urea; high temperatures cause urea to hydrolyse to isocyanate, which modifies protein by **carbamoylation**.

Carbamoylation
The transfer of the carbamoyl from a carbamoyl-containing molecule to an acceptor such as an amino group.

1.4.1.2 Lysis of cells

In order to fully analyse all cellular content all cells must be effectively disrupted. This can be achieved using a variety of lysis methods (Table 1.3). The main aim is to extract as much protein as possible from the lysate, while minimizing contamination from other molecules, e.g. lipids and nucleic acid. This can be achieved with the help of:

- detergents (e.g. sodium dodecyl sulfate (SDS), 3-([3-cholamidopropyl]dimethylammonio)-1-propane sulfonate (CHAPS), Tween), which helps to solubilize membrane proteins;
- reducing agents (e.g. dithiothreitol [DTT], mercaptoethanol, thiourea), which reduce disulfide bonds or prevent protein oxidation;
- denaturing agents (e.g. urea and acids), which disrupt protein–protein interactions, secondary, and tertiary structures by altering solution ionic strength and pH;

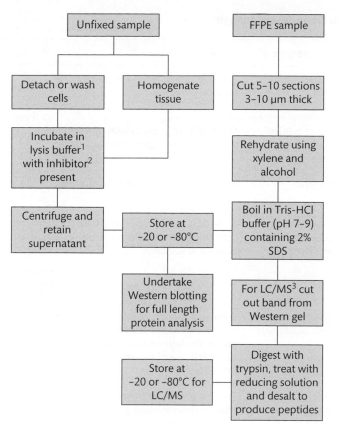

FIGURE 1.7

Preparation of homogenate samples for proteomic analysis.

Key: FFPE, formalin-fixed paraffin wax embedded; LC/MS, liquid chromatography/mass spectrometry; SDS, sodium dodecyl sulfate.

[1] Lysis buffers will differ according to subcellular fraction required. Those containing SDS and other ionic detergents can give highest yields.

[2] Appropriate protease and phosphatase inhibitors should be added to slow proteolysis and dephosphorylation.

[3] Direct preparation for LC/MS can also be undertaken.

- enzymes (e.g. DNAse, RNAse), which digest contaminating nucleic acids, carbohydrates, and lipids.

1.4.1.3 Sample purification/enrichment

The protein content of any given sample is very complex after the initial lysis. One of the key processes required for ensuring that all protein content is analysed equally is to reduce complexity. Protein purification and enrichments are workflows intended to isolate one or a few proteins from a complex mixture.

Protein purification is vital for the characterization of the function, structure, and interactions of the protein of interest. Purification takes advantage of the differences of protein size, physio-chemical properties, binding affinity, and biological activity to ensure specificity and selectivity. Enrichment is used to help concentrate low-abundance proteins to improve their downstream analysis.

Protein fractionation is routinely used in any proteomics approach that requires the need to create a less complex mixture that is suitable for detection by MS or other quantitative tool. In simple terms, this is chromatography with a collection of methods for separation complex mixtures. The process requires two phases—a mobile (i.e. sample) and a stationary phase (static holding material). Essentially, a sample matrix is loaded onto the stationary phase, either pumped or gravity-fed. The matrix is separated according to the physical properties of the stationary phase and eluted.

TABLE 1.3 Methodologies available for the lysis of samples for proteomic analysis

Lysis method	Process	Sample type	Procedure
Osmotic	A gentle method suited for when lysate is used for subsequent fractionation	Tissue culture cells	Cells are suspended in a hypo-osmotic solution
Freeze–thaw cycles	Cells or tissue can be lysed by subjecting material to one or more cycles of quick freezing and thawing	Tissue culture and bacterial cells	Material frozen rapidly with nitrogen then thawed. Can be repeated several times
Enzymatic	Lysis is performed by cell-specific enzymes, e.g. lysozymes	Bacteria	Cells are treated with specific enzymes
Detergent-based	Detergents are very good at solubilizing the cellular membrane and expelling the internal material. There are a number of detergents that can be used, e.g.: Triton-X100, Tween-20, SDS, CHAPS, NP-40,	Tissue culture cells, solid tissue, and biofluids	Material is usually mixed in a base solution containing any one of the detergents and allowed to solubilize.
Sonication	Sonication is performed using a 'sonicator' an instrument that delivers ultrasound (high-frequency) energy to samples to agitate and disrupt the cell membranes	Tissue culture cells, solid tissue	Sonication is most commonly performed using an ultrasonic bath or an ultrasonic probe. The latter is better known as a 'sonica-tor,' and is one of the most preferred methods for cell lysis
Grinding	Cells or tissue is ground into powder form using a pestle and mortar	Solid tissue and bacterial cells	Tissue is preferably snap-frozen, using liquid nitrogen before being ground up
Mechanical/liquid homogenization	Mechanical and liquid-based homog-enization are the most widely used cell disruption techniques for either solid tis-sue or cell culture	Solid tissue or cell culture	Cells are lysed by forcing the cell or tissue suspension through a narrow space, thereby shearing the cell membranes

There are a number of approaches that exist, for example, by molecular weight (gel filtration or size-exclusion), protein charge (ion exchange), hydrophobicity (hydrophobic chromatogra-phy), and specific ligand interaction (affinity chromatography).

Size-exclusion chromatography (SEC) or gel filtration chromatography is a method in which molecules are separated by size through a gel matrix. Principally, the gel is comprised of small spherical beads that contain pores of differing size. Essentially, once a sample is passed through these beads, the smaller molecules get trapped in the pores, while the larger molecules pass through. Subsequently, the smaller molecules are eluted based on their size.

Alternatively, separation can be performed based on either hydrophobic or ionic properties. Hydrophobic separation relies on the separation of molecules based on surface hydropho-bicity. On the other hand, ion exchange chromatography separates molecules based on the degree of their charge. These come in two varieties—anion resins, which retain negatively charged molecules, and cation columns, which retain positively charged molecules.

Finally, affinity chromatography is an approach based on the molecular properties of the pro-teins. The stationary phase resin is bound to a specific ligand, e.g. nucleic acid, antibodies, antigens, metal ions.

Protein enrichment is a fundamental step in the identifications of protein modifications. The biggest problem with mapping modifications is simply obtaining MS fragmentation spectra of the modified peptides. Of all the copies of any particular protein in a cell, only a small fraction may bear any specific modification. For example, many protein kinase substrates are rapidly phosphorylated and dephosphorylated, such that only a few phosphorylated copies of a protein may be present at a particular time. The process of enrichment makes use of the purification procedures mentioned previously, especially affinity chromatography. Enrichment of specific proteins or protein complexes can most easily be accomplished using target-specific immobilized antibody bound to the stationary phase in a column. One of the widely used approaches is immobilized metal affinity chromatography (IMAC) to enrich phosphor-specific peptides or proteins. Briefly, the process works by allowing either proteins or peptides to be retained in the column containing immobilized metal ions, such as cobalt, copper, gallium, iron, nickel, or zinc.

1.4.2 Genomic methods

1.4.2.1 Principles

The extraction and purification of nucleic acids must be carefully undertaken if optimal analysis is to be obtained. As a result, investigators may be specific about protocols or kits to be used. Protection from **nucleases** is one of the most important considerations during nucleic acid preparation, particularly when isolating RNA. If RNA is not protected from nucleases, it degrades and vital transcripts may be lost, potentially leading to false negative results. Overcoming these issues involves working in dedicated nucleic acid isolation areas, wearing gloves at all times, using pipettes with filter tips, and maintaining a clean work area. Poor sample preparation can also result in contaminants being present that can inhibit polymerase chain reaction (PCR) reactions, in amplification of non-specific bands or no amplification being obtained. As the range of downstream applications has expanded, so has the requirement for optimal nucleic acid preparation.

Nucleases

Enzymes that degrade DNA (DNAses) and RNA (RNAses).

The key goals for all methodologies are good yield, quality, and purity. In addition, scalability and throughput may be added as practical requirements. External factors, such as sample age, type, previous handling, and storage may all affect yield and quality. Samples that have been stored correctly generally produce better quality nucleic acid. The amount of starting material is also a consideration. Small samples, such as biopsies and chorionic villus sampling (CVS) are challenging, partly because there is usually no option to retain a reserve for extraction. While a small amount of material is a limitation, overloading during extraction with more than recommended input will also affect yield. This is because overloading results in clogging of the system and leads to a lower yield that can also affect quality. With respect to RNA extraction a DNAse digestion step is required to remove contaminating DNA.

Cell disruption is critical to release the DNA and RNA. Incomplete cell disruption will lead to decreased nucleic acid yield. The method used to disrupt cells is dependent upon the starting material. Whole blood, bone marrow, and cultured cells generally require little or no disruption, whereas tissue requires sufficient homogenization into small enough pieces to enable the cells to be released, thereby allowing lysis buffer access to the cells. Nucleic acid extraction from FFPE material is challenging as the fixation process that has already taken place causes cross-linking between the nucleic acids and the protein. As a result, the molecules are very tightly bound and highly susceptible to degradation. All FFPE tissue require a deparaffinization step in xylene at 37°C to solubilize and remove the paraffin wax. The sample is then washed a number of times in ethanol to remove the xylene. A protease digestion step, e.g. proteinase K or pepsin, is then required and this is generally longer than required from an extraction from 'fresh' samples. Digestion time can typically be overnight compared with 2–3 hours for a fresh tissue sample.

1.4.2.2 Methodologies

Identifying the right method, whether kit-based, manual, or automated, is key to obtaining the desired result. In recent years, the use of kits, as well as automation has increased. There are a wide range of DNA extraction kits on the market, including a number that offer small-scale extraction. Which one to choose can seem daunting as they all seemingly do the same thing. However, depending on the sample type some provide better results than others and validation/verification will therefore be required. Laboratories or individuals need to select the kit of choice to fit their requirements.

Cross reference

See Chapter 2, Section 2.4 for further discussion of validification and verification.

Considerations when selecting the right kit-based methodology may include the use of silica membranes, anion exchange, or magnetic particle technology for the isolation step. Silica membranes yield highly pure nucleic acids suitable for most applications. As nucleic acids are highly negatively charged, they can also be separated from other components by anion-exchange chromatography. This allows separation of nucleic acids from proteins, polysaccharides, and metabolites. This technology is equivalent to phenol/chloroform purification and the highest quality DNA is obtained using this method without the associated health risks of the older procedure.

Cross reference

Barcode scanning and audit control are discussed further in Chapter 2.

Automated extraction solutions have become increasingly available and are commonly used as the requirement for routine genetic testing, both germline and somatic, has increased significantly over recent years. Small-scale technologies generally accommodate 12–14 samples per run, and are suitable for research or small- to medium-sized diagnostic laboratories. Larger scale platforms, such as QIAsymphony (QIAGEN Inc., Germantown, MD), which utilizes a silica-based purification protocol, can process up to 96 samples at once utilizing four individual loading cartridges of 24 samples each as illustrated in Figure 1.8. These platforms generally allow more hands-off time, which is desirable in a busy diagnostic laboratory. Automated platforms also aid laboratories in streamlining workflow, as well as providing reliability and uniformity. Additionally, the majority offer barcode scanning and audit control—an essential requirement for diagnostic laboratories. The ability to process multiple sample types and with several distinct methods simultaneously is an additional advantage. The QIAsymphony is equipped with a cooling platform, which provides the ability to elute DNA and RNA, and maintain this at 4°C.

DNA extracts can be stored in Tris-EDTA buffer at 4°C in the short term (weeks to a few months). This prevents repeated freeze–thaw cycles, which can result in shearing of the DNA. Storage at –20°C or –80°C is suitable for long-term storage of DNA, RNA, or cell pellets. This preserves the integrity and prevents degradation.

Key Point

An optimal extraction method will conserve the constituents at high purity, maximize recovery, and provide sufficient quantity for subsequent molecular analysis.

1.3 SELF-CHECK QUESTION

What factors may determine the quality of protein and nucleic acid extracts?

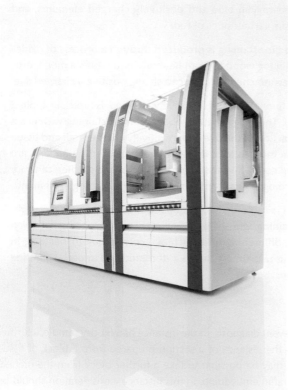

FIGURE 1.8
Automated high throughput nucleic acid platform. The
QIAsymphony is a fully automated nucleic acid extraction platform
enabling extracts of DNA and RNA to be made from a wide range
of sample types. It has a continuous loading ability, barcode
reading, and sample tracking. It has the ability to both import
sample lists, as well as export sample sheets.

1.5 **Assessment of sample quality**

Before using any sample, its suitability for analysis should be established. Failure to do this
could lead to inaccurate diagnostic interpretation—a false negative or skew of research data
obtained in a cohort of samples. In short, the purpose of pre-assessment of sample quality

ensures a level playing field for subsequent analysis. This section considers the types of sample assessment methods that can be used and their application.

1.5.1 Intact sample assessment

Morphological assessment of haematoxylin and eosin (Box 1.5) stained tissue sections will be undertaken as part of the routine diagnostic process with preservation and representation being assessed.

BOX 1.5 Haematoxylin and eosin staining

This represents the standard architectural stain employed for histopathological diagnosis. Negatively charged elements of cells and tissues, principally nucleic acids (nuclei), are stained with haematoxylin blue and positively charged elements, and supporting structures, with eosin, various hues of red.

The haematoxylin solution used for staining is produced through a process of oxidation of the natural haematoxylin log wood to form haematein to which a metal, usually aluminium, 'mordant' is then attached to produce a strong positively charged dye.

Eosin, a derivative of synthetic fluorescein and negatively charged, is typically applied diluted as an aqueous solution. The range of hues obtained after staining with eosin reflects the density of the interaction of the dye with positively charged cell and tissue components. Accordingly, from most to least dense interactions, red blood cells are stained orange, muscle red, and connective tissue pink. Cytoplasmic staining can vary in colour from red to pink according to density of available positively charged elements.

While the staining mechanism is simple, the technique requires some finesse in application with deliberate overstaining with haematoxylin then eosin being followed after each dye application by differentiation to achieve correct colour balance. In diagnostic situations the technique is almost always automated and undertaken using commercially supplied solutions.

Inadequate preservation could prevent diagnosis. In the instance of a biopsy sample, a further sample may need to be taken. In the instance of a section prepared from a block of tissue taken from a resection sample, it may be possible to take a further block from the original sample and to process this for paraffin embedding. In this instance, consideration should be given to alteration of the processing schedule if this is thought to have contributed to the poor preservation of the initial block.

Providing the sample has been correctly embedded and sectioned carefully to avoid loss of tissue, then inadequate representation would normally be addressed by cutting sections at a deeper level in the tissue block.

For cytological preparations, the opportunity to redress issues of preservation and representation disclosed in a stained sample will often mean that a new cellular collection will be required. Exceptions to this are the possibility of using spare Cytospin™ (ThermoFisher, Santa Clara, CA) slides, preparation of additional thin layer slides of gynaecological cases and the cutting of sections deeper into a cell block preparation.

Assessment of molecular preservation of proteins and nucleic acids in intact cells and tissues can be undertaken using antibodies and nucleic acid probes to ubiquitously expressed **housekeeping genes**.

Using a panel of a few antibodies (Table 1.4 and Figure 1.9) almost all cell types can be verified as suitable, or otherwise, for use for immunocytochemistry. When labile targets, such as phosphorylated proteins, are to be demonstrated then a pAKT antibody can be used, but if precise target conservation is required then this can be confirmed in extracts of the preparations using Western blotting.

For the assessment of conservation of mRNA *in situ* hybridization of a poly(d)T probe that hybridizes to the poly A tails that are present on about 90% of mRNA transcripts (Figure 1.10) can provide a ready means of assessing the global conservation of gene expression. This may be supplemented by establishing a ribosome integrity number (RIN) using parallel homogenate extracts (Schroeder et al., 2006) or the paraffin-embedded RNA metric (PERM), an adaptation of this that is claimed to add finesse to the former (Chung et al., 2016). For assessment of the conservation of microRNA *in situ* hybridization to miR-126 that is expressed in blood vessel endothelia can be used (Jørgensen et al., 2010).

At the DNA level test and control probe, *in situ* hybridization is often undertaken simultaneously. The control probe hybridizes to centromeric or peri-centromeric sequences present on the same chromosome in which the gene of interest is to be demonstrated. A diagnostic example is the assessment of HER2 copy number change in breast and gastric carcinoma (Bartlett et al., 2011), where the control probe hybridizes to centromeric sequences of chromosome 17 (Figure 1.11).

1.5.2 Homogenate sample assessment

Assessment of sample quality following nucleic acid extraction is critical and integral to any laboratory protocol. It is required for a number of reasons including:

- estimation of DNA or RNA concentration;
- estimation of purity;
- determination of suitability for downstream assay;
- ensuring that the correct quantity of nucleic acid is included in an experiment, such as a PCR reaction;
- assessment of degradation, particularly applicable for FFPE DNA and RNA extractions.

TABLE 1.4 Antibodies for assessment of antigenic preservation in cells and tissues

Cell/tissue type	Antibodies targeting
Epithelial	Low molecular weight cytokeratin for simple epithelia (LP34) Hepatocyte-specific antigen for liver High molecular weight cytokeratin for stratified epithelium (AE1/3)
Haemopoietic	Leucocyte common antigen (CD45)
Muscle	Desmin and vimentin
Nervous	Neurofilament
Soft tissue	Vimentin

Housekeeping gene
A DNA gene or RNA transcript, which is required for basic cellular function, therefore, widely expressed and can be used as a comparison with DNA/RNA analytical targets.

(A)

(B)

(C)

(D)

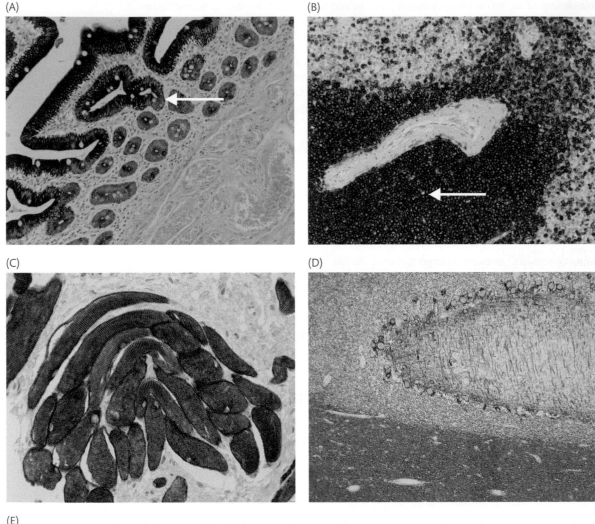

(E)

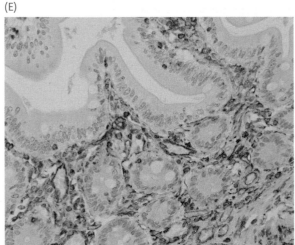

FIGURE 1.9

Immunocytochemical staining verifying protein conservation in FFPE tissue sections. (A) Demonstration of cytokeratin using AE1/3 antibody in the epithelium (arrow) of small intestine. (B) Demonstration of lymphoid cells using anti-CD45 antibody in the white pulp (arrow) and surrounding red pulp in spleen. (C) Demonstration of striated muscle fibres using an anti-desmin antibody. (D) Demonstration of neurofilaments in cerebellar tissue. (E) Demonstration of capillaries and basal membranes in lamina propria of small intestine using an anti-vimentin antibody. Immunoperoxidase immunocytochemistry with production of brown polymerized diaminobenzidine precipitate marking sites of antibody/antigen interaction counterstained with haematoxylin to demonstrate nuclei (blue).

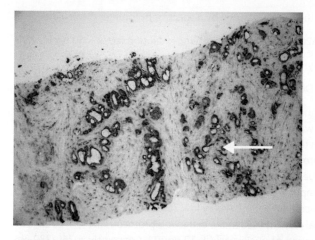

FIGURE 1.10
Demonstration of total mRNA in formalin fixed paraffin embedded tissue. Poly A-tailed mRNA demonstrated using a poly(d)T probe by *in situ* hybridization in Gleeson stage 7 prostate cancer. Note the intense reaction in the tumour glands (arrow).

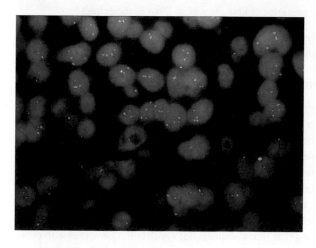

FIGURE 1.11
Dual *in situ* hybridization staining of breast carcinoma tissue to demonstrate HER2 gene copy (red) and centromeric sequence (green) on chromosome 17.
Note: In the majority of nuclei a normal gene copy number for the two hybridization signals for each probe is present.
Image kindly provided by S Forrest, Royal Liverpool University Hospital, UK.

Assessment of sample quality can be categorized according to methodology:

- Absorbance measurement, which includes quantitation, nucleic acid concentration, and purity analysis. The methodologies include the use of a UV spectrophotometer, Nanodrop, and QIAexpert technologies.

- Agarose gel electrophoresis to provide semi-quantitative, purity, and quality analysis.

- Fluorescent DNA-binding dyes.

- Amplification of a housekeeping gene by PCR to assess integrity.

1.5.2.1 Absorbance measurement

The most common method used to measure DNA and RNA yield and purity is by absorbance using a **spectrophotometer**. It is simple and generally requires commonly available laboratory equipment; a spectrophotometer with a UV lamp, a UV transparent cuvette, and the sample to be measured.

The wavelength for measurement of DNA is absorbance at 260 nm. This reading needs to be adjusted for the turbidity of the sample and is measured by absorbance at 320 nm. The DNA concentration can then be calculated by using the relationship:

A260 of 1.0 = 50 ng/μL of pure double-stranded DNA

Spectrophotometer
An instrument used to measure the absorbance of light through a sample.

Therefore, the concentration of the DNA measured by a spectrophotometer is calculated as:

$$\text{Concentration (ng/μL)} = (\text{A260 reading} - \text{A320 reading}) \times \text{dilution factor} \times 50 \text{ ng/μL}$$

The yield of DNA is calculated by:

$$\text{DNA yield (ng)} = \text{DNA concentration} \times \text{total sample volume (μL)}$$

One of the drawbacks of measuring absorbance at 260 nm is that DNA is not the only molecule to absorb UV light at this wavelength. RNA has a marked absorbance at 260 nm and, if present in the extract, will distort the reading by contributing to it and provide an overestimate of the DNA yield. Additionally, protein absorbance at 280 nm can distort the DNA reading. The ideal 260/280 nm ratio should be 1.8, indicating a highly purified DNA sample. Slight movement from this ratio indicates impurity, e.g. a 260/280 ratio of 1.73 indicates the presence of only 30% DNA and 70% protein contamination. A very low 260/280 ratio can also be an indication of the presence of residual reagent used in the extraction protocol. Finally, it could also be a result of very low concentration of nucleic acid (<10 ng/μL). Conversely, a 260/280 nm ratio approaching 2 indicates the presence of RNA, as well as DNA in the extract.

Absorbance at 230 nm is indicative of the presence of organic compounds or salts. As a guide, the ratio of 260/230 should be greater than 1.5; therefore, the lower the ratio the greater the amount of contaminant present. This could be in the form of residual phenol or guanidine, which is often used in the column-based kits or glycogen.

Nanodrop

A type of spectrophotometer designed to measure the absorbance of DNA and RNA.

A commonly used spectrophotometer in molecular laboratories is the **Nanodrop** (Figure 1.12). It is a compact, bench-top instrument, which is simple to use. No dilution of DNA is required and measurement can be obtained in less than 10 seconds. It uses the surface tension of the aqueous solution containing the DNA to form a sample column between two pedestals through which light is directed. This allows very small volumes of sample to be used. As an example, only 1 μL of extracted undiluted DNA is required to measure the concentration using a Nanodrop. The drawback of a Nanodrop is its inability to measure low concentrations, i.e. <0.2 ng/μL of nucleic acid accurately. Consecutive readings of the same sample are not always reproducible, suggesting that repeatability is not consistent. Compared with other

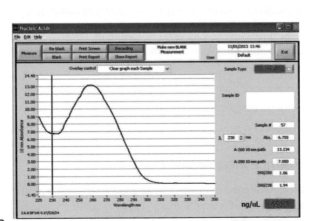

FIGURE 1.12

Absorbance spectrum of a DNA sample using a Nanodrop spectrophotometer. Absorbance is measured at 230, 260, and 280 nm. Sample purity is determined by the 260/280 ratio with a ratio of 1.8–2.0 being considered as good quality DNA and RNA, respectively. The 260/230 ratio can be used as an additional purity measure. Good quality nucleic acids will generally have a slightly higher 260/230 ratio than the 260/280 ratio. Lower than expected ratios may be indicative of contaminants that absorb at 230 nm. The concentration in ng/μL is shown in the green box.

quantification methods, the Nanodrop also tends to produce higher concentration values, which could be an issue if accurate quantification is required.

The QIAexpert system (QIAGEN Inc., Germantown, MD) is UV spectrophotometer-based, allows for the measurement of up to 16 samples at once, and is suited to a medium through-put laboratory. It distinguishes between DNA and RNA, and also measures sample impurities. Samples are pipetted onto a QIAexpert slide and placed in the reader. A pre-programmed method is then selected and the results are available in less than 2 minutes for all samples.

1.5.2.2 Agarose gel electrophoresis

The use of agarose gel electrophoresis is another quick and relatively straightforward method to estimate and assess DNA and RNA quantity and quality. It is not widely employed, but as it requires no specialist equipment, it is an attractive option for a laboratory with limited resources.

To perform this method, a horizontal gel electrophoresis tank with an external power sup-ply is required. Reagent requirements include analytical grade agarose and running buffer of Tris acetate EDTA (TAE) together with an intercalating dye, such as SYBY green, to stain the DNA. Ethidium bromide, another intercalating dye, is a known carcinogen, and is, increas-ingly, being replaced by dyes such as SYBR green. An appropriately DNA-sized marker ladder is also required to estimate integrity of the sample DNA. Usually a 1% agarose gel is cast with the samples and ladder run alongside of each other. To assess the quality of RNA on a gel, a denaturing agarose gel is recommended, as RNA forms secondary structures via intramolecu-lar base-pairing, and this prevents it from migrating according to size. This comprises a 1% agarose gel, plus 37% formaldehyde and a MOPS buffer.

To assess the quality and quantity of the sample, a 1- or 2-µL DNA sample is diluted in a total volume of 10 µL of loading buffer. The latter is typically composed of Tris-EDTA to which a dye is added to show the migration of the samples during electrophoresis. This is then loaded into a well made in the agarose gel and then an electrical current is switched on. As DNA is negatively charged, it will migrate towards the anode. Smaller DNA fragments migrate faster and, hence, the DNA is separated by size. Any contaminating protein or RNA will migrate at different rates and can be visualized on the gel (Figure 1.13). The concentration and yield of the DNA can be estimated by comparing the migration of the sample with a DNA ladder.

1.5.2.3 Fluorescence methods

The availability of fluorometers and fluorescent DNA-binding dyes has made the measurement of DNA concentration using this methodology an attractive option. When DNA concentration

L 1 2 3 4 5 6 L

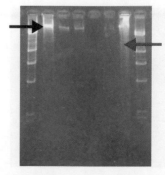

FIGURE 1.13
DNA quality assessment using agarose gel electrophoresis.
1–3 cell samples immersed in saline for 30 min before protease digestion for 2, 24, and 48 hours, respectively. 4–6 cell samples immersed in formol saline for 30 min before protease digestion for 2, 24, and 48 hours, respectively.

L, lambda DNA size ladder of 23, 9.5, 6.5, 4.3, 2.3, and 2 kbp.
Note the presence of high molecular weight DNA at the exclusion zone (red arrow) in the saline samples, lower yield of DNA in formal saline samples 4 and 5, and presence of degraded DNA in lane 6 (green arrow)—probably revealed due to extended protease digestion.

or yield is low, this option is more accurate than absorbance methods. Examples of fluorescent dyes include PicoGreen, QuantiFluor, and SYBR green. The availability of single-tube or microplate fluorometers allows flexibility and the ability to measure multiple samples simultaneously. The dye selected will determine the fluorescence excitation and emission spectra. The concentration of unknown samples is calculated based on comparison with a standard curve generated from samples of known DNA concentration.

A popular choice for fluorometric quantitation is the Qubit™ (ThermoFisher, Waltham, MA). It is sensitive, does not overestimate the quantities of the nucleic acid, and is quick and simple to use. It requires as little as 1 µL of sample and has a small footprint.

1.5.2.4 Amplification of a housekeeping gene to assess integrity

This methodology is widely used to assess the integrity of both DNA and cDNA, either as a separate assay or as a multiplex (Figure 1.14). A suitable housekeeping gene needs to be one that is either expressed in all cells or is widely expressed in many. The amplification of the selected gene acts as an internal control for the quality of the DNA or cDNA. A key example of this is assessing the integrity of RNA extracted from FFPE tissues and its amplification by PCR. This can be affected by a number of factors, such as length of storage before analysis, fixation time, fixation type, composition, and thickness of the tissue. A key consideration is that the PCR products of the housekeeping genes must be in the size range of the specific target amplicons for the different targets. Detection of *GAPDH* for DNA extraction and *G6PD* for RNA extraction will confirm that the quality of the sample is appropriate for testing. In a diagnostic laboratory, the requirement to demonstrate successful amplification of a housekeeping gene is critical for reporting cases where the test result is 'negative'.

Key Point

Before molecular analysis can proceed, it is essential that the quality and, in the instance of extracts, the quantity of preparations is assessed.

1.4 SELF-CHECK QUESTION

What methods are available to assess the quality of DNA in intact cellular and extract preparations?

FIGURE 1.14

PCR of a housekeeping gene to assess cDNA integrity.
Amplification using a common *G6PD* forward primer and four different reverse primers produces PCR products of 100, 200, 300, and 400 bp. Lanes 1 and 7 show the 1 kb size ladder. Lanes 2, 3, and 5 demonstrate samples that have amplified 100, 200, 300, and 400 bp amplicons. Lanes 4 and 6 show two samples that failed to amplify the housekeeping gene.

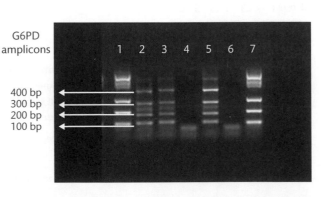

1.6 **Future directions**

At the beginning of this chapter, the need to minimize or at least standardize the ischaemic interval to ensure optimal and consistent sample preparation was examined. For cultured cell preparations and for samples, such as blood or bone marrow, collected from patients under controlled conditions, there should be no issue in this regard. However, when storage and transportation are required, then due care and attention should be paid to possible degradative effects. With these factors in view, an ideal would be to have a universal medium into which a cellular sample could be placed that would preserve cytology and molecular constituents for a period of time before laboratory procedures commence. Such a universal medium would be temperature-insensitive and preserve samples without degradation for up to 72 hours. No such medium has been described to date, but the quest to attain at least partial success in this area would greatly assist in the standardization of quality and simplify sample preparation.

The positive and negative aspects of the FFPE sample preparation process have been highlighted several times in this chapter. With combined morphological and molecular analysis playing an increasingly vital role in patient diagnosis, can a molecular fixative replace formalin? The answer at this time is a qualified 'yes' and a firm 'no'. The 'yes' applies to the description of fixatives (Cox et al., 2006; Klopfleisch et al., 2011; Moelans et al., 2011; Craft et al., 2014; Howat and Wilson 2014) that conserve morphology and molecular integrity for biopsy-sized samples that are up to 1 cm^3. The 'no' applies to larger samples, where there is no data at present indicating that a molecular fixative can replace formalin's ability to penetrate and stabilize bulky samples. Even if such a fixative were devised, there would still be a need for diagnostic pathologists to be convinced that they could safely change from standard procedures using formalin fixation.

Until these present constraints are overcome, it is recommended that close attention is always paid to sample preparation and storage before molecular analysis.

CASE STUDY 1.1

A heparinized whole-bone marrow aspirate sample was received in the laboratory on a Monday morning with a request for T- and B-cell gene rearrangement analysis. When the date on the form and sample were checked it became clear that the sample was collected from the patient on the previous Friday morning and there was no indication that it had been stored at other than ambient temperature over the weekend. After enquiry with the clinical team looking after the patient the option of taking a repeat sample was ruled out. Accordingly, sample preparation was undertaken with special emphasis on checking for its integrity for meaningful analysis.

As a first step, the peripheral blood mononuclear cells were separated using the Lymphoprep™ method (see Section 1.3.1 and Figure 1.5). The cell concentrate was pipetted off, resuspended in PBS and stored at –20°C. DNA was extracted and purified from an aliquot of the stored sample using a suitable commercial DNA extraction kit. The concentration of the DNA was determined using a Qubit™ instrument.

As the sample had been initially left at ambient temperature for 72 hours, it was considered necessary to undertake a PCR assessment of potential DNA fragmentation using a housekeeping gene. In this instance, this was undertaken using primers for *G6PD* gene that could generate PCR products of 100, 250, and 500 bp. Following the PCR, a 2% agarose gel was run to visualize amplified products. This revealed amplified products at 100 and 250 bp only, with the former providing a stronger band than the latter.

Both T- and B-cell primers were available to amplify fragments below 250bp and these were then utilized in PCR reactions to assess for gene rearrangement. The results provided conclusive evidence of a B-cell rearrangement indicating the presence of a B-cell lymphoproliferation in the bone marrow.

Chapter summary

This chapter has provided guidelines for optimal sample preparation ahead of molecular analysis. These are given in the context of the handling of cells and tissue for subsequent protein and nucleic acid analysis.

- All samples degrade due to the effects of ischaemia leading to autolysis and, in a non-sterile environment, putrefaction. Sample integrity can also be compromised by chemical interactions, such as formalin fixation and physical handling, as can be caused by inappropriate freezing of tissue. Consequently, to avoid compromising subsequent analysis sample preparation needs to be carefully controlled.

- Pathways for sample preparation vary according to the sample type and its analytical destination. Samples such as blood and cultured cells are amenable to standardized preparation. By contrast, surgical resection samples can present a challenge due to their bulk and the non-uniform effects of ischaemia exacerbated by the subsequent influence of FFPE.

- Once samples have been stabilized via an appropriate initial handling procedure they should be assessed for quality and, as appropriate, for concentration before analysis. A variety of methods by which this may be done ahead of intact or homogenate analysis are presented.

- When samples are not to be immediately analysed or residual material is available then these must be appropriately stored. Options are given that, again, vary according to the sample type.

- By combining an understanding of the underlying principles that influence sample integrity and adopting correct preparation pathways as outlined in this chapter samples should be ready for optimal molecular analysis.

Further reading

- Shambayati B (ed.) *Cytopathology*, Fundamentals of Biomedical Science series, 2nd edn Oxford, Oxford University Press, 2018.

- Orchard G, Nation B (eds) *Histopathology*, Fundamentals of Biomedical Science series, 2nd edn. Oxford, Oxford University Press, 2018.

References

- Atkin G, Daley FM, Bourne S, Glynne-Jones R, Northover JMA, Wilson GD. The effect of surgically induced ischaemia on gene expression in a colorectal cancer xenograft model. British Journal of Cancer 2006; 94: 121–127.

● Baker AF, Dragovich T, Ihle NT, Williams R, Fenoglio-Preiser C, Powis G. Stability of phosphoprotein as a biological marker of tumor signaling. Clinical Cancer Research 2005; 11: 4338–4340.

● Bartlett JM, Starczynski J, Atkey N, et al. HER2 testing in the UK: recommendations for breast and gastric in-situ hybridisation methods. Journal of Clinical Pathology 2011; 64: 649–653.

● Chung JY, Cho H, Hewitt SM. The paraffin-embedded RNA metric (PERM) for RNA isolated from formalin-fixed, paraffin-embedded tissue. Biotechniques 2016; 60: 239–244.

● Cox ML, Schray CL, Luster CN, et al. Assessment of fixatives, fixation, and tissue processing on morphology and RNA integrity. Experimental Molecular Pathology 2006; 80: 183–191.

● Craft WF, Conway JA, Dark MJ. Comparison of histomorphology and DNA preservation produced by fixatives in the veterinary diagnostic laboratory setting. Peer Journal 2014; 2: e377.

● Do H, Dobrovic A. Sequence artifacts in DNA from formalin-fixed tissues: causes and strategies for minimization. Clinical Chemistry 2015; 61: 64–71.

● Hall JS, Taylor J, Valentine HR, et al. Enhanced stability of microRNA expression facilitates classification of FFPE tumour samples exhibiting near total mRNA degradation. British Journal of Cancer 2012; 107: 684–694.

● Hammond ME, Hayes DF, Dowsett M, et al. American Society of Clinical Oncology/College of American Pathologists guideline recommendations for immunohistochemical testing of estrogen and progesterone receptors in breast cancer (unabridged version). Archives of Pathology & Laboratory Medicine 2010; 134: 48–72.

● Howat WJ, Wilson BA. Tissue fixation and the effect of molecular fixatives on downstream staining procedures. Methods 2014; 70: 12–19.

● Jørgensen S, Baker A, Møller S, Nielsen BS. Robust one-day in situ hybridization protocol for detection of microRNAs in paraffin samples using LNA probes. Methods 2010; 52: 375–381.

● Klopfleisch R, Weiss AT, Gruber AD. Excavation of a buried treasure—DNA, mRNA, miRNA and protein analysis in formalin fixed, paraffin embedded tissues. Histology & Histopathology 2011; 26: 797–810.

● Moelans CB, ter Hoeve N, van Ginkel JW, ten Kate FJ, van Diest PJ. Formaldehyde substitute fixatives. Analysis of macroscopy, morphologic analysis, and immunohistochemical analysis. American Journal of Clinical Pathology 2011; 136: 548–556.

● Orchard GE, Shams M, Nwokie T, et al. Development of new and accurate measurement devices (TruSlice and TruSlice Digital) for use in histological dissection: an attempt to improve specimen dissection precision. British Journal of Biomedical Science 2015; 72: 140–145.

● Rakha EA, Pinder SE, Bartlett JM, et al. Updated UK Recommendations for HER2 assessment in breast cancer. Journal of Clinical Pathology 2015; 68: 93–99.

● Schroeder A, Mueller O, Stocker S, et al. The RIN: an RNA integrity number for assigning integrity values to RNA measurements. BMC Molecular Biology 2006; 7: 3.

- Shi SR, Liu C, Balgley BM, Lee C, Taylor CR. Protein extraction from formalin-fixed, paraffin-embedded tissue sections: quality evaluation by mass spectrometry. Journal of Histochemistry and Cytochemistry 2006; 54: 739–743.

- Sprung RW Jr, Brock JW, Tanksley JP, et al. Equivalence of protein inventories obtained from formalin-fixed paraffin-embedded and frozen tissue in multidimensional liquid chromatography-tandem mass spectrometry shotgun proteomic analysis. Molecular & Cellular Proteomics 2009; 8: 1988–1998.

- Strander B, Andersson-Ellström A, Milsom I, Rådberg T, Ryd W. Liquid-based cytology versus conventional Papanicolaou smear in an organized screening program: a prospective randomized study. Cancer 2007; 111: 285–291.

- Strober W. Trypan blue exclusion test of cell viability, Appendix 3. Current Protocols in Immunology 2015; 111(1): A3.B.1–A3.B.3.

- Tanca A, Uzzau S, Addis MF. Full-length protein extraction protocols and gel-based downstream applications in formalin-fixed tissue proteomics. Methods in Molecular Biology 2015; 1295: 117–134.

- Vincenti DC, Murray GI. The proteomics of formalin-fixed wax-embedded tissue. Clinical Biochemistry 2013; 46: 546–551.

- Wolff AC, Hammond ME, Hicks DG, et al. Recommendations for human epidermal growth factor receptor 2 testing in breast cancer: American Society of Clinical Oncology/College of American Pathologists clinical practice guideline update. Archives of Pathology & Laboratory Medicine 2014; 138: 241–256.

Useful websites

- For summary of cervical cytology screening programmes:
  https://healthcaredelivery.cancer.gov/icsn/cervical/screening.html

- For Western blotting:
  http://www.abcam.com/protocols/sample-preparation-for-western-blot

Discussion question

1.1 What procedures should be in place to ensure the optimal preparation of intact surgical resection tissue samples for protein analysis using immunocytochemistry?

Acknowledgements

Box 1.2, Chromosome preparations for cytogenetic analysis, was written by Emanuela Volpi, Reader in Biomedical Sciences, Life Sciences, University of Westminster, London, UK.

Sections 1.4.1.1–1.4.1.3 and Table 1.3 on methods for protein extraction were written by Abdul Hye, Senior Research Scientist at Institute of Psychiatry, Psychology and Neuroscience, King's College London, UK, and Nicholas Ashton, Post- doctoral Scientist, Department of Psychiatry and Neurochemistry at Institute of Neuroscience and Physiology, University of Gothenburg, Sweden.

Answers to the self-check questions and tips for responding to the discussion question are provided on the book's accompanying website:

 Visit: www.oup.com/uk/warford

- Pre- and post-PCR areas.
- Dedicated quiet office area.

The dedicated sample reception area should preferably be situated near to the entrance of the laboratory to promote smooth workflow. The area should be furnished with sufficient personal computers (PCs) to enable staff to enter patient details onto a laboratory information management system (LIMS). Ideally, this system should include the use of a barcode system to identify and track patients, and their tests (See Box 2.1). There should be sufficient racking to minimize sample mix-up, as well as to create an orderly system for the samples to move through to the processing laboratory for downstream procedures, such as DNA or RNA extraction, Lymphoprep TM superscript or other preprocessing steps (Figure 1.5). An appropriate checking system must be in place in the sample reception area, together with standard operating procedures (SOP) to investigate any questions relating to patient identification and/or requested tests. No sample should leave the reception area unless there is absolute confidence that there is match between the patient request form, identifiers on the sample, and that the latter is appropriate for testing.

Cross reference

More information about cell-based preparations can be found in Chapter 1, Section 1.3.1.

BOX 2.1 Laboratory information systems

Advances in information technology, together with the availability of high-performance computers and servers, have made it possible for laboratories to have LIMS that allow:

- e-requests for laboratory testing to be made;
- sample tracking from reception to completion of testing;
- monitoring of equipment with automatic 'paging', when these are not performing optimally;
- monitoring of reagent stocks and automatic ordering;
- issuing of e-reports of tests.

Increasingly, laboratories provide services across a network of hospitals and health care providers, such as GP surgeries. LIMS, as far as requesting tests and the issuing of reports, are being integrated into these.

The set-up of a LIMS is no longer a challenge in terms of availability of hardware, software, or connectivity. However, it does need considerable forward planning if it is to meet present requirements, and to be able to be future-proofed as the scope, volume, and diversity of the laboratory work increases. While the appointment of an Information Technology (IT) manager will inevitably be made and responsibilities allocated within the disciplines that make up the laboratory, this will be insufficient for a LIMS set-up. In this critical phase. The laboratory IT manager will need to work closely with colleagues within the laboratory and those external to it, who require its services. LIMS set-up can involve the use of external consultants or, when a completely new laboratory is being designed, its provision can be specified via the lead equipment/reagent provider.

The sample processing area must be a clean area and should be large enough to suit the workload without undue cluttering. Ideally, it should be directly accessible to sample reception so as to enable samples to move smoothly through. The processing area will include laminar flow hoods, centrifuges, nucleic acid extraction platforms if available, or alternatively adequate bench space to perform the manual preparation. The processing areas should include dedicated pipettes, racking systems, centrifuges, and other equipment as necessary. Adequate fridge and freezer space should also be available with distinct separation of reagents from samples.

A most important consideration is the design of the pre-PCR (clean) and post-PCR areas to minimize the risk of cross-contamination between samples. Avoidance of cross-contamination is essential as, by its very nature, the PCR turns nucleic acid sequence 'needles' into 'haystacks', so that they can be identified. There are several options that are considered good practice, and this is dependent upon availability of space and resources. The pre- and post-areas must not share any equipment, including pipettes and racks. Movement of materials, such as notepads and stationery, should be kept to an absolute minimum or, better still, avoided altogether by using PC-accessed LIMS tracking.

In summary, laboratory layout options include:

- clean area comprising sample reception, sample processing and PCR set up;
- clean rooms comprising sample reception and processing in one area with a separate room for PCR set-up;
- three rooms with one each for sample reception, sample processing, and PCR set-up;
- four rooms with one each for sample reception, sample processing, PCR reagent preparation, and finally, PCR set-up.

The area where PCR and downstream analysis is undertaken must be separated from the pre-PCR areas above. There must be no movement of material from these post-PCR preparation areas to the pre-PCR areas. Traffic must only be one way, i.e. from clean/pre- to post- areas. To help partition activities, different coloured laboratory coats can be used for each area.

The PCR and analysis area can, again, be subdivided. If multiple rooms are an option these can be configured as follows:

- Amplification area specifically housing the PCR thermocyclers.
- Detection area, such as agarose gel electrophoresis equipment, QIAxcel® (QIAGEN, Valencia, CA), UV equipment.
- Downstream assay area including, for example, pyrosequencer and other sequencers.

When nucleic acid sequencing is undertaken, then the laboratory layout should be similar to that for a dedicated PCR laboratory. Special care must be taken to prevent cross-contamination when cutting sections for nucleic acid analysis using PCR and sequencing methods—see Box 1.4. However, when protein or nucleic acid analysis is being undertaken using intact samples, there is no risk of cross-contamination and these technologies can be incorporated into a standard laboratory layout.

Cross reference

More information on intact sample analysis can be found in Chapter 3.

Key Point

Matching the patient request with the sample is always an absolute requirement for any analysis. Stringent measures need to be in place to avoid the possibility of cross-contamination of samples being analysed by PCR or sequencing methods.

Assuming no resource or space limitations, how would you set up a laboratory dedicated to PCR analysis?

2.3 Laboratory accreditation and quality management systems

To undertake diagnostic molecular work, formal accreditation is normally required. UK laboratories offering these services must demonstrate conformance to ISO 15189:2012 (https://www.iso.org/standard/56115.html) and be accredited as meeting this international standard by Clinical Pathology Accreditation (UK) Ltd (CPA), which operates as part of the United Kingdom Accreditation Service (UKAS–https://www.ukas.com/services/accreditation-services/clinical-pathology-accreditation/). In other countries, different accreditation bodies operate and different standards may be in place.

Accreditation provides formal external recognition that an organization is competent to perform specific activities or processes, in a reliable and accurate manner. Accreditation also allows laboratories to operate with confidence and ensures the validity of all work produced.

To gain and maintain accreditation, a quality management system (QMS) must be in place and be shown to be operating according to the standard as it fulfils 'customer' requirements. Two core elements define a successful QMS:

- Recognition of customer requirements and, therefore, providing customer satisfaction in terms of tangible outcomes.
- The presence of a dynamic quality management cycle (Figure 2.2) that constantly responds to customer requirements by making sure internal processes align with these.

In the context of the diagnostic molecular laboratory the 'customer' will usually be the clinician. Requests for tests may submitted from the hospital where the laboratory is situated or as referrals made from other pathology laboratories. Service agreements will be in place with individual parties requiring the laboratory tests.

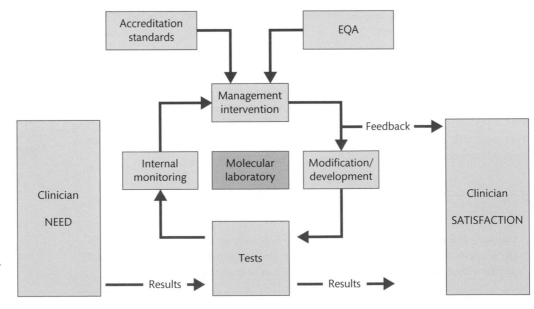

FIGURE 2.2
The quality management cycle.
EQA, external quality assurance.

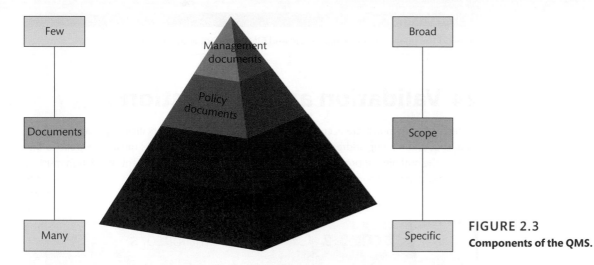

FIGURE 2.3
Components of the QMS.

Structurally, the QMS is often visualized as a hierarchical pyramid with overarching management documents at its apex, supported at its base by forms, lists, and records. Above the base sit SOPs that precisely guide all processes and, linking these with the management documents, are policy documents. The pyramid image is used as there will be relatively few management documents compared with the large number of forms, lists, and records at its base (Figure 2.3).

Management documents will cover organizational aspects of a laboratory, of which the molecular laboratory may be a component. They will emphasize the overall responsibilities, inclusive of the appointment of a Laboratory Director and Quality Manager, together with interrelationships of key personnel. These documents will also state goals and objectives, and define lines of accountability.

At the policy level, the documents will be written to cover such aspects as staff training, risk assessment, and internal audit. In these individual areas, responsibility will be clearly stated.

SOPs cover the day-to-day running of laboratory procedures from sample reception to the issuing of results. Good SOPs will focus on individual processes, and be clearly and concisely written. They must be followed without deviation, and consequently, may be formally reviewed at any time and modified within a controlled document management framework. A stable SOP will need to be reviewed at set periods of time even if it is just to make sure that any cross-references to other documents included in it are up to date.

Finally, there is the array of forms, lists and records that are at the operational end of the QMS. These are almost always linked to SOPs and provide evidence that the QMS is working, for example, in recording the temperature of a freezer on a continuous basis.

All components of the QMS must be operated within a controlled document system, the framework for which is often based on commercially available software.

Key Point

To provide diagnostic molecular services, the laboratory must be accredited by an external regulatory body. To maintain accreditation and to provide an optimal service a quality management system (QMS) must be in place and a responsive quality management cycle in operation.

2.2. SELF-CHECK QUESTION

How do the quality management cycle and QMS interact with each other?

2.4 **Validation and verification**

All tests must demonstrate accuracy, precision, sensitivity, and specificity appropriate to the clinical question being addressed (Box 2.2). Accordingly, laboratories must have policies that include the outlining of procedures for validation of tests and equipment. Validation must be performed prior to the introduction of any new laboratory procedure.

 **BOX 2.2** *Critical test indicators*

Sensitivity

The ability of a test to correctly identify individuals who have a given disease or condition.

Specificity

The ability of a test to correctly exclude individuals who do not have a given disease or condition.

Precision

The treatment of a patient with a targeted therapy based on the molecular profile of a disease.

Accuracy

The ability of a test to measure what it is meant to measure in true amounts.

Two indicators—sensitivity and specificity—are used to determine how well a test is able to distinguish between the presence or absence of a disease or condition. **Sensitivity** is the ability of a test to correctly identify individuals who have a given disease or condition and **specificity** is the ability of a test to correctly exclude individuals who do not have a given disease or condition. Data for sensitivity and specificity are usually available from the scientific publications and must be considered when adopting or refining a test.

In addition, two further indicators are used to determine the reliability of a test. These are precision and accuracy. **Precision** is the ability of repeated analyses on the same sample to give similar results. **Accuracy** is the ability of a test to measure what it is meant to measure in true amounts. Precision and accuracy should be monitored on a day-to-day basis for each test.

The following link provides a practical explanation of the use of these four critical test indicators: http://labtestsonline.org.uk/understanding/features/reliability/start/1/.

Validation may be defined as the 'confirmation through the provision of objective evidence that the requirements for a specific intended use or application have been fulfilled (doing correct test)' (International Organization for Standardization, 2005).

Once the requirement for a new test has been established, the process of implementation follows a number of steps that include:

- intended use or application;
- work plan, overall aim, requirements, and scope;
- outline methodology;
- sample numbers to test;
- parameters to be measured;
- results and conclusion, including any limitation;
- defined acceptance criteria and procedures in place for recording these.

During the validation phase, comparability measures must be performed with any new equipment to ensure consistency between platforms.

If the procedure is an adaptation or development to an existing test or technology, then assessment is more straightforward and is described as a verification process.

Verification is defined as the 'confirmation through the provision of objective evidence, that specified requirements have been fulfilled (doing test correctly)' (International Organization for Standardization, 2005).

In vitro diagnostic (IVD) reagent kits, equipment, and commercial software (including version updates) will all require verification. Examples of verification would include the sequencing of a new gene or exon in a laboratory that already undertakes validated sequencing or introduction, and the introduction of a new FISH probe, where the procedure is already performed.

For a detailed discussion of validation and verification in the context of diagnostic molecular genetic testing, please refer to guidelines from the EuroGentest Validation Group (Mattocks et al., 2010).

Key Point

Validation of equipment and reagents must be done before their use for diagnostic testing. Verification is required when either is changed to ensure that output quality is equivalent or enhanced.

2.3 SELF-CHECK QUESTION

What performance criteria must be addressed to validate a new diagnostic test?

2.5 **Best practice**

In aiming to provide a first class service and in meeting accreditation standards, the laboratory should address the 'nuts and bolts' issues of its day-to-day operation and underlying aspects that relate to its organization. In the following, individual areas for consideration have been grouped into these categories.

2.5.1 Operation

2.5.1.1 *Sample acceptance criteria*

This will include defining a minimum set of identifiers that need to be matched on the request form and the sample. The majority of laboratories require a minimum of three identifiers to be present, including:

- given name and family name;
- date of birth;
- hospital number.

Clinical details must be clearly stated on the referral form and test selection must be relevant. The condition of the sample must be acceptable and the laboratory must have procedures in place to follow up those received in a suboptimal condition, e.g. leaking or clotted samples. The laboratory must have a rejection procedure in place to deal with this situation, which includes contacting the clinician. For example, in instances where it may be difficult to collect another sample, e.g. bone marrow, then it is essential that the clinician is contacted. Reflex (repeat) testing procedures should be in place and a policy that covers outsourcing to another accredited laboratory.

2.5.1.2 Sample transfer checkpoints

Transfer of samples between tubes should be minimized, but where necessary an independent check must be performed. This can involve checking that the order and corresponding laboratory numbers are correct. It may be supplemented by a witness check during the transfer process. These records must be kept.

2.5.1.3 Data transfer

Laboratories should minimize the use of manual transcription, particularly of patient information. Barcoded tracking systems as part of a LIMS will minimize the effort required during transfer and greatly facilitate accuracy of this. All stored data must be backed up daily. This is generally provided by organization-wide IT services. Any instrument not networked must be backed up regularly, and policies need to be in place to ensure this is done and documented.

2.5.1.4 Reagents and consumables

Records must be kept of all reagents and consumables received. These should include the date of receipt, date of opening, and date finished. Unless revalidated, reagents and consumables must not be used beyond their expiry date. In practice, the time and effort required to revalidate these may outweigh the prospective financial benefit of so doing.

Acceptance testing records must be kept and facilities must be in place to store those tested separately from those not tested. Until acceptance testing has taken place, reagents should not be used for testing. Reagents and consumables must not be stored in the same place as nucleic acid test samples. A system for temperature monitoring must be in place, including a means of recording ambient (room) temperature. Many laboratories have remote monitors, which do not require daily physical recording of temperatures. Information is accessed online, and alerts sent by email and/or text message.

2.5.1.5 Equipment

All equipment used for service delivery must be calibrated, maintained according to manufacturing requirements, and documented in accordance with accreditation body requirements. Any instrumentation used in another facility, even if used infrequently, must be maintained according to the same standards. All staff must be appropriately trained, and their competency assessed before using equipment on their own. Daily/weekly and annual cleaning and maintenance records must also be recorded and kept.

Factors that should be considered when purchasing equipment are discussed in Box 2.3.

BOX 2.3 Equipment

This will inevitably be defined by the diversity of services offered by the molecular laboratory. Indeed, the type of equipment required may well shape the type of laboratory and its relationship with those who use its services. For example, consider a laboratory offering nucleic acid analysis of sample homogenates only. In this situation, equipment may include qualitative and quantitative (real time) PCR machines, together with automated sample preparation and sequencing equipment. The cost and footprint of these, together with the expertise required to run them, could be accommodated into a segregated area within an existing laboratory. However, the provision of a mass spectroscopy service will require a considerable capital and infrastructure outlay, and specialist expertise to run the equipment and ancillary items. In this situation, a stand-alone facility is required and one that, to be viable, will almost certainly need to take referrals from external institutions, and supplement the service provision with research usage.

In selecting individual equipment several questions need to be asked that will shape the decision process. Among these are:

- *Will it match present and future demand?* Forward planning for a minimum of 5 years is required.

- *Is it 'closed' or 'open' in functionality?* Open equipment allows users to choose their own reagents and protocols, and therefore, this option permits flexibility. Closed systems use reagent kits and defined protocols. These apparent restrictions are usually balanced by the provision of regulatory approval for the integrated use of equipment and reagents. Semi-closed equipment options may also be available and can offer the best of both worlds.

- *What are the reagent costs and should these be integrated into the purchase package?* This point is closely allied to the choice of an open or closed system.

- *How easy is the software to use, will upgrades be automatically supplied and how easy is it for the equipment to be integrated into the LIMS?*

- *What is the 'footprint' of the equipment?* Free-standing equipment may require reconfiguration of a laboratory, and its physical delivery may need to be considered very carefully if it is a large and heavy unit.

- *What will be the overall cost of the equipment, reagents, and servicing over a minimum of 5 years?* An ancillary question should be asked; should the equipment be bought outright or supplied under an inclusive 'pay as you go' equipment/reagent/servicing agreement?

2.5.1.6 Control material

All assays must include appropriate controls. These should be assessed during the validation phase of any new procedure (see Section 2.4), and should include the use of positive and negative assay controls. The availability of control material can be challenging. It is generally preferable to use commercially available material when possible. However, where this not possible, in-house substitutes such as previously tested patient samples are acceptable, providing appropriate consent is in place.

Cross reference

The types of controls that should be used for intact and homogenate analysis are covered in Chapters 3 and 4 respectively.

2.5.1.7 Reporting

The post-analytical process comprises the analysis and reporting of results. Reports should be clear in their presentation, checked for clerical accuracy, such as patient information, date of request, and date of reporting, together with scientific accuracy, including the use of correct nomenclature, e.g., the Human Genome Variation Society or International Society of Cytogenetic Nomenclature. Reports will also require authorization by an appropriately trained and qualified clinical scientist or pathologist. Any amendment or addendum to a report must be versioned and controlled. Reports should be issued securely and, if sent externally, the transmission route for the report must be checked to ensure confidentiality.

2.5.2 Organization

2.5.2.1 *Staff competency and training*

Work should only be undertaken by appropriately qualified and trained staff. Training would normally include a period of observation, working under supervision and, finally, the signing-off by a member of staff competent to undertake an activity on their own. Competency records need to be kept, and maintained and updated on an agreed time scale.

Implicit in this is the need for staff to have up-to-date job descriptions, training records, and provision for **continued professional development (CPD)**.

Continued professional development

A formal process of tracking and documenting the skills, knowledge and experience that gained in the workplace.

Adverse incident

An event that causes or has the potential to cause unexpected or unwanted effects involving safety.

2.5.2.2 *Risk assessment*

This covers equipment and processes with the aim of minimizing the possibility of an **adverse incident**. Risk assessments should be prospective, rather than retrospective, and the aim should be to eliminate or minimize risk.

2.5.2.3 *Audit and improvement*

All laboratories are required as part of their QMS to perform audits that monitor effectiveness of their processes. Vertical and horizontal audits will need to be undertaken. The former will trace an item through the several linked processes; the latter will investigate several items in one process. The item is often a sample and these audits rely on the availability of records for inspection.

When non-conformities are identified they must be recorded, their importance determined, and defined action taken. If there is a clinical consequence, the clinician must be informed. Allied to this is the need for corrective and preventative action plans to be in place.

2.5.2.4 *Good communication*

Good communication is essential and staff should be encouraged to make suggestions. Laboratories should hold regular meetings for all staff, as well as quality and training meetings.

2.5.2.5 *Laboratory information managements system*

It is a formal requirement that a LIMS system should be in place for accreditation to ISO 15189: 2012. This should comprise a software-driven package that will link together all the operational aspects of the laboratory, and include an information conduit for customers(s) (Box 2.1). As with QMS, commercial software is available that can be adapted for use to meet individual laboratory needs.

Key Point

An integrated approach to operation and organization is required if a molecular laboratory is to meet diagnostic needs. Both should be subject to regular review and development.

How should equipment, reagents, and consumables be confirmed as suitable for use?

2.6 External quality assessment

Participation of laboratories in external quality assessment (EQA) schemes is a fundamental requirement for laboratory accreditation.

The UK National External Quality Assessment Service (UK NEQAS, http://www.ukneqas.org.uk/) has been in operation since 1969. There are over 390 schemes operating from 26 centres in the UK. These centres are based across hospitals and research institutions, as well as universities. Services cover quantitative and qualitative tests, and also include interpretive skills, as well as clerical accuracy for some schemes. All schemes are directed by a Scheme Director and include experts in the relevant field who form an advisory group. In the UK, molecular pathology and genetics schemes operate primarily from the UK NEQAS for Molecular Genetics and Molecular Pathology, as well as the UK NEQAS for Leukocyte Immunophenotyping. Other UK NEQAS schemes applying to intact sample analysis include those for immunocytochemistry and *in situ* hybridization, cellular pathology techniques, and four schemes that involve flow cytometry. In other countries, similar EQA schemes are operational and participation in these, like UK NEQAS, is not limited by national boundaries.

Operationally, all participating laboratories are identified only by a code. This code is known only to the Scheme Director and other key personnel involved. Samples are distributed on a regular basis to laboratories with frequency varying from quarterly to twice a year. The frequency of distribution is intended to reflect the frequency of testing in laboratories. Participating laboratories are required to perform the testing as they would do for diagnostic clinical specimens and return results, including laboratory reports, for assessment. Participating laboratories receive objective reports on their results and their performance. Any laboratory where performance becomes unsatisfactory or critical is advised to seek help from the relevant NEQAS scheme and work collaboratively with them to identify any weakness within the testing strategy. Appropriate action must be taken and reported to the NEQAS scheme. Laboratories that consistently underperform can be prohibited from offering a test. The advent of **precision** medicine, using targeted therapies for a number of cancers, has given both patients and clinicians treatment options not previously available, and in these situations the time interval to demonstrate improvement in the test will, of necessity, be short.

Due to the circulation of the same materials to many laboratories, the results obtained are very useful in guiding best practice. Consequently, the results published after each assessment can be used to modify practice or to provide evidence for the adoption of a new system inclusive of reagents and equipment.

Key Point

EQA schemes provide objective evidence that tests are being performed to an acceptable standard for diagnostic use.

How can the published EQA results for individual schemes assist in promoting the performance and development of a participating laboratory?

2.7 **Future directions**

The understanding of disease has moved from the macroscopic to the microscopic, and now to the molecular. With this, there has come increasing finesse in the categorization of diseases and, with the advent of targeted molecular therapies, the availability of precise treatment for individual patients. Accordingly, in this era of precision medicine it is often essential for the patient and the health care provider to know the exact molecular status of the disease under investigation. The patient should benefit from precise treatment, while the provider can allocate limited resources more appropriately. For these reasons, the number of diagnostic molecular tests will increase and, with each, the need to combine these into a dedicated laboratory is reinforced.

Unlike the discipline-specific laboratories of haematology and clinical chemistry, for example, the sample types received into the molecular laboratory are more diverse, and the range of tests required are of greater technical variety. Perhaps the biggest challenge ahead for the molecular laboratory is the choice between one laboratory attempting to do everything and the alternative of specialization. The equipment required to run a molecular laboratory dedicated to nucleic acid analysis only is not particularly expensive nor does it require a large footprint (see Box 2.3). This could be seen as attractive to keeping all such tests under one roof. However, due to the diversity of these tests and patient impact of the analyses, a strong argument can also be made for having specialized laboratories, not only to undertake the tests, but also to refine and introduce new technologies. In support of this model are the demands of laboratory accreditation and the meeting of EQA standards. Accordingly, it is suggested that specialized laboratories may well be the way forward; indeed, these are already emerging under the appellation of 'referral laboratories'.

An exception to this potential paradigm is the need to provide precise molecular analysis in situations where there is no possibility of the construction of a molecular laboratory. This could apply to the identification of viruses and other infective agents. The outbreaks of Ebola and Zika virus epidemics point to such a need and, here, the design of molecular tests into a self-contained, self-calibrating portable device could be of considerable value.

In summary, in whichever form the testing takes place, the need for diagnostic molecular analysis is firmly established and the continued investigation of the complex cellular pathways that are perturbed in disease will only reinforce and expand this requirement.

CASE STUDY 2.1 Identification of potential PCR cross-contamination

Following a PCR assay for X, and subsequent direct sequencing, two patients were identified with the same mutation. The mutation, expected to be present in 10% of patients with this disease, appeared to be present at a lower level in one of the patients. The initial gel electrophoresis demonstrated no evidence of general contamination as, in the no-template control, no PCR product was present. A scientist was asked to investigate the likelihood that both results were genuine.

The following investigations were undertaken:

1.   The paperwork for the PCR assay was checked to establish that all was in order. This included:

a. The name of the scientist and of the 'witness check' and that both had signed off the work. Competency records of the staff involved was then undertaken to ensure that all training was up to date.

b. Expiry date and lot numbers of reagents as recorded in the complete audit trail. This was done to ensure all reagents used in the assay were up-to-date and this could be excluded as a potential source of error.

c. The order of samples on the worksheet. This was done to establish that, as far as possible, the samples were added in the order specified. This backed up the 'witness check' evidence.

d. Laboratory numbers, dates of DNA extraction, and in particular, if these were consecutive. If they were not, it would be unlikely that DNA was mixed at this stage. However, if consecutive then this would raise the possibility that the samples were contaminated during DNA extraction. In the investigation it was established that the samples were non-consecutive.

2. **The PCR assay set-up record was then checked. Note that, as the two samples in question were next to each other, this raised a suspicion that one may have become contaminated by the other. This could have occurred in a number of ways, for example, good laboratory practice was not followed, and the tube was not closed properly following addition of the first DNA, or the scientist may have inadvertently added the same DNA to both tubes.**

3. **The PCR assay was repeated, ensuring that the two samples in question were not next to each other. In this assay, a commercial control sample was included between these as a precaution against potential contamination. The control was a sample known to be wild type, i.e. with no mutations.**

4. **Following the repeat PCR, direct sequencing was undertaken. This time, the mutation was only detected in one test sample, confirming that there had been contamination between them. This result also excluded the possibility of any issues of cross-contamination at the DNA extraction stage.**

 # Chapter summary

- The need for diagnostic molecular testing is increasing in line with the understanding of cellular pathway perturbations in disease, together with the emergence of therapies that target specific elements of these.

- Options available for the physical construction of a molecular laboratory are outlined. Special emphasis is placed on the need to avoid cross-contamination of samples requiring nucleic acid analysis.

- Best practice for the operation and organization of an accredited laboratory is considered from practical viewpoints.

- The necessity for diagnostic laboratories to have formal external accreditation is noted. To obtain this the need for a dynamic QMS to be in place that is, in turn, driven by a Quality Management Cycle is emphasized.

- Validation and verification are defined in the context of ensuring that tests meet acceptable levels of specificity, sensitivity, and reproducibility.

- EQA is defined. The necessity for participation in EQA schemes and their benefit to the laboratory in improving its service is discussed.

 # References

- International Organization for Standardization: General requirements for the competence of testing and calibration laboratories. ISO/IEC 17025, 2005.

- Mattocks CJ, Morris MA, Matthijs G, et al. A standardized framework for the validation and verification of clinical molecular genetic tests. European Journal of Human Genetics 2010; 18: 1276–1288.

 # Useful websites

- Critical test indicators: http://labtestsonline.org.uk/understanding/features/reliability/start/1/.

- ISO 15189:2012 Medical laboratories—requirements for quality and competence: https://www.iso.org/standard/56115.html.

- United Kingdom Accreditation Service (UKAS): https://www.ukas.com/services/accreditation-services/clinical-pathology-accreditation/.

- UK National External Quality assessment service: UK NEQAS: http://www.ukneqas.org.uk/.

 # Discussion question

2.1 As a member of the scientific team delivering a molecular-based diagnostic laboratory service you have been asked to introduce a new test. How would you approach this challenge in the context of meeting relevant standards?

Answers to the self-check questions and tips for responding to the discussion question are provided on the book's accompanying website:

◯ Visit: www.oup.com/uk/warford

3

Intact Sample Analysis

Anthony Warford and Emanuela V. Volpi

Learning objectives

This chapter will provide the reader with:

- An understanding of the technologies used for the analysis of nucleic acids and proteins in intact samples, either as single cells or tissues.

- Guidance on the validation of antibodies for protein analysis and nucleic acid probes for the visualization of DNA and RNA.

- Principles and key technical considerations of immunocytochemistry and flow cytometry for the demonstration of proteins.

- Principles and key technical considerations of *in situ* hybridization for the demonstration of nucleic acids.

3.1 **Introduction**

This chapter explains the principles and key technical considerations for the successful demonstration of proteins and nucleic acids in intact cell and tissue samples. It links back to Chapter 1, where optimal sample preparation was covered, and is complementary to Chapter 4, in which technologies associated with the analysis of proteins and nucleic acids in homogenized samples are considered.

The technologies described here of **immunocytochemistry (ICC)**, **flow cytometry (FC)** and **in situ hybridization (ISH)** allow for the specific identification of proteins and nucleic acids, either for semi-quantitative microscopic analysis—ICC and ISH—or, in the case of FC, for the quantitative assessment and sorting of cells. As such they provide a means of precise identification of cell types in normal and pathological samples, and are used either as primary tools or as 'follow-ons' after initial cytological or morphological staining. As will become evident in Part 2 of this book each of these technologies, used either individually or in combination, provide valuable diagnostic information. It should also be recognized that these techniques also make a valuable contribution to furthering research into the understanding of disease processes.

Immunocytochemistry, also known as *immunohistochemistry*, An analytical method employing the use of antibodies that bind to antigens in intact cell and tissue preparations and are then visualized to demonstrate the location and distribution of proteins using a microscope.

***In situ* hybridization** An analytical method for the visualization of nucleic acids in intact cell and tissue preparations. It is based on the hybridization of labelled nucleic acids to the target sequence in the preparation. Visualization of the label allows the target sequence to be observed using a microscope.

3.2 **Technologies**

For the microscopic localization of proteins within and on the surface of intact single cells, and in sections of tissue samples immunocytochemistry is used. Antibodies are used to bind to target molecules and, often, elaborate detection methods are applied to provide for the microscopic visualization of this interaction. Strictly speaking, the title 'immunocytochemistry' should be applied when the technology is used with cell preparations. Again, applying strict terminology, the word **immunohistochemistry** should be used to describe the same technology when applied to tissue sections. In reality, this strict separation is seldom applied and a literature search will reveal that about equal numbers of publications use either term, irrespective of the exact nature of the sample. In this chapter, to avoid confusion, immunocytochemistry (ICC) will be used throughout.

Another prominent technique that similarly to ICC employs antibodies is flow cytometry (FC). With this technology, labelled antibodies directed against cell surface antigens are used to categorize intact single cells in suspension or flow. In contrast to ICC, FC does not normally entail microscopic visualization and in a further point of distinction the FC 'read out' is quantitative, rather than semi-quantitative. The technology may be further refined by sorting the labelled cells to enrich populations for further analysis.

For the localization of nucleic acids in intact cells—either single cells or tissue samples—the technique of *in situ hybridization* (ISH) is applied. In this technology, labelled nucleic acid probes hybridize by complementary base pairing to the nucleic acid sequences to be demonstrated. After this, the label is visualized to allow for microscopic analysis. The final 'read out' can be semi-quantitative or quantitative. ISH can be applied to demonstrate chromosomal territories and DNA gene sequences within nuclei together with the products of gene expression, mRNA, and microRNA, in the cytoplasm.

Key Point

Technologies for the demonstration of proteins and nucleic acids in intact cell and tissue samples include immunocytochemistry, flow cytometry, and in situ hybridization.

3.1 SELF-CHECK QUESTION

What are the similarities and distinctions between the techniques of immunocytochemistry, flow cytometry, and *in situ* hybridization?

3.3 **Immunocytochemistry**

3.3.1 Development

It is difficult to precisely determine a time point for the first description for ICC, but certainly by the early 1940s the concept of using a labelled antibody to demonstrate a protein in a cellular preparation had been translated into practice (Coons et al., 1941). The technique was gradually applied in research, but it was about two decades later that it was first adopted for diagnostic use for the demonstration of auto-antibodies using frozen section preparations. The difficulty of applying ICC to formalin-fixed paraffin embedded (FFPE) tissue sections was described in the early 1960s (Sainte-Marie, 1962) and this held back the application of the technology for over a decade. The breakthrough came when it became clear that the majority of epitopes—specific parts of the antigens to which the antibodies bind—were masked, rather than destroyed, through the cross-linking effects of formalin fixation in the FFPE process. At this time, it was also shown that unmasking could be achieved using controlled proteolytic digestion prior to application of the primary antibody (Huang et al., 1976). In the same decade, monoclonal antibodies were introduced offering greater specificity than polyclonal antibodies. Together, these developments heralded the widespread introduction of ICC for use in diagnostic cellular pathology. Later developments have seen the introduction of heat-induced epitope retrieval (HIER; Shi et al., 1991) and a plethora of sensitive methods to detect the binding of the primary antibody with the target antigen. Today, ICC is routinely used to confirm or facilitate cytological and morphological diagnosis, and in several instances as a companion diagnostic to guide targeted therapy in the area of precision medicine.

3.3.2 Antibody validation

The diagnostic utility of ICC pivots on the specificity and sensitivity (affinity) of the primary antibody interaction with the antigen it is to demonstrate. That the use of a primary antibody, coupled with a suitably sensitive ICC method, gives an intense reaction that is free from background reaction—a high signal/noise ratio result—does not in itself mean that the primary antibody is specific for an antigen. The fundamental reason for this is that an antibody simply recognizes a small three-dimensional 'cloud' of amino acids (the epitope) and, if this shape is shared by more than one protein, it will demonstrate those antigens as well. Even when a partial shape similarity occurs, a primary antibody may also interact with other proteins. When such events occur, this is known as cross-reactivity. Furthermore, due to the milieu of proteins present in the matrix of cells and tissues, even the localization of staining to the apparently correct cell type or extracellular substance does not guarantee a specific interaction of the primary antibody with a target antigen. Behind these potential pitfalls is the process by which antibodies are produced.

Polyclonal antibodies are produced by harnessing the natural immune response mechanisms that result in antibody production when a foreign antigen is introduced into an animal. The output is a mixture of antibodies that interact with all epitopes present on the antigen. When monoclonal antibodies are produced, then each one interacts with only one epitope, but if this is shared by more than one protein then cross-reactivity will occur. Taken together, these variables lead to the necessity for the specificity and sensitivity of each primary antibody to be assured (Bordeaux et al., 2010). The process by which this is done is called antibody validation.

For many diagnostic antibodies and ICC kits, validation will have been done by the commercial supplier; users should look out for 'approved for diagnostic use' and, in the case of companion diagnostics, check that the product is FDA approved (http://www.fda.gov/medicaldevices/productsandmedicalprocedures/invitrodiagnostics/ucm301431.htm).   However, when antibodies are sold for 'research use only' then they cannot be assured as suitable for diagnostic use. In these situations, guidance is available for antibody validation. This can be very complex (Smith and Womack, 2014) or tiered, and relatively straightforward (Howat et al., 2014). The key steps for the latter are outlined in Box 3.1.

BOX 3.1 Steps for antibody validation

The following is based on Howat et al. (2014). For full consideration of antibody validation please refer to the paper.

1. Undertake a full literature review to establish the expected distribution of the target antigen in the cell/tissue type that is to be demonstrated.

2. Choose the most appropriate tier of validation:

   a. Well-known antibody with high quality literature evidence. The requirement is to reproduce the same staining pattern.

   b. Well-known antibody used in a non-validated cell/tissue preparation. Run similar 'validated' preparations alongside the new cell/tissue preparations and ask: 'is the new staining pattern sensible?'

   c. Unknown antibody with inconsistent or no literature evidence of ICC staining. Test antibody in at least one other non-ICC method, for example, by Western blotting. Assessment can include ICC undertaken with 'knock in' and 'knock down' cell lines.

3. Identify cell lines and tissues to serve as positive and negative controls.

4. Choose an appropriately sensitive ICC method and use a commercially validated kit if possible.

5. Test multiple epitope retrieval conditions to optimize the staining conditions.

Key Point

Antibodies for use in diagnostic ICC must be validated for specificity and sensitivity.

3.2 SELF-CHECK QUESTION

What factors will affect the specificity of a primary antibody?

3.3.3 Sample preparation

Cross reference

Please refer to Chapter 1 for full details on sample preparation.

In summary, ischaemic time should be kept to a minimum or at least standardized. Fixation follows to halt cell/tissue degradation due to autolysis and/or putrefaction. This 'one-way' process will stabilize proteins, and may also fix nucleic acids, and conserve carbohydrates and lipids. Fixation for cell preparations is usually undertaken using an alcohol-based protein precipitating fixative. This type of fixation or brief formaldehyde (cross-linking) fixation is used for frozen sections before ICC. Neutral buffered formalin is most commonly used to fix tissue samples prior to paraffin wax embedding.

When sections are cut for ICC these should normally be 2–4 μm thick and mounted on adhesive-coated slides to ensure retention during the staining procedure. Suitable adhesives include poly-L-lysine and silane. Alternatively, positively charged coated slides are commercially available. There has been some debate over the course of development of ICC technique as to whether the availability of antigens for demonstration is adversely affected by the storage of FFPE sections at ambient temperature before staining. It would seem that some antigens, such as CD15, are affected even after only a few days' storage, but no general rule can be suggested. A prudent approach is to standardize the interval between section cutting and staining, as is recommended for the **predictive** use of ICC for the demonstration of HER2 expression in cases of invasive breast cancer samples (Wolff et al., 2014; Rakha et al., 2015).

3.3.4 ICC procedure

In order, the key steps of an ICC procedure are:

1. epitope retrieval when formalin-fixed preparations are used;

2. blocking non-specific staining;

3. primary antibody incubation;

4. detection of antibody/antigen interaction. The procedure is outlined in Figure 3.1 and discussed below.

Prior to the ICC staining of FFPE sections the paraffin wax must be completely removed and the sections rehydrated to water. This is accomplished by sequential immersion of slides in a wax solvent, such as xylene, followed by immersion in alcohol before immersion in water. Cell and frozen section preparations should be fixed, if not already done so, just before staining. When using protein precipitating fixatives, slides are often dried after fixation. In contrast, when cross-linking fixation is used, the preparations are usually immersed in water or a slightly alkaline pH buffer (TRIS-buffered or phosphate-buffered saline) before staining. Once the ICC procedure has commenced, it is important to keep preparations wet at all times to avoid background caused by non-specific adherence of reagents to the preparations.

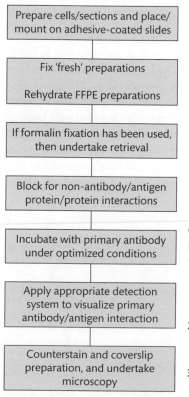

FIGURE 3.1

Outline of immunocytochemistry procedural steps.
Notes
1. Blocking steps to suppress the demonstration of endogenous components that would otherwise be demonstrated in the ICC procedure should be inserted into the procedure as appropriate.
2. Between each step slides should be washed in buffer. An exception is after the protein/protein blocking step.
3. To avoid background staining preparations should not be allowed to dry during the procedure.

3.3.4.1 Epitope retrieval

When ICC is undertaken on FFPE preparations, epitope retrieval is almost always required, with exceptions being noted amongst the peptide hormone family. Even when formalin is briefly applied to cell or frozen section preparations epitope masking will occur. Consequently, the application of appropriate and carefully controlled epitope retrieval methods is required.

Proteolytic-induced epitope retrieval (PIER) was introduced by Huang et al. (1976) who used trypsin to enhance ICC sensitivity. Pepsin and proteinase K may be used as alternative enzymes for this purpose. The concept behind the use of these enzymes is that they expose epitopes that have become hidden by the additive and cross-linking effects of formalin fixation. PIER needs to be carefully controlled, as under-digestion will not optimally expose epitopes on the target antigen and over-digestion may liberate the antigen from the formalin-fixed protein matrix, and can destroy cell or tissue integrity. As a general rule, the shorter the formalin fixation time, the milder the conditions for PIER need to be. Controlled PIER is often used for cell and fixed frozen tissue preparations, but for FFPE sections it has been largely replaced by HIER.

HIER involves the super-heating of tissue sections and was first described by Shi et al. in 1991. The underlying mechanism by which epitopes are retrieved has been subject to some investigations (Morgan et al., 1997; Sompuram et al., 2004), but it is not fully understood how this is achieved. Remarkably, destruction of tissue morphology is rare and standard treatment times can often be used irrespective of the duration of fixation time, accommodating a range from hours to months. This contrasts with PIER, where formalin fixation beyond a few days may render epitope retrieval using this method ineffective. Finally, in comparison with proteolytic digestion, HIER has been shown to be effective in revealing epitopes on a far greater range of antigen targets (Shi et al., 2011). For a few antigens, optimal demonstration may only be obtained when HIER and PIER are combined (D'Amico et al., 2009).

In its initial description, HIER employed microwave heating with slides immersed in buffer solutions that included metal ions (Shi et al., 1991). When using microwaves, the power output and consistency of heat distribution over the slides needs to be considered. The use of pressure cookers (Norton et al., 1994) as an alternative equipment option will often allow a greater number of slides to be treated at any one time, together with more consistent slide heating. Metal ions are no longer incorporated in HIER buffers and the use of two buffer solutions predominates. These buffers are 10 mmol/L citrate at pH 6 and 10 mmol/L TRIS, 1 mmol/L EDTA at pH 9. Often the data sheet supplied with a primary antibody will state which one to use, but when this information is not available then the choice should be determined empirically beforehand.

3.3.4.2 Blocking non-specific reactions

This falls into two main categories:

- Suppression of antibody interaction caused by non-epitope driven protein–protein interactions.
- Blocking of reaction of endogenous substances present in the preparations that could otherwise be detected by the ICC procedure.

Examples of blocking methods and technical notes are provided in Table 3.1.

TABLE 3.1 Examples of blocking methods used in immunocytochemistry

Blocking method	Application	Insertion into ICC method	Comment
0.3% hydrogen peroxide in methanol	Blocks endogenous peroxidase activity	Prior to or post-epitope retrieval	Can destroy antigens in cell and frozen section preparations
1 mM levamisole	Blocks endogenous alkaline phosphatase activity	Incorporated into the enzyme substrate solution	Will not suppress intestinal alkaline phosphatase activity
Avidin saturation of endogenous biotin	Blocks endogenous binding activity prevalent in liver and kidney	After epitope retrieval, before application of primary antibody	Required only when avidin/biotin detection systems are used
Serum, immunoglobulins, and/or albumin	Reduces non-epitope protein–protein interactions of primary and detection antibodies	Usually immediately before application of the primary antibody. Can also be used as components of antibody diluents	Care necessary to ensure blocking proteins do not interact with detection systems
0.1% Tween or 0.1% Triton X 100	Reduces non-epitope protein–protein interactions of primary and detection antibodies	Used in detection systems as constituent of reagent diluent and wash buffers	Can reduce sensitivity of ICC detection

Adapted from Warford et al., (2014).

3.3.4.3 Primary antibody incubation

During validation of a primary antibody, the optimal conditions for its incubation with a cell and/or tissue preparation should have been determined. It is important to keep in mind that, if the preparation is changed or any of the aspects of the ICC procedure are altered, then these conditions may need to be re-optimized. Furthermore, when a primary antibody is sourced from an alternative supplier or there is a great number of changes in the supply of an existing antibody, then ICC results should be monitored carefully and, if necessary, adjusted to ensure that optimal staining is maintained.

Cross reference

See Chapter 2, Section 2.4 for details of validation and verification.

For diagnostic applications, ready-to-use primary antibodies are often available. They may form part of an ICC kit or be stand-alone reagents. Their main advantage is that they are easy to use and should give optimal results, providing the data sheet instructions are followed precisely.

The most common mode of manual application of a primary antibody is to dilute it in a blocking solution; to add it after tipping off, but not washing off the latter; and then incubating at ambient temperature for 30–60 minutes. Incubation at 37°C for less than 30 minutes may be used in closed-system automated immunostainers, while overnight incubation at 4°C can be used in manual methods to ensure that the antibody/antigen reaction is driven to completion. The latter can be especially useful when the target antigen is present in low abundance or if, of necessity, the antibody is of low affinity.

3.3.4.4 Detection systems

The endpoint of an ICC procedure is the microscopic examination of the interaction of the primary antibody with the relevant epitope. This interaction needs to be precisely localized and 'coloured'. On occasions that the primary antibody has been labelled with a fluorescent compound or active enzyme, then this can be readily undertaken. This is known as direct detection. However, in most instances the primary antibody is applied unlabelled and, therefore, a detection system is required to produce the necessary visualization of the interaction with the antigen. In this situation, the many and various detection systems that are available are grouped together under the heading of indirect detection. Although indirect detection systems introduce complexity into the ICC procedure, they also provide the means to amplify the initial primary antibody/antigen reaction and introduce a diversity of endpoints suited for examination. Examples of detection systems are provided in Table 3.2.

TABLE 3.2 Alternative immunocytochemical detection methods

Sensitivity	Method	Comment	Reference
Low	*Direct*: The primary antibody is labelled	Simplest method, but each antibody must be labelled. In all other methods the primary antibody is unlabelled	
	Indirect: The primary antibody is unlabelled. It is visualized using a labelled secondary antibody to the species of the primary antibody	It is suggested that between 7 and 10 secondary antibodies can interact with one primary antibody, therefore providing some amplification of staining	
Medium	*Peroxidase anti-peroxidase (PAP)*: Three-step method with last step being a peroxidase antibody complex	Due to the use of the peroxidase antibody complex and saturated interaction of the linking secondary antibody, three-stage amplification is achieved without the possibility of non-specific antibody interactions with the preparation	Sternberger et al. (1970)
	Alkaline phosphatase anti alkaline phosphatase (APAAP)	As for PAP, but using alkaline phosphatase. Often used for blood cell ICC.	Mason and Sammons (1978)
High	*Avidin/biotin complex (ABC)*: The primary antibody is detected with a biotin-labelled antibody that then interacts with an avidin/biotin complex containing many labels	The endogenous presence of biotin in liver and kidney, and to a lesser extent, in other tissues needs to be suppressed using special blocking procedures	Hsu et al. (1981)
	Labelled polymer: Antibodies reactive with the species of the primary antibody and labels are attached to a polymer backbone	A simpler method in comparison with ABC and not requiring special blocking procedures. Often used in automated ICC kits	Sabattini et al. (1998)
Ultra	*Tyramide amplification*: Relies on the tyramide catalysed deposition of a label via the action of tyramide with horseradish peroxidase	In kit form, whether manual or automated, is reliable	Bobrow et al. (1992)
	Rolling circle amplification: Relies on the interaction of two oligonucleotide-labelled antibodies. Once the oligonucleotides are ligated a circularized amplification product is generated using a polymerase	Complex manual procedure. Can allow two antigens that are in close proximity to each other to be demonstrated simultaneously thus showing their biological association	Gusev et al. (2001); Soderberg et al. (2008)

Two alternative endpoints are possible with light microscopic ICC. These are fluorescent and chromogenic. The former provides 'bright lights' on a dark background and—given the ability of different fluorophores used for fluorescent labelling to emit light of different wavelengths—these are especially useful for the simultaneous localization of multiple antigens in a preparation or multiplexing (Figure 3.2). For observation, a fluorescent microscope is required. In contrast, chromogenic endpoints allow, after suitable counterstaining, for the cytology and morphology of ICC to be examined simultaneously. This is crucial for diagnostic assessment of cancer samples, where the tumour cells need to be precisely identified for interpretation of staining (Figure 3.3). Most chromogenic methods are based on the use of active enzymes that catalyse substrates to produce insoluble precipitates in the vicinity of the antibody/antigen reaction. The main distinguishing features of the most commonly used fluorescent and chromogenic labels are provided in Table 3.3.

3.3.4.5 Counterstaining and mounting

These represent the last stages of an ICC procedure. Counterstaining is important to allow cell and tissue features to be readily distinguished. It usually is accomplished using a nuclear counterstain; 4',6-diamidino-2-phenylindole (DAPI) for fluorescent systems and haematoxylin for the horseradish peroxidase/diaminobenzidene chromogenic system.

FIGURE 3.2

Dual label indirect immunofluorescence. Mouse small intestine demonstrating laminin in red (Sigma L9393 with AF647 visualization) in the basal lamina, capillaries, and underlying submucosa, together with proliferating cells in the villi in green (SP6—Neomarkers RM-9106 with AF488 visualization). The blue nuclei are DAPI-counterstained.

Image provided by Dr Will Howat, Wellcome Trust Sanger Institute.

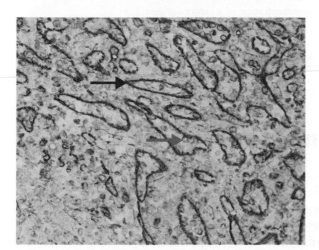

FIGURE 3.3

Chromogenic immunocytochemistry. Vimentin demonstration of sinusoids (red arrow) and lymphocytes (green arrow) in the red pulp of human spleen. Horseradish peroxidase polymer-based detection with diaminobenzene substrate visualization, together with haematoxylin counterstaining of nuclei.

TABLE 3.3 Commonly used fluorescent and chromogenic labels

Type	Colour	Permanence	Comment
Fluorescent			
Fluorescein-isothiocyanate (FITC)	Green	Non-permanent	Quenched by repeated excitation and emission of fluorophores
Rhodamine	Red	Non-permanent	
Cyan dyes	Various, according to dye type	Semi-permanent	Good for multiple detection of targets
Quantum dots	Various, depending on diameter of dot	Permanent	Not quenched under fluorescent examination
Chromogenic			
Horseradish peroxidase Diaminobenzidene (HRP/DAB)	Brown	Permanent with resinous mounting	Short reaction time (10 min). Most commonly used label for ICC
Horseradish peroxidase/ 3-Amino-9-ethylcarbazole	Red	Permanent if aqueously mounted	Better localization than HRP/DAB
Alkaline phosphatase/nitroblue tetrazolium, 5-bromo-4- chloro-3-indolyl phosphate (NBT-BCIP)	Blue-black	Permanent if aqueously mounted	Can be cycled for over 24 hours to increase sensitivity. Commonly used label for ISH
Alkaline phosphatase/fast red	Red	Permanent with resinous mounting	Short reaction time (1 hour)

Ideally, the completed ICC preparation should be permanent. For some endpoints this is possible by dehydrating the counterstained preparation in alcohol, immersing in xylene (or equivalent solvent), then mounting under a coverslip using a resinous mount with a refractive index near that of glass. Others require aqueous mounting, resulting in a lower refractive index.

Key Point

The ICC procedure is a multistep process, in which each step needs to be optimized to achieve specific and sensitive demonstration of a target antigen.

3.3 SELF-CHECK QUESTION

What options are available for epitope retrieval for preparations that have been formalin fixed?

3.4 Flow cytometry

3.4.1 Principles

Similarly to ICC, flow cytometry can be applied for the detection of macromolecules within intact cells by means of antigen recognition-based techniques. However, in distinction to ICC, FC analysis is carried out on cells in liquid suspension (in *flow*) and no microscopic visualization of tissue sections is involved. Tissue architecture is lost as cells from solid tissues require disaggregation before they can be analysed by FC.

FC entails the use of fluorescent antibodies for the recognition and labelling of proteinaceous cellular components and is amenable to 'multiplexing' or the concurrent detection of different targets through the simultaneous use of multiple, spectrally compatible fluorescent labels. FC can also be applied to assess ploidy levels, and for studies of cell cycle and gene expression by using fluorescent affinity dyes for DNA and RNA detection (Ashraf and Hisham, 2011; Hanley et al., 2013). As an analytical approach, the advantage proffered by FC versus traditional microscopy-based techniques resides with its high-throughput nature, ensuing from the elevated acquisition rate in combination with the fluorescence sensitivity.

FC analysis is undertaken by using a flow cytometer, which is essentially comprised of a fluidic system to transport the fluorescently labelled cells in a thin stream through a laser beam, coupled with an optics-to-electronics system consisting of a set of filters and sensitive detectors (photomultipliers) to record the light scattering and fluorescence, and transmit and process the data (see Figure 3.4).

Following laser excitation of the labelled cells in the fluid stream, emitted light can be precisely measured from a single observation point on a cell-by-cell basis. Emitted fluorescence of different wavelengths is indicative of which of the different antibodies are attached to the cells and, in addition, light *scattering* or deflection is used as an indication of the size and granularity of the cells. By measuring the intensity of a fluorescent signal or multiple fluorescent signals within tens of thousands of cells per sample, FC generates large datasets amenable to statistical analysis, and allows for accurate, quantitative assessment of multiple cellular properties, including intracellular complexity and intercellular heterogeneity.

FC provides an example of a powerful analytical tool that has been successfully translated from research onto the clinical diagnostic scenario. As a technology, the principle of flow cytometry was introduced in the 1950s by Wallace Coulter and a prototypical form of the technique was initially applied to assess the number and size of cells. FC as a method for counting and evaluating cells has now become an important tool for the diagnosis, **prognosis** and monitoring of

Prognosis

The probable outcome of a disease. This is based on clinical presentation, surgical intervention, precise diagnosis and monitoring of the disease.

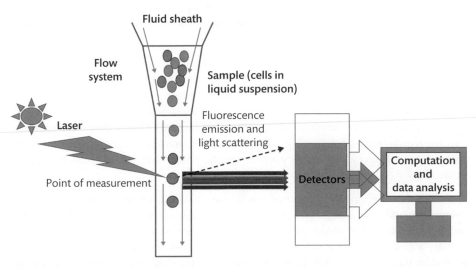

FIGURE 3.4
Schematic representation of a flow cytometer design and function.

Immunophenotyping
Determination of cell or tissue type using antibodies as markers.

haematological and immunological disorders. **Immunophenotyping** by FC or the process of classification of immune cells on the basis of cluster of differentiation (CD) antigens present on their plasma membranes allows the identification of cellular subpopulations, and can be used to help diagnose and classify haematological cancers, and monitor some forms of immunodeficiencies and autoimmune disease. For a contemporary, comprehensive review of principles and applications of FC please refer to Adan et al. (2017).

Key Point

Flow cytometry relies on the use of fluorescent antibodies or affinity dyes for the detection of macromolecules (proteins and nucleic acids) within intact cells in suspension or in *flow*.

3.4 SELF-CHECK QUESTION

What are the key functional components of a flow cytometer?

3.4.2 Procedure

The successful design of a FC test presupposes an understanding of the biology of the system under investigation, as well as knowledge of the physical principles governing the methodological approach.

A key aspect of the sample preparation process is the preservation of the native cellular characteristics as artefacts can be inadvertently elicited during sample manipulation and potentially affect the analytical outcomes (Filby, 2016). When preparing suspensions of cells from biopsies by tissue enzymatic digestion or mechanical disassociation the principal risk is the stripping of surface receptors or other determining cellular components. On the other hand, with peripheral blood, the main technical challenge derives from the preponderance of red blood cells over white blood cells, and the artefacts that can be potentially introduced during the process of cell separation. Common approaches to cell separation from whole blood rely on the addition of red blood cells lysing agents to the cell suspension or—alternatively—centrifugation of unlysed blood on ready-to-use density gradient centrifugation media (see Figure 1.5). In fact, both separation approaches can compromise cell viability and retention, and require optimization and analytical adjustments.

Designing an effective and informative panel of antibodies is also fundamental for the success of a FC experiment. Given the criticality of this step, and the complexity of the technical and commercial landscape, a number of online resources have been created for the purpose of assisting with the antibody panel design, for example, Chromocyte (http://www.chromocyte.com) and Fluorofinder (https://fluorofinder.com). From a diagnostic perspective, the requirement is determining combinations of FC markers for screening approaches that balance efficiency and accuracy within specific clinical settings (Williams-Voorbeijtel et al., 2017).

As in ICC, the immunostaining method can follow either a direct route (use of fluorescent primary antibodies in a single-step staining procedure) or indirect route (use of fluorescent secondary antibodies to detect primary antibodies). Panels normally consist of two or three primary antibodies (and—for indirect immunostaining—a set of species-of-origin-compatible secondary antibodies). For multi-parametric analysis, antibodies will be expected to be conjugated to spectrally distinguishable fluorochromes (e.g. green fluorescein isothiocyanate (FITC), yellow phycoerythrin (PE), and red allophycocyanin (APC)), although initially the choice of fluorochromes is normally dictated by the laser availability and configuration of the cytometer to be used for the sample analysis. Within each experiment, control tubes are typically set up to test individual fluorescent dyes, and a negative or 'no-antibody' control is also normally included. The latter type of control is essential to exclude the presence of *autofluorescence*—naturally occurring fluorescence in unlabelled cells—that could affect the emission readings.

During sample processing for FC of whole blood, monoclonal antibodies aimed at cell surface antigens and/or other cellular markers are added to the heparinized blood (100 µl) at the required concentration, either as recommended by the manufacturer or as determined by prior optimization through titration. The samples are then briefly incubated in the dark at room temperature. The erythrocyte lysis buffer—consisting of a mix of ammonium chloride to selectively lyse red blood cells and formaldehyde for chemical cross-linking of proteins—is then added to the samples. After a couple of cycles of centrifugation and washing in phosphate buffered saline, the pellet of each sample is finally resuspended in buffer at the final concentration of 1×10^7 cells/ml and ready to be analysed.

Fluidic systems of FCs are designed to ensure that cells are delivered to the laser in a very thin fluid stream. Samples are injected into the stream of sheath fluid (water or buffer) passing through the fluidic chamber. The sample stream is narrowed to a diameter of a few microns, which results in the cells passing through the laser beam one at a time, ensuring fluorescence and light scattering measurements on a cell-by-cell basis.

Key Point

Key procedural aspects in flow cytometry are the preservation of native cellular characteristics during sample preparation and the selection of an effective and informative set of antibodies.

3.5 SELF-CHECK QUESTION

What is the main technical challenge encountered when undertaking flow cytometry of whole blood?

3.4.3 Data analysis

FC allows the rapid analysis of large populations of cells, with a typical analytical throughput of 5000 cells per second. The combination of high acquisition rates with fluorescence sensitivity (less than 100 molecules per cell) means that even rare cell subpopulations (e.g. less than 1 cell in 10,000) can be detected in statistically significant numbers in a reasonable period of time.

In addition to fluorescence emission, most instruments can measure light scattered by the cells at right angles to the laser beam (side scatter, SS) and light scattered in a forward direction (forward scatter, FS). The amount of light scattered is affected by the size, shape, and optical homogeneity of the cells, with the magnitude of FS being proportional to the size of the cell, while the magnitude of SS bis proportional to granularity and structural complexity in the cell.

Data collection consists of numerical values obtained by the conversion of light pulse into voltage pulse and, finally, numerical values. The processed data are usually rendered as single parameter histograms or graphs, or as two-dimensional plots of correlated parameters, often presented as a cytogram or dot plot, wherein each cell is recorded as a single dot (see Figure 3.5).

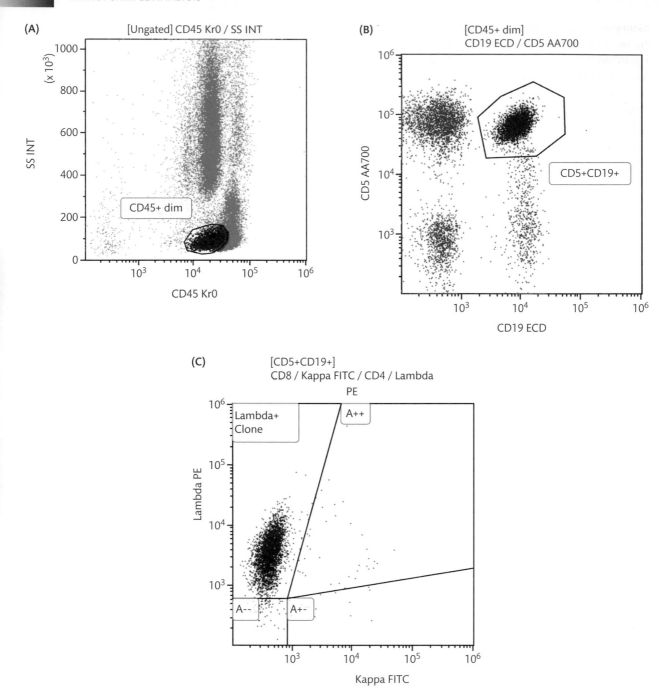

FIGURE 3.5

Flow cytometry plots from a case of chronic lymphocytic leukaemia (CLL). (A) An abnormal lymphocyte population isolated by side scatter (SS) with diminished CD45+ intensity (dim CD45+), compared with normal CD45+ lymphocytes. This feature may be seen in lymphoid leukaemia. (B) Gated abnormal cells with CD5+CD19+ phenotype. (C) Gated from the CD5+/19+ cells shows lambda immunoglobulin monoclonality indicating a B-cell CLL. For further information on CLL please refer to Chapter 8.
Plots kindly provided by Stephen Darwood, Applications Specialist Flow Cytometry, Beckman Coulter.

'Gating' in FC analysis is the process of defining a selected region of a single or a two-dimensional cytogram. Only cells within the 'gate' can progress to the next stage of analysis. Gating is normally based on the use of antibodies against highly expressed antigens, which normally define distinct cellular subpopulations of interest.

Key Point

By measuring the intensity of a fluorescent signal or multiple fluorescent signals within tens of thousands of cells per sample, flow cytometry creates large, computable datasets, which allow for accurate quantitative assessment.

3.6 SELF-CHECK QUESTION

What affects fluorescence intensity and light scattering?

3.4.4 Fluorescence-activated cell sorting (FACS)

Some FCs have a 'sorting' feature in the electronics system that can be used to deflect particles so that certain cell populations can be physically separated or sorted for further analysis. This specific application is called Fluorescence Activated Cell Sorting (FACS) analysis. Although the acronym FACS is a trademark, the term is now generically used to indicate this specialized type of FC.

The physical separation of the interrogated and identified cells happens through a vibration process and the ensuing breaking of the cell suspension flow followed by the formation of thousands of individual droplets, a fraction of which contain a cell. Those droplets that contain a cell of interest (as detected by the presence of a certain fluorescence) are electrically charged as they pass into an electrical field. The charged droplets are then diverted through an electrostatic deflecting system into a final receptacle. The applications of cell sorting include protein engineering and disease characterization through nucleic acid extraction, protein expression, and study of cellular functions (Abraham and Maliekal, 2017).

Key Point

Fluorescence-activated cell sorting is a specialized form of flow cytometry that allows physical separation of cell sub-populations.

3.7 SELF-CHECK QUESTION

How does the physical separation of cellular sub-populations happen?

3.5 *In situ* hybridization

3.5.1 Development

In intact samples, either single cells or tissue samples, the technique of ISH is applied to localize nucleic acids. In this technology, labelled nucleic acid probes hybridize by complementary base pairing to the nucleic acid sequences to be demonstrated. After this, the label is visualized by light microscopy. The final 'read out' can be semi-quantitative or quantitative.

The technique was first described by Gall and Pardue (1969) and John et al., also in 1969, and, initially, only applied for research. In these descriptions, and those that followed throughout the 1970s, the nucleic acid probes were labelled radioactively. Tritium (^{3}H) labelled probes gave fine resolution and sensitivity, but required prolonged auto-radiographic exposure (weeks and sometimes months). ^{32}P-labelled probes gave results in 24–48 hours, but with poor resolution, while ^{35}S probes provided results within a few days with the deposits of silver grains being closely localized to the target nucleic acid sequences. A significant breakthrough was described by Langer et al. (1981), who labelled nucleic acid probes with biotin, a water-soluble vitamin also called vitamin B7. Using such probes, it now became possible to employ chromogenic detection systems, as used in ICC, to demonstrate target gene sequences within 1 or 2 days with precise localization and associated morphology. Subsequently, non-radioactive ISH was applied to FFPE sections and biotin, as a label, was largely replaced by the hapten digoxigenin due to the fact that—unlike biotin, a naturally occurring vitamin—digoxigenin is not endogenously present in tissues (Herrington et al., 1989). This technology has been applied to the demonstration of infectious agents together with changes in mRNA and microRNA associated with various disease states (Warford, 2016).

With the advent of precision medicine in the late 1990s, and the need to assess DNA gene amplification, copy number change, or gene fusion and translocation, ISH took on a diagnostic priority. For these applications fluorescent-labelled probes are often used, but for some applications non-fluorescent endpoints are available. These may replace the former as they provide for the simultaneous examination of cell and tissue morphology (Warford, 2016).

3.5.2 Nucleic acid probes

All probes used in ISH have two components:

- a nucleic acid sequence complementary in base sequence to the target nucleic acid;
- an attached label to provide for visualization of the formed hybrid.

There are several options available when choosing a nucleic acid sequence for probe preparation, and the principal advantage and disadvantage of a selection of these is provided in Table 3.4. While the majority of probe sequences are composed of standard nucleic acid bases, locked nucleic acids (LNA; Koshkin et al., 1998) and peptide nucleic acids (PNA; Nielsen and Egholm, 1999) provide synthetic alternatives. Evidence has been presented that these enhance hybridization efficiency and stability.

As already noted, the most frequently used label for chromogenic ISH is digoxigenin, while the choice of fluorescent labels is the same as for ICC (Table 3.3). When possible their inclusion during probe synthesis is preferable as this ensures maximum incorporation. Furthermore,

TABLE 3.4 Examples of nucleic acid probe types used in ISH

	Recombinant sequences		Oligonucleotides	PCR-generated
	Double-stranded DNA	Single-stranded RNA	Single-stranded DNA	Double-stranded DNA
Application	Demonstration of DNA and RNA targets	Demonstration of mRNA targets	Demonstration of DNA, mRNA, and microRNA targets	Demonstration of DNA and mRNA targets
Principal advantage	Probes can be very long and incorporate many labels. Ideal for gene visualization	Very high hybridization efficiency	High hybridization efficiency due to short length. Must be used for micro-RNA visualization	Length can be precisely controlled and produced when needed
Principal disadvantage	Probe requires denaturation before hybridization	Difficult to produce and store without degradation	Multiple sequences may be needed to raise sensitivity levels	Potential batch variation

From Warford and Volpi (2018), reproduced with permission of the publishers.

a variety of purification procedures are available to remove any unincorporated labels or incomplete probe nucleic acid sequences.

3.5.3 Probe validation

The specificity of nucleic acid probes should be assured by the use of search engines such as BLAST (https://blast.ncbi.nlm.nih.gov/Blast.cgi) that compare the proposed probe nucleic acid sequence against possibly similar DNA sequences present in the genome. Subsequently, it is important that each probe is checked for sensitivity and specificity on the cell or tissue preparations it is to be used on. Factors that may affect the former include the influence of secondary structure of the probe and target sequences on hybridization efficiency and, like ICC, the availability of the target gene sequences for hybridization due to sample preparation. Consequently, unless a commercially validated probe is used as specified by the supplier, in-house validation should be undertaken. This can proceed using a tiered approach as already discussed for ICC in Section 3.3.2.

3.5.4 ISH procedure

In order and principle there are close similarities to the steps used in ISH and ICC. Accordingly, the main distinctions are discussed here.

To guard against the ubiquitous presence of nucleases, solutions used for ISH up to and including the hybridization step should be nuclease free and gloves used to handle slides. Post-hybridization steps concern the demonstration of the nucleic acid probe label and do not require the use of any special protective procedures.

3.5.4.1 Sample preparation and exposure of target nucleic acid sequences

The native stability of nucleic acids differs; DNA is more stable than mRNA, but there is evidence suggesting that microRNA sequences are more stable than mRNA (Hall et al., 2012).

Cross reference

Formalin fixation is fully discussed in Chapter 1, Box 1.3.

When precipitating fixation methods are used then nucleic acids are readily available for hybridization. Formalin fixation introduces cross-links between nucleic acids and associated proteins. These are most abundant between histones and DNA. Furthermore, in a time-dependent manner formalin fixation shears nucleic acids. This may have severe effects on the homogenate analysis of such samples, but for ISH it tends only to reduce the sensitivity of the procedure, as fragmented targets are held together in the fixed matrix and probe sequences are often short enough to hybridize to them.

To achieve optimal hybridization the fixation-induced 'mesh' needs to be opened up. Although PIER and HIER methods can be used for this the former is most often employed. As with ICC, this step needs to be carefully controlled, balancing exposure of target with conservation of cytology or morphology. Proteinase K is commonly employed for target exposure and suggested concentrations for use to demonstrate DNA and RNA targets in formalin-fixed preparations are provided in Table 3.5.

3.5.4.2 Hybridization

This step represents the pivotal procedural step in an ISH procedure. The goal is to maximize complementary base pairing between the nucleic acid probe and the target gene sequence, and at the same time, to minimize partial or non-specific base pairing.

There are a number of chemical and physical factors that need to be taken into account to ensure specific hybridization. These are normally discussed under the heading of 'stringency'. In contrast to ICC, where the primary antibody is applied in a simple blocking buffer, the chemical constitution of a hybridization solution is more complex. The constituents of a typical hybridization solution and their effects are provided in Table 3.6.

Physical factors are the temperature used and time of incubation. Higher temperatures will favour specific hybridization, but these may compromise cell/tissue preservation. Consequently, formamide has traditionally been included as a constituent of hybridization solutions. Formamide acts as a helix-destabilizing reagent and, thereby, promotes the

TABLE 3.5 Proteinase K pretreatment concentrations according to preparation and target nucleic acid following formaldehyde-based fixation

Target	Fixation time	Preparation	Proteinase K range (μg/mL)[1]
DNA	2–5 min	Cell culture, Thinpreps[2], fine needle aspirations, frozen sections	0.5–10
	6–168 hours	FFPE	25–50
RNA[3]	2–15 min	Cell culture, Thinpreps[2], fine needle aspirations, frozen sections	0.5–5
	6–168 hours	FFPE	2–15

Key:
[1]Assumes incubation at 37°C for 30 minutes.
[2]Applies to all similar preparations.
[3]Applies to mRNA and microRNA.
Adapted from Warford and Volpi (2018), used with permission of the publishers.

TABLE 3.6 Hybridization solution components and their effects

Component/condition	Purpose	Comment
Probe	Hybridization to target sequence	Concentration influences sensitivity of hybrid formation
Monovalent cation (sodium chloride)	Regulation of specificity and sensitivity of hybrid formation	Concentration regulates charge repulsion between nucleic acid probe and target sequence. Higher concentration negates charge repulsion, enhancing hybrid formation, but potentially allowing non-specific hybridization
Formamide	Regulation of specificity and sensitivity of hybrid formation together with sample preservation	As a helix destabilizing reagent, it introduces repulsion between probe and target sequence, thereby allowing hybridization to be undertaken at lower temperatures enhancing cell/tissue preservation
Macromolecules (dextran sulfate)	Rate of hybrid formation and regulation of non-specific hybridization	Negates charge repulsion between probe and target, and locally concentrates the hybridization solution, thereby increasing rate of hybrid formation
Divalent anions	Inhibition of nuclease activity	Should not be used as a substitute for nuclease-free preparation of the hybridization solution
Buffer	Maintains pH	

Adapted from Warford and Volpi (2018), used with permission of the publishers.

formation of only fully complementary hybrids at lower temperatures. When this reagent is included, hybridization can often be undertaken at and about physiological temperatures. An important safety consideration is that formamide is a teratogen and, therefore, the use of LNA probes, where it is not required, is advantageous in this respect (Jørgensen et al., 2010).

ISH procedures often include a pre-hybridization step. In composition identical to the hybridization solution, but lacking the nucleic acid probe, its purpose is to equilibrate the preparations and 'prime' the sample before hybridization. Post-hybridization washes are also undertaken to remove any partially formed hybrids. These typically include the use of a series of salt solutions, sometimes with formamide present, into which the preparations are immersed at different temperatures. It is possible to avoid the use of these washes if the conditions of hybridization are precisely controlled.

3.5.4.3 Detection

The underlying principles for signal detection in ISH are the same as for ICC; see Section 3.3.4.4. Simple indirect procedures are often used. An example is the detection of the hapten digoxigenin using an antibody labelled with alkaline phosphatase followed by nitroblue tetrazolium, 5-bromo-4-chloro-3-indolyl phosphate (NBT/BCIP) substrate incubation (Figure 3.6). Sensitivity is achieved in this instance by enzyme cycling of the substrate, which can, when required, proceed for over 24 hours. Silver-enhanced ISH (SISH; Powell et al., 2007) detection together with tyramide signal and adapted rolling circle amplification detection procedures may also be used for enhanced detection sensitivity (Cassidy and Jones 2014).

FIGURE 3.6

Chromogenic *in situ* hybridization. Epstein–Barr virus RNA present in Hodgkin's lymphoma. Note the nuclear localization of the viral RNA. EBER-targeted oligonucleotide probe with indirect detection and nitroblue tetrazolium, 5-bromo-4-chloro-3-indolyl phosphate (NBT/BCIP) substrate incubation.

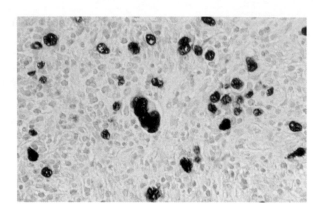

Key Point

The ISH procedure shares many steps that are also present in ICC methods, such as pre-treatment and detection. However, the hybridization solution containing the nucleic acid probe is more complex than that of the antibody diluent used for the incubation of a primary antibody in ICC. This reflects the need to precisely control the specificity and sensitivity of complementary base pairing formation between the probe and the target gene sequence.

3.8 SELF-CHECK QUESTION

What are the principal chemical components and physical factors that will affect the formation of base pairing between a nucleic acid probe and a target gene sequence in an ISH procedure?

3.6 **Controls**

In addition to the validation of antibodies and nucleic acid probes prior to their use, controls should always be included in ICC, FC, and ISH runs. Two types of control should be included:

- *Positive*: This can be a separate preparation known to contain the target. When it contains moderate amounts of the target antigen or nucleic acid sequence then the result can be used to assess run-to-run variations and act, when a weak result is obtained, as an early warning of potential procedural issues.

  When chromosomal DNA gene sequences are demonstrated using ISH in interphase nuclei, the simultaneous or sequential hybridization to a repetitive gene sequence present on the same chromosome in the same preparation may be employed. This not only acts as a positive control, but it may also be used to calculate gene copy number ratio change between the target gene and repetitive sequence. For example, in the assessment of HER2 amplification in invasive breast cancer a ratio of HER2/CEP17 (centromeric repetitive sequence) ratio of ≥2.0 with an average HER2 copy number ≥4.0 signals per cell is interpreted as positive gene amplification (Wolff et al., 2014; Rakha et al., 2015).

- *Negative*: These can be simple or complex. In its simplest form it can be the use of antibody diluent or hybridization solution without, respectively, the addition of the primary antibody or nucleic acid probe. More complex negative controls include the use of isotype-matched antibodies for ICC and FC, and in RNA-ISH scrambled (nonsense) and sense nucleic acid probes.

Scrambled sequence probes are identical in construction to the labelled probe, except that they have no complementarity with the target sequence. Sense probes have the identical orientation and nucleic acid sequence to the target. Accordingly, they cannot bind by complementary base pairing to the target nucleic acid. When these controls are used, the expected negative ISH result provides confirmation that the hybridization of the test probe is specific. If staining occurs, then these controls indicate that non-specific interactions are occurring with the sample.

Ideally, a negative control should accompany each test sample to reveal any non-specific staining. When multiple antigens or genes are being demonstrated across several slides then comparison across these can reveal non-specific reactions without the need to include a separate negative control. The adoption of this approach is particularly useful when minimal quantities of sample preparations are available.

As discussed in Chapter 2, Section 2.3, participation in an external quality assurance scheme will be required for ISO15189 accreditation, and this is obligatory for several ICC and ISH precision medicine-linked procedures.

3.7 Future directions

ICC, FC, and ISH may be regarded as technologies that stand or fall according to the specificity and sensitivity of their primary antibodies or nucleic acid probes. FC is reliant on automated instrumentation, while in the diagnostic environment, ICC is often automated. With ISH, semi-automated and fully automated platforms are available. In the increasingly regulated practice of diagnostic pathology, it is almost certain that all of these technologies will be used on fully automated instruments, where all reagents will be supplied validated and in ready-to-use format.

For the foreseeable future, formalin fixation will continue to be used as the initial step before paraffin wax embedding and as an optional fixative for cell preparations prior to ICC and ISH. As noted here and in Chapter 1, the fixative has known molecular deficiencies. Alternative fixatives have been proposed to address this issue, but none are, so far, able to adequately fix samples that are larger than small biopsies. The development and introduction of a 'one size fits all' fixative that preserves cytology and morphology, together with molecular sample integrity, is awaited with anticipation.

The demonstration of gene expression at the protein level by ICC is dominant in diagnostic molecular pathology. In some instances, where protein product is secreted this can give rise to significant 'background' staining that may make diagnostic interpretation difficult. An example is the secretion of immunoglobulin in B-cell lymphomas, which can make assessment for clonality difficult. As mRNA is never secreted from its cell of origin, the use of ISH suggests itself as an alternative demonstration technology. Sensitivity issues have hampered its use until recently, but with the description of the branched DNA ISH method, which combines specificity via couplet oligonucleotide hybridization and an ultra-sensitive detection procedure, this issue has now been addressed (Cassidy and Jones, 2014; Figure 3.7). Accordingly, ISH for gene expression may well become a more frequently used technology in diagnostic molecular pathology.

Cross reference
See Chapter 1, Section 1.6 for details of alternative fixatives.

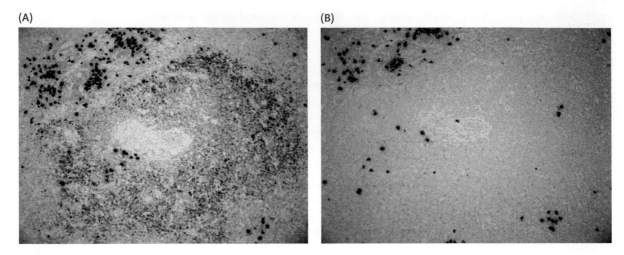

FIGURE 3.7
Branched DNA demonstration of light chain immunoglobulin mRNA restriction in follicular lymphoma. (A) Kappa light chain mRNA. (B) Lambda light chain mRNA. Note presence of stained plasma cells in both preparations, but staining of malignant follicle (centre) for kappa mRNA only.
RNAScope®, peroxidase detection (Advanced Cell Diagnostics Inc).

The last few years have seen significant developments in the field of FC (reviewed in Adan et al., 2017). The introduction of quantum dots (QDs)—characterized by wider excitation spectra, narrower emission peaks and increased photo-stability, then traditional fluorophores—has led to increased sensitivity and design flexibility. Another significant advance has been the introduction of microfluidics, enabling very small amount of fluids (down to 10^{-18} L) to be processed in micro-channels; a development ideal for analysis of small cell populations, which has also led to the introduction of microchip-based, portable FCs. The establishment of 'imaging flow cytometry' with the introduction of techniques combining the statistical power and fluorescence sensitivity of FC with the strengths of digital microscopy has raised great expectations in terms of potential enhancement of diagnostic capabilities and clinical applicability (Basiji et al., 2007; Fuller et al., 2016). Efforts will continue to address concerns about consistency (inter- and intra-laboratories variations) and the need for improved standardization of FC tests in both clinic and research (Mizrahi et al., 2018).

Finally, it is inevitable that, as the scope of precision medicine broadens, new diagnostic tests involving ICC, FC, and ISH will be introduced to supplement those already in existence. These will reinforce the need for these technologies to be undertaken in a regulated environment using only validated reagents and procedures.

CASE STUDY 3.1

Antibody validation for a companion diagnostic biomarker

During the development of a targeted therapeutic drug it is now a requirement to have a biomarker or biomarkers that can be used to segregate patient groups during clinical trials. The use of such biomarkers enables the patient groups who will receive the drug to be segregated, according to their molecular profile, into those who may benefit from receiving the targeted treatment and those where a positive effect cannot be anticipated. Antibodies that recognize either the drug target or a component of its downstream signalling pathway may be used for this purpose. If clinical trials are successful and segregation of patients for targeted therapy is indicated then the same antibody may be developed into a companion diagnostic. Accordingly, selection of an antibody that is sensitive and specific for its target is essential. In the following the steps, the means by which this validation process can be undertaken is outlined. For more detailed information, please refer to Smith and Womack, 2014.

• *Step 1:* Before undertaking any practical work the drug/target interaction needs to be thoroughly understood. This will be based on both literature reviews and preclinical testing, with the latter including work with cell line and animal models. The results of this work will point to potential biomarkers that could be developed for use in clinical trials. Once this has been completed, then a further literature review can commence to identify which antibody target would be best suited for development, and potential cell and tissue preparations that could be used for antibody validation.

• *Step 2:* In preparation for testing antibodies the development of knock-in or knock-down cell lines should be attempted. These are very valuable as they provide immediate evidence of antibody sensitivity and specificity. They should be prepared as formalin-fixed cell blocks (see Chapter 1 so as to be representative of the FFPE clinical material that will be used to segregate patients. Similarly, consideration should be given to the selection of normal and pathological FFPE tissue that can be used for antibody validation. For these, both positive and negative material should be identified, and with respect to the former samples that express the antigen at weak to strong levels.

• *Step 3:* Search for suppliers of the antibody target chosen as the potential biomarker. Exceptionally, antibodies that have been approved for diagnostic use will be available and their evaluation should take precedence over those sold as for research only. Whatever their category, more than one antibody should be evaluated and, at this stage, consideration of how the successful antibody might be used in a sensitive ICC procedure under manual or automated conditions should be considered.

• *Step 4:* Commence practical work using the knock-in and knock-down cell lines. This could be done by making a small cell array so that all samples are cut at the same time and mounted on the same slides. Using data sheets as a guide, concentrate on the effects of epitope retrieval and antibody concentration/dilution on staining. At the same time, employ at least one other technique to evaluate for specificity. This should include Western blotting.

• *Step 5:* Evaluate results and narrow down the range of antibodies for further validation. Any antibodies that are unreactive or do not correctly discriminate between the cell lines should be eliminated. Furthermore, when multiple bands are revealed in Western blots then this should (allowing for phosphorylation and glycosylation) be taken as a warning that these antibodies may be cross-reactive with other antigens. Antibodies that should be taken forward should show appropriate staining discrimination between the cell lines, appropriate localization (nuclear, cytoplasmic, membrane), a single Western blot band, and sensitive staining.

• *Step 6:* Using the range of FFPE tissues identified in Step 2 evaluate the selected antibodies from Step 5 on either individual sections or using a tissue microarray approach. Assessment of the effect of epitope retrieval and varying antibody concentration/dilution should be undertaken. Results should be analysed as in Step 5, and a primary and back-up antibody recommended for use as a biomarker for clinical trials.

This procedure is not a trivial exercise and will involve several weeks of 'person hours' to complete. Its downstream consequence can be critical for clinical trials and for the development of a companion diagnostic. Accordingly, painstaking attention to the validation process is essential.

Chapter summary

This chapter has covered the principles and practice of three techniques that can be used for the detection of protein and nucleic acid macromolecules in intact samples, either as single cells or tissue preparations.

■ The techniques for the microscopic visualization of proteins are immunocytochemistry (ICC), and for nucleic acids *in situ* hybridization (ISH). ICC relies on the application of primary antibodies to bind to epitopes on antigens to identify proteins and provides a semi-quantitative 'readout'. ISH relies on the use of nucleic acid probes binding to target gene sequences by complementary base pairing. The analysis for ISH may be semi-quantitative or quantitative.

■ ICC and ISH may be undertaken manually or using automated platforms, and they share many procedural similarities. These include sample preparation, unmasking of target sequences, and detection of bound primary antibodies or probe/target hybrids using fluorescent or chromogenic detection methods.

■ In contrast, flow cytometry (FC) is a quantitative automated technique that is used to identify cells that have bound labelled antibodies to antigens present on their cell membranes. FC may also be used as a precursor to fluorescence-automated cell sorting (FACS) for further analysis.

■ The need to ensure that primary antibodies and nucleic acid probes are validated as specific and sensitive for use in these technologies before operational use together, with the incorporation of appropriate controls when these reagents are subsequently used, is emphasized.

 Further reading

● Adan A, Alizada G, Kiraz Y, Baran Y, Nalbant A. Flow cytometry: basic principles and applications. Critical Reviews in Biotechnology 2017; 37: 163–176.

● Suvarna KS, Layton C, Bancroft JD (Eds). *Bancroft's Theory and Practice of Histological Techniques*, 7th edn. London, Churchill Livingstone, 2013.

● Taylor CR, Rudbeck L (Eds). *Immunohistochemical Staining Methods*, 6th edn. Santa Clara, CA, Agilent, 2013. Available at: https://www.agilent.com/cs/library/technical-overviews/public/08002_ihc_staining_methods.pdf (accessed 25 August 2018).

● Renshaw S (Ed.). *Immunohistochemistry and Immunocytochemistry: Essential Methods*, 2nd edn. Oxford, Wiley Blackwell, 2017.

● Hauptmann G (Ed.). *In Situ Hybridization Methods*. Berlin, Springer, 2015.

● Nielsen B (Ed.). *In Situ Hybridization Protocols*. Berlin, Springer, 2014.

● Shapiro HM (Ed.). *Practical Flow Cytometry*, 4th edn. Wilmington, DE, Wiley Liss, 2003. Available at: https://www.beckman.com/resources/reading-material/ebooks/practical-flow-cytometry (accessed 25 August 2018).

 References

● Abraham P, Maliekal TT. Single cell biology beyond the era of antibodies: relevance, challenges, and promises in biomedical research. Cellular and Molecular Life Sciences 2017; 74: 1177–1189.

- Adan A, Alizada G, Kiraz Y, Baran Y, Nalbant A. Flow cytometry: basic principles and applications. Critical Reviews in Biotechnology 2017; 37: 163–176.

- Ashraf T, Hisham I. The use of flow cytometric DNA ploidy analysis of liver biopsies in liver cirrhosis and hepatocellular carcinoma. In: *Liver Biopsy*, Takahashi H (Ed.). London, InTech, 2011.

- Basiji DA, Ortyn WE, Liang L, Venkatachalam V, Morrissey P. Cellular image analysis and imaging by flow cytometry. Clinical and Laboratory Medicine 2007; 27: 653–670.

- Bobrow, MN, Litt GJ, Shaughnessy KJ, Mayer PC, Conlon J. The use of catalyzed reporter deposition as a means of signal amplification in a variety of formats. Journal of Immunological Methods 1992; 150: 145–149.

- Bordeaux J, Welsh A, Agarwal S, et al. Antibody validation. Biotechniques 2010; 48: 197–209.

- Cassidy A, Jones J. Developments in *in situ* hybridisation. Methods 2014; 70: 39–45.

- Coons AH, Creech HJ, Jones RN. Immunological properties of an antibody containing a fluorescent group. Proceedings of the Society for Experimental Biology 1941; 47: 200–202.

- D'Amico F, Skarmoutsou E, Stivala F. State of the art in antigen retrieval for immunohistochemistry. Journal of Immunological Methods 2009; 341: 1–18.

- Filby A. Sample preparation for flow cytometry benefits from some lateral thinking. Cytometry A 2016; 89: 1054–1056.

- Fuller KA, Bennett S, Hui H, Chakera A, Erber NW. Development of a robust immuno-S-FISH protocol using imaging flow cytometry. Cytometry A 2016; 89: 720–730.

- Gall JG, Pardue ML. Formation and detection of RNA-DNA hybrid molecules in cytological preparations. Proceedings of the National Academy of Sciences USA 1969; 63: 378–383.

- Gusev Y, Sparkowski J, Raghunathan A, et al. Rolling circle amplification: a new approach to increase sensitivity for immunohistochemistry and flow cytometry. American Journal of Pathology 2001; 159: 63–69.

- Hall JS, Taylor J, Valentine HR, et al. Enhanced stability of microRNA expression facilitates classification of FFPE tumour samples exhibiting near total mRNA degradation. British Journal of Cancer 2012; 107: 684–694.

- Hanley MB, Lomas W, Mittar D, Maino V, Park E. Detection of low abundance RNA molecules in individual cells by flow cytometry. PLoS One 2013; 8(2): e57002.

- Herrington CS, Burns J, Graham AK, Evans MF, McGee JO'D. Interphase cytogenetics using biotin and digoxigenin-labelled probes I: relative sensitivity of both reporter molecules for HPV16 detection in CaSki cells. Journal of Clinical Pathology 1989; 41: 592–600.

- Howat WJ, Lewis A, Jones P, et al. Antibody validation of immunohistochemistry for biomarker discovery: recommendations of a consortium of academic and pharmaceutical based histopathology researchers. Methods 2014; 70: 34–38.

- Hsu SM, Raine L, Fanger H. Use of avidin-biotin-peroxidase complex (ABC) in immunoperoxidase techniques: a comparison between ABC and unlabelled antibody (PAP) procedures. Journal of Histochemistry and Cytochemistry 1981; 29: 577–580.

Discussion question

3.1 During microscopic quality control following an ICC run using several primary antibodies, one positive control for a primary antibody gives weaker results than expected. What practical steps would you take to address this observation?

Answers to the self-check questions and tips for responding to the discussion question are provided on the book's accompanying website:

⬤ Visit www.oup.com/uk/warford

4

Homogenate Sample Analysis

Nadège Presneau, Abdul Hye, and Nicholas Ashton

Learning objectives

After studying this chapter, you should be able to:

- Describe the different technologies used for the analysis of nucleic acids.
- Discuss the principles and key technical considerations of the PCR-based technologies.
- Understand the major proteomic technologies.
- Appreciate the importance of protein modifications.

4.1 **Introduction**

This chapter provides an overview of the current landscape for molecular analysis of nucleic acids and proteins from homogenate samples. It focuses on new technologies that have revolutionized nucleic acids and proteins analysis. Techniques such as Southern, Northern and Western blotting, which have been used widely in clinical diagnostics and biological/medical research since the 1970s, are now being complemented or replaced by techniques based on the polymerase chain reaction (PCR) and quantitative single/multi-analyte immunocapture assays.

This chapter will, therefore, explain the principles and key technical considerations for the use of these newer technologies for the analysis of nucleic acids and proteins in homogenate samples in a clinical setting.

The chapter has been divided into two major parts with the first one dedicated to the analysis of nucleic acids—DNA and RNA—followed by the analyses of proteins. The application of the technologies in molecular diagnosis, particularly for DNA and RNA, are mentioned here, but

> **Cross reference**
> This chapter links back to Chapter 1, where optimal sample preparation was covered.

these are covered extensively in Part 2 of this book in the context of application to diagnosis of specific diseases.

This chapter will provide the reader with an idea of the choice of appropriate technique in context of the type of analyte under investigation.

4.2 **Nucleic acid analysis**

The preparation of nucleic acids using homogenate methods, together with quality checks is covered in Chapter 1, and it should be emphasized again that samples of high integrity are most likely to provide good analytical results.

4.2.1 DNA

4.2.1.1 PCR-based methods

PCR was first described as a method of DNA amplification in 1985 by Kary Mullis and co-workers (Mullis et al., 1986), for which he won the Nobel Prize for Chemistry in 1993. PCR can produce large amounts of a specific DNA fragment from small starting amounts of a complex template. It is, therefore, distinctive among diagnostic molecular methods in turning the 'needle' of the target into a 'haystack' so that it can be easily detected. Due to this, special care must be taken to avoid non-specific amplification due to sub-optimal technique and possible cross-contamination between samples. For steps that need to be taken to avoid the latter please refer to Chapter 2, Section 2.2.

PCR represents a form of *in vitro* cloning that can generate, as well as modify, DNA fragments of defined length and sequence in a simple automated reaction. It has revolutionized the detection of genetic disorders, cancers, and infectious diseases, as well as being employed in paternity and forensic analysis.

4.2.1.1.1 The principles of PCR

PCR is based on the biological process of DNA replication and is performed in a test tube by using an enzyme called the *Taq* DNA polymerase (from *Thermus aquaticus*). This enzyme is unusual as it is resistant to degradation at high temperature up to 95°C and is relatively unaffected by rapid changes of temperatures. *Taq* polymerase will synthesize a complementary new strand of DNA in the presence of other reagents contained in a reaction buffer. The constituents of a reaction buffer typically include $MgCl_2$, a co-factor and catalyser for the polymerase, KCl, and tris-HCl for pH control, deoxynucleotide triphosphates (dNTPs: dATP, dCTP, dGTP, and dTTP) for incorporation in the copied template and a set of primers (forward and reverse) that flank the region to amplify. A primer is a short sequence of DNA (oligonucleotide) and two of them are needed to flank the region of DNA that needs to be amplified. Primers are necessary for the DNA polymerase to start DNA synthesis. Good primer design is essential to ensure specific amplification of the target DNA sequence—see Box 4.1. Several software programs exist for primer design with one of the most popular being Primer3 (Koressaar and Remm, 2007; Untergasser et al., 2012).

Standard PCR involves repeated cycling of three steps as described below and illustrated in Figure 4.1.

1. **Denaturation** of the DNA at 94–95°C. This breaks the weak hydrogen bonds that hold DNA strands together in a helix, allowing the strands to separate creating single-stranded

BOX 4.1 *Primer design essentials*

Below are listed the essential parameters to consider when designing primers. Often, software such as Primer3 or software on companies websites (e.g. Integrated DNA technologies (IDT), ThermoFisher, Roche, Eurofins Genomics) have default settings that would cover some of these parameters. Nonetheless, it is good to be aware of them.

The oligonucleotide primers should be a minimum of 18 bp to a maximum 25–30 bp with an ideal size at 20 bp. This may vary depending on the application, but for a conventional PCR, 20 bp is good.

GC clamp at the 3′ end of the primer to allow stability of the primer annealing to the target DNA sequence as the GC pair is bound by three hydrogen bonds, while the AT pair is bound by two hydrogen bonds.

The GC-content of the primer should be between a minimum of 30% and maximum of 65%. Ideally, a GC-content around 50% is good.

No self-complementarity should be present to avoid the formation of hairpin structures. There should also be no complementarity between the primers as this will create primer-dimers.

Check for presence of SNPs using a SNP check either by looking at primer sequence on a genome browser such as UCSC (https://genome.ucsc.edu) or using the online software SNPCheck v3 (https://secure.ngrl.org.uk/SNPCheck/snpcheck.htm). Pay attention to the allele frequency and validity status of the SNP, as well as the position of the SNP in the primer. For instance, a primer with a SNP at 3′ end with high allele frequency should be redesigned. When there is a perfect match between primer and target sequence, there is 100% annealing and, therefore, full efficiency of amplification, but if there is a mismatch due to a SNP at 3′ end of the primer it will result in low-level background amplification.

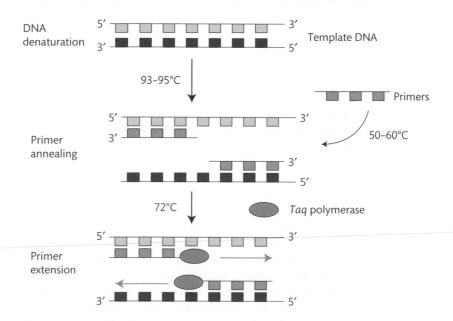

FIGURE 4.1

Template DNA is denatured into single strands by heating to 93–95°C. Primers in the reaction mix anneal at their complementary sites at various temperature depending on the Tm of the primers that would depend on the number of AT and GC nucleotides. This may vary between 50 and 65 degrees. *Taq* polymerase begins to work at 72°C, adding nucleotides from the reaction mixture to produce a duplicate copy of the original template. This sequence of events is repeated, doubling the amount of DNA in every cycle. Reproduced from O'Sullivan and Taniere, chapter 11 in *Histopathology*, Orchard and Nation eds, 2nd edn, Oxford University Press, 2017.

DNA. Initial denaturation at 95°C for 2 minutes is recommended prior to PCR cycling to fully denature the DNA. Longer or higher temperature incubations, unless required due to high-GC content of template, should be avoided. Typically, a 15–30-second denaturation at 95°C should be utilized during subsequent thermocycles.

2. **Annealing** to permit the primers to bind to their complementary sequence in the denatured template and so allow for extension. The polymerase begins synthesizing new DNA from the 3' end of the primer in the 5' to 3' end direction. Typical primer annealing temperatures are 5°C below the lowest primer's **melting temperature (Tm)** and often fall in the range of 50–60°C. If spurious amplification products occur, higher annealing temperatures may be tried as the kinetics will favour only precisely complementary base pairing between the primers and template DNA. Typical annealing times are 15–30 seconds.

3. **Extension** that allows the *Taq* polymerase to synthesize DNA in order to produce a double-strand DNA molecule. As *Taq* polymerase is stable at high temperatures extension is usually undertaken at 72°C. As a general rule, extension times of 1 minute per 1000 base pairs (e.g. 3 minutes for a 3 kb product) are used. For products less than 1 kb, 45–60-second extension times can be employed. Products greater than 3 kb, or reactions using more than 30 cycles, may require longer extension times.

These steps are repeated 30–40 times until the reaction reaches a 'plateau' or endpoint, where the *Taq* polymerase stops amplifying the target sequence. PCR amplification is exponential because, for each cycle, the DNA made in the previous cycles can also serve as a template; therefore, after each successive cycle, an exponential increase of 2^n occurs, where n is the total number of cycles. For instance, a PCR of 30 cycles, 2^{30} will give 1,073,741,824 copies of the DNA (amplicons).

The PCR usually finishes after the last cycle with an additional extension of 5–15 minutes. This last step allows synthesis of many uncompleted fragments to finish and the terminal transferase activity of *Taq* DNA polymerase permits the addition of an adenine residue to the 3' ends of all PCR products.

Key Point

Standard PCR is based on three steps that consist of denaturation, annealing, and extension steps. This is repeated for up to 35–40 cycles during which exponential amplification of the target DNA occurs.

The DNA template for a PCR can be from various sources as either genomic DNA (gDNA) or cDNA (*in vitro* reverse transcription of mRNA). gDNA can be extracted from various clinical specimens obtained from bodily fluids, for example, blood, buccal swabs, saliva, urine, semen, and amniotic fluid. Microbacterial samples and viral clinical samples can also be analysed. Frozen and FFPE tissue from biopsy and resected surgical samples are also suitable for PCR. However, as highlighted in Chapter 1, Section 1.3.2, due to the degrading effect of the FFPE process, amplification is normally restricted to shorter length fragments compared with the use of DNA from fresh or frozen samples. Therefore, care should be taken in primer design to ensure that amplicons are less than 150–200 bp in length when using such material for PCR.

When amplifying longer fragments of DNA or where there is a need for high fidelity amplification, other types of DNA polymerases have been introduced. One example is the *Pfu* polymerase (from *Pyrococcus furiosus*) that displays a lower error rate of 1.5×10^{-6} errors/bp as opposed to the *Taq* polymerase that has an error rate of approximately 1×10^{-4} to $2 \times$

Melting temperature (Tm)

The point at which 50% of a complementary sequence of DNA is disassociated.

10^{-5} errors/bp. These differences are inherent to the 3′–5′ exonuclease proofreading activities that the *Taq* polymerase lacks. DNA polymerase enzymes have also been modified to improve their specificity by adding an activation step before the start of the amplification, so-called **hot-start DNA polymerases**. They can be either chemically- or antibody-based, modified to allow them to start the polymerization only after a first step of activation in the thermocycler. This has the advantage of avoiding non-specific amplification due to the primers annealing at a lower temperature (mispriming). Indeed, at lower temperatures, the DNA polymerase can start to extend while the operator is still assembling the reagents on the bench and, therefore, can create spurious PCR products. Other enzymes such as *Phire*™ hot start and quick hot start have been engineered to have no reactivation step and work four times faster than conventional hot-start DNA *Taq* polymerase. They allow amplification of significantly longer DNA fragments as they are less sensitive to inhibitors that can be present in the DNA template.

Hot-start DNA-polymerase
DNA polymerase that would have its activity activated at high temperature.

4.2.1.1.2 Variations around standard PCR

Many variations to standard PCR steps have been made to increase the sensitivity and specificity, particularly for applications in the diagnostic laboratory. Among these are:

- *Touch-down PCR*: This approach is based on the start of the PCR at a higher annealing temperature from the calculated Tm of the primers. In the next 5–10 cycles, a decrease of 1°C to the annealing temperature at each cycle is employed to reach the optimum annealing temperature for the primers, which is then kept for the subsequent cycles. The objective is to enhance specificity, and avoid non-specific primer binding and the creation of spurious PCR products.

- *COLD-PCR (COamplification at Lower Denaturation temperature PCR)*: This approach has been applied to the detection of low-abundance DNA variants or mutations. This form of PCR selectively amplifies low-abundance DNA mutant-containing (or variant-containing) sequences present in a mixture with abundant wild type sequence. By using a critical denaturation temperature, it allows amplification of the low abundant variant DNA, irrespective of the mutation type or position on the amplicon. The use of a lower denaturation temperature in COLD-PCR results in selective denaturation of amplicons with mutation-containing molecules due to slight changes in the Tm of these compared with the wild type sequences (Milbury et al., 2011). COLD-PCR is used to assist in cancer prognosis coupled with targeted therapy.

- *Nested PCR*: This method was developed to enhance specificity of amplification by introducing to the conventional PCR, a second round of PCR using, as a template, the PCR product from the first PCR round. The second PCR will have primers designed within the DNA sequence of the first PCR product (amplicon) to augment specificity. The probability that the second set of primers will bind to a non-specific region is very low. The drawback of this method is that prior knowledge of the DNA sequence is necessary in order to design a set of primers within the first amplified DNA segment. Nested PCR has been used in several areas of molecular diagnostics, but particularly in microbiology and viral testing.

- *Multiplex PCR*: Multiplex PCR involves the amplification of more than one target sequence using multiple primers in a single reaction. The optimization of multiplex PCRs can pose several difficulties, including poor sensitivity or specificity, and/or preferential amplification of certain DNA sequences. Therefore, special attention to primer design parameters, such as homology of primers with their target DNA sequence, their length, GC content (affecting Tm), and their concentration in the PCR reaction mix is required. Ideally, all the primer pairs in a multiplex PCR should enable similar amplification efficiencies for their respective target. This may be achieved through the utilization of primers with nearly identical optimum annealing temperatures (primer length of 18–25 bp or more and a GC content of 35–60% may prove satisfactory), and should not display significant homology, either internally or to one another to avoid primer-dimer formation (primers that hybridize to each other) or self-complementary complexes to form.

Since its introduction, multiplex PCR has successfully been applied in many areas of nucleic acid diagnostics, including gene deletion analysis [e.g. mutations in the *DMD* gene that cause Duchenne and Becker muscular dystrophy (DMD/BMD); Chamberlain et al., 1992], mutation and polymorphism analysis [e.g. mutations in the cystic fibrosis transmembrane conductance regulator gene, *CFTR*, responsible for cystic fibrosis (Moutou et al., 2002)], and quantitative analysis (HIV/HBV/HCV testing). In the field of infectious diseases, the technique has been shown to be a valuable method for the identification of viruses, bacteria, fungi, and/or parasites.

Loop-mediated isothermal amplification (LAMP). This method employs a DNA polymerase with high strand displacement activity opening double-stranded DNA in order to extend from the primer in addition to a replication activity. Typically, a *Bst DNA polymerase* (large fragment of *Bacillus stearothermophilus*), and a set of *four specially designed primers* that recognize a total of six distinct sequences on the target DNA are used in this form of PCR. An inner primer complementary to the target DNA sequence initiates the LAMP. This is followed by strand displacement DNA synthesis primed by an outer primer that releases a single-stranded DNA. This serves as template for DNA synthesis primed by the second inner and outer primers, which hybridize to the other end of the target, producing a stem–loop DNA structure (see https://www.neb.com/applications/dna-amplification-pcr-and-qpcr/isothermal-amplification). In subsequent LAMP cycling, one inner primer hybridizes to the loop on the product and initiates displacement DNA synthesis, yielding the original stem–loop DNA and a new stem–loop DNA with a stem twice as long. The cycling reaction continues with accumulation of 10^9 copies of target in less than an hour. The final products are stem–loop DNAs with several inverted repeats of the target and cauliflower-like structures with multiple loops formed by annealing between alternately inverted repeats of the target in the same strand. Because LAMP recognizes the target by six distinct sequences initially and by four distinct sequences afterwards, it is expected to amplify the target sequence with high selectivity. LAMP results can be accurately observed using a real-time turbidimeter, but less expensive and simplified methods have recently been developed, which include using fluorescent dyes contained in a solid bead of wax that are released at higher temperature following the cessation of the reaction (Notomi et al., 2000; Tao et al., 2011).

The distinct advantage of the LAMP reaction is that it takes place at a single temperature (isothermal amplification between 60–67°C). Therefore, it does not require a thermocycler. In addition, the synthesis is not inhibited by the secondary structure of DNA or by inhibitors that may be present in the sample. It is, therefore, a method of choice for point of care (POC) diagnostic applications in developing countries.

4.1 SELF-CHECK QUESTION

Explain how the PCR reaction amplifies a target nucleic acid so that it can be readily detected using electrophoresis.

4.2.1.1.3 Downstream analysis of the PCR-amplified fragments

Once the PCR products are generated they can be submitted to different type of analysis:

(a) Electrophoresis and determination of amplicon size

Agarose gel electrophoresis provides for the direct visualization of PCR products. Smaller molecules travel further, so increasing the agarose concentration of a gel will reduce the migration speed and enable separation of smaller DNA molecules. A 0.7% agarose gel will allow good separation of large DNA fragments between 5–10 kb and a 2% agarose gel will have a good resolution for small fragment around 0.2–1 kb.

Alternatively, polyacrylamide gels can be used to provide for higher resolution of PCR products. Thus, the product of a deletion or insertion can be visualized in this way [e.g. DMD and cystic fibrosis (CF)].

(b) Mutation-specific restriction enzyme digestion (MSRED)

Point mutations can change restriction sites in DNA, causing alteration in cleavage by **restriction endonucleases** which produce fragments of various sizes.

PCR products that contain the potential site of the somatic mutation are submitted to a restriction digestion using restriction endonucleases to reveal, for instance, a mutation that changes the recognition of the nucleotide sequence for a specific restriction enzyme. The enzyme will cut the DNA depending if a mutation is present or not. The product of restriction digestion is submitted to electrophoresis to identify a change in the band pattern compared with controls.

The MSRED approach has been successfully used for the detection of β-catenin mutations in FFPE samples of sporadic desmoid-type fibromatosis (Amary et al., 2007; see also Chapter 13). The digestion of DNA with restriction enzymes is also employed to reveal restriction fragment length polymorphisms (RFLP). These genetic markers are used as a method of DNA profiling for genotyping (Dausset et al., 1990), forensics, and paternity testing.

> **Restriction endonucleases (restriction enzymes)**
> Enzymes that cut DNA into short pieces. Each restriction endonuclease targets different nucleotide sequences in a DNA strand (generally from 4 to 6 base pairs in length) and cuts at different sites.

(c) Heteroduplexes analysis HDA

HDA is based on conformational differences in double-stranded DNA caused by the formation of a heteroduplex molecule. Often in mutation screening, a normal control PCR product (homoduplex wild type control) is used and mixed with a test sample PCR product, which potentially contains a mutation. The denaturing and renaturing of the mixture of wild type fragments with mutated fragments will generate both homoduplex and heteroduplex molecules. Heteroduplex molecules will have a mismatch in the double strand, due to the presence of a point mutation, for example, which will cause a distortion in its usual conformation. The same 'feature' can also be generated by PCR of a heterozygous sample or by adding mutant and wild type DNA in the same PCR reaction. The distortion or altered conformation can be detected on polyacrylamide gels, where the heteroduplex molecule will migrate more slowly than the corresponding homoduplex molecule. Hence, by running both a wild type control and a potentially mutated test sample a different mobility in a polyacrylamide gel can be observed. This approach has been successful for the analysis of mutations in cystic fibrosis samples, for instance (Wang et al., 1992).

(d) Single-strand conformational polymorphism analysis (SSCP) and denaturing gradient gel electrophoresis (DGGE)

Very similar to HDA, SSCP and DGGE analyses are sensitive techniques for mutation detection and are used in molecular diagnostic laboratories. The principle of SSCP analysis is based on the fact that relatively short single-stranded DNA fragments have defined conformations that can be altered due to a single base change. Accordingly, such single-stranded DNA will migrate differently under native (non-denaturing) electrophoresis conditions. Therefore, wild type and mutant DNA samples will display different band patterns. SSCP analysis involves the following steps:

- polymerase chain reaction (PCR) amplification of any given DNA sequence of interest;
- denaturation of double-stranded PCR products with either heat or chemical agents, such as formamide;
- cooling of the denatured DNA (single-stranded) to maximize self-annealing;
- detection of mobility difference of the single-stranded DNAs by electrophoresis through a native polyacrylamide gel. (For review, read Kakavas et al., 2008.)

Similarly, DGGE is a technique used to separate double-stranded DNA fragments that are identical in length, but which differ in sequence. DGGE is based on the fact that a single-stranded DNA molecule migrates more slowly than the equivalent double-stranded molecule during electrophoresis, due to increased interaction of the unbound nucleotides in the single-stranded molecule with the polyacrylamide gel. The polyacrylamide gel contains an increasing gradient of chemical denaturants (urea and formamide are often used) through which the DNA molecules pass by electrophoresis. As double-stranded DNA passes through the gradient, each molecule will begin to denature at a particular concentration of denaturant dependent upon its %GC content and the exact arrangement of bases in the sequence. Therefore, the separation of molecules is based on GC content and DNA sequence, rather than size. The method has been used frequently for identifying single-nucleotide polymorphisms without the need for DNA sequencing and as a molecular fingerprinting method (for review read Fodde and Losekoot, 1994).

Cross reference

See Chapter 5 for more information about Sanger sequencing and Next Generation sequencing.

A disadvantage of these techniques is that they need to be followed by Sanger sequencing to determine the precise mutation description. With the advent of Next Generation sequencing, these methods are progressively being replaced in the clinical laboratory setting.

Key Point

PCR products can be analysed downstream by various methods. The most used and most straightforward is gel electrophoresis. Agarose gels are frequently used, but for finer resolution and certain specialized applications, polyacrylamide gels in denaturing or non-denaturing conditions can be used.

4.2.1.1.4 Consideration for PCR

There are several considerations that should be taken into account before performing a PCR and they are summarized next.

As considered in detail in Chapter 2, Section 2.2, particular care should be taken to avoid sources of contaminants. Contaminant DNA usually finds its way into PCR assays through:

- *Working environment*: e.g. rooms and equipment, laboratory benches and other work surfaces.
- *Consumable reagents and supplies*: e.g. oligonucleotides, media for sample collection and transport, plasticware.
- *Laboratory staff and their work habits*: e.g. contaminants from skin, hair, gloves, production of aerosols during pipetting.

PCR inhibitors can also present issues. These can include the presence of EDTA, phenol, ethanol, and isopropanol. In forensics, methods for extracting DNA that limit inhibitors have been developed. The choice of *Taq* polymerase might also need to be considered (e.g. AmpliTaq Gold polymerase is widely used, but it is the most sensitive to inhibition. Therefore, the choice of DNA extraction method, the type of *Taq* DNA polymerase, the quantity of template DNA added to the reaction, when less DNA can improve performance, are all factors to consider for optimum amplification. To detect PCR inhibitors, it is highly recommended to use an internal positive control (IPC).

4.2 SELF-CHECK QUESTION

What are the main sources of contamination in a PCR and how can these be avoided?

4.2.1.2 Real-time PCR

Real-time PCR is based on the same principle as conventional PCR, but the amplification of the target DNA (PCR products or amplicons) is analysed simultaneously as the reaction proceeds. Real-time PCR is used for sensitive, specific qualitative detection of a target and quantification of the number of copies of a DNA sequence of interest. To be able to monitor the progress of the PCR in 'real time', thermocycler technology has been designed to integrate an excitation light source, e.g. a lamp, a laser, and a camera or detector to detect fluorescence emissions. The fluorescence detected during the exponential growth (log) phase of PCR reflects the amount of amplified product in each cycle and is proportional to the amount of template DNA.

Different detection chemistries can be used, the two main ones being the use of an **intercalating agent** or dye, e.g. SYBR green or Eva Green, or the use of a sequence-specific probe-based detection, e.g. TaqMan^{TM} (Applied Biosystems, Waltham, MA), Scorpions® (Sigma-Aldrich, Gillingham), or amplification of the set of primers (ARMS), tagged with fluorescence molecules added to the reaction mixture. For a comprehensive list of different real-time PCR detection chemistry, the reader is referred to a review from Navarro et al. (2015).

> **DNA intercalating agent**
> Often also called a fluorescent dye; is a molecule that will insert itself in the DNA and will bind to it. These are used to visualize DNA under ultraviolet, for example, in agarose gel or to monitor the progress of a real-time PCR.

If an intercalating dye is used, it is based on the principle that it will increase its fluorescence when integrated in the minor groove of double-stranded DNA (dsDNA). Consequently, fluorescence can be measured in the extension phase of each cycle of qPCR. The advantage of this chemistry is that it is cheap and does not require any extra design steps in addition to the specific primers for the target DNA. The main downside of this method of detection is that it is not always specific to the target DNA of interest. If the PCR generates primer-dimers or non-specific amplification, all will contribute to the production of the fluorescence; therefore, it is highly recommended to add a melting curve analysis at the end of the qPCR to check the specificity of the amplified fragments (see Section 4.2.1.3).

The use of a fluorescently labelled probe, as opposed to an intercalating dye, will enhance specificity (Box 4.2 and Figure 4.2). The principle is that an oligonucleotide (probe) is designed to hybridize with the central region of the target DNA that is amplified. The probe is tagged with a fluorescent reporter molecule, such as FAM, VIC, JOE, or CY5 at the 5′ end, and on the opposite 3′ end with a quencher. Due to the close proximity of the quencher and the reporter, no fluorescence will be emitted until the PCR process starts and release of the quencher that will allow the detection of fluorescence from the probe reporter after the double-strand synthesis. There are a wide variety of reporters and quenchers with different excitation and emission spectra that can be used in qPCR. The commonly used probes are the TaqMan® probes and Scorpions probes. For more information see Didenko (2006).

BOX 4.2 Different type of probes for real-time PCR

TaqMan® probe: This probe is a complementary oligonucleotide to a target DNA sequence; that is, dual-labelled covalently with a fluorophore on the 5′ end (there are different types of fluorophores, e.g. FAM: 6-carboxy-fluorescein or TET: tetrachlorofluorescein) and with a quencher molecule on the 3′ end, e.g. tetramethylrhodamine (TAMRA). Due to the small distance between the fluorophore and the quencher, there is no emission of light after excitation if the probe has not bound to its complementary sequence. However, as soon as the probe binds to its DNA sequence and the *Taq* polymerase starts extension and incorporating nucleotides on the complementary strand, when it reaches the TaqMan probe the 5′ end fluorophore is released due to the 3′–5′ exonucleases activity of the polymerase allowing fluorescence upon excitation. Therefore, fluorescence can be detected during the exponential phase of the amplification and will be used to determine the quantitative accumulation of the PCR product.

Molecular Beacon probe: Similarly to the TaqMan® probe, this probe has two molecules attached at both 5′ and 3′ extremities of the probe, one being a fluorophore and the other a quencher. The distinctive feature of the Molecular Beacon probe is that it forms a hairpin structure with the loop containing the specific sequence complementary to the DNA sequence to amplify. When the probe is not hybridized to the target sequence, it keeps the quencher and the fluorophore in close proximity and there is no fluorescence emission after excitation. When annealing of the probe occurs, it causes the separation of the stem and, therefore, the separation of the fluorophore and the quencher. This allows emission of light from the fluorophore on excitation during the exponential phase of the PCR, thus allowing the quantification of the PCR product.

Scorpions® primer and probe: Scorpions probes are bi-functional molecules in which a PCR primer is covalently linked to a bi-labelled fluorescent probe sequence held in a hairpin-loop conformation with a 5′ reporter (fluorophore such as FAM) and a quencher (methyl red monomer) linked to the extremity end of the PCR primer via a non-amplifiable monomer blocker (hexethylene glycol). Similarly to the TaqMan® and Molecular Beacon probes, in the absence of a target to amplify, fluorescence is absorbed due to the presence of the quencher. During the Scorpions PCR reaction, in the presence of the DNA sequence target, the extension of primer causes the hairpin to unfold, and to separate the fluorophore and the quencher, which leads to fluorescence being emitted upon excitation with a laser source. Accordingly, the fluorescence can be detected and measured in the reaction tube during the exponential phase of the amplification.

ARMS™ (Amplification Refractory Mutation System™) primer: The technique works on the principle that *Taq* polymerase will distinguish between a match and a mismatch at the 3′ end of a PCR primer. When there is a perfect match, full efficiency of amplification will occur. When mismatches are present low-level background amplification will occur. Therefore, an ARMS primer can be designed with a mismatch base corresponding to a SNP or mutation allowing amplification only when the mutation is present, wild type sequence would not amplify. The ARMS primers can be used in combination with TaqMan® probes and Scorpions® probes in diagnostic kits—see Box 4.3 for examples.

Real-time PCR has many benefits over conventional PCR. These include:

- The reaction is monitored live, and so permits identification of reactions that have worked well and those that have failed.
- The efficiency of the reaction can be precisely calculated.
- There is no need to run the PCR product out on a gel after the reaction, and a melting curve analysis can be done at the end to assess if one or more fragments have been amplified or if primer-dimers are present.
- Real-time PCR data can be used to perform truly quantitative analysis of gene expression. In comparison, standard PCR is only semi-quantitative at best.

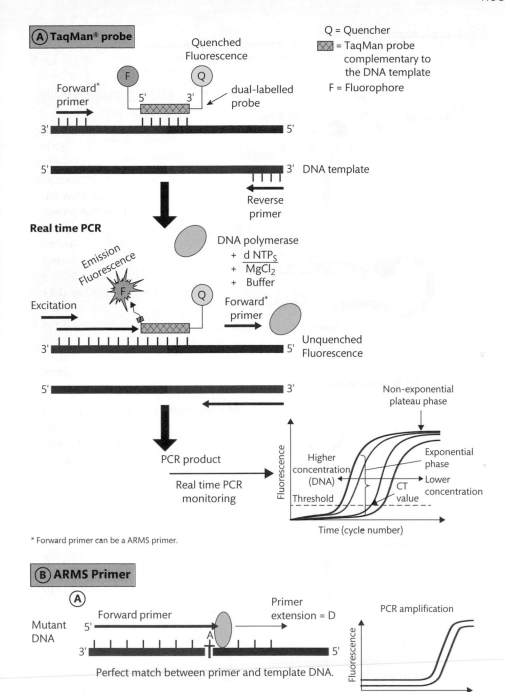

FIGURE 4.2
Probes used for qPCR.

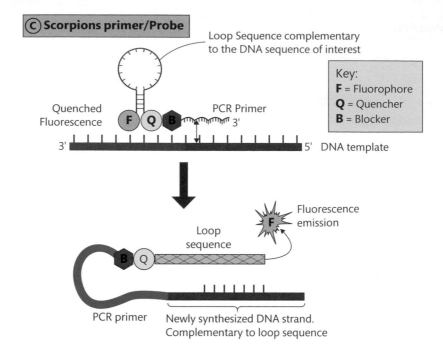

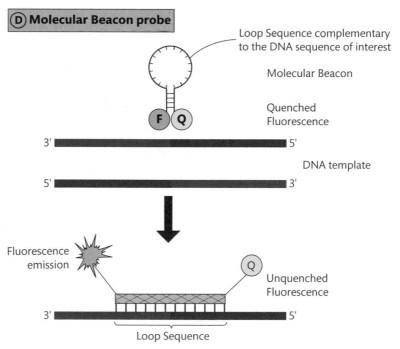

FIGURE 4.2 *continued*
Probes used for qPCR.

What are the different chemistries used in real-time PCR to detect the result of the amplification?

Real-time PCR technology has proven its versatility and usefulness in different research areas, including veterinary science, agriculture, pharmacology, biotechnology, and toxicology, but also in molecular clinical diagnostic for cancer (see Box 4.3) and in microbiology. It can be used for the quantification and genotyping of pathogens, gene expression, methylated DNA, and microRNA analysis, validation of micro-array data, allelic discrimination and genotyping, analysis of SNPs and microsatellites, identification of chromosomal alterations, validation of drug therapy efficacy, forensic studies, and quantification of genetically modified organisms (GMOs).

BOX 4.3 Example of a real-time PCR kit for oncology

Qiagen (Therascreen®) and other companies such as Roche Molecular Systems Inc. (COBAS®) have developed CE-IVD and FDA approved kits for testing for mutation in genes in solid tumours using qPCR for the qualitative detection of mutations. The following examples are of Therascreen® kits.

Three Therascreen® assays have been developed that target hot-spot mutations in KRAS (metastatic colorectal cancer), EGFR (non-small cell lung cancer NSCLC), and BRAF (melanoma).

The Therascreen® KRAS RGQ PCR Kit was FDA approved as companion diagnostic for Vectibix® (panitumumab) in treatment of metastatic colorectal cancer in May 2014. The kit detects seven KRAS hot-spot mutations.

Therascreen® EGFR RGQ PCR Kit was FDA-Approved as companion diagnostic for afatinib and gefitinib for the treatment of NSCLC in July 2013. The kit detects 29 EGFR mutations.

Therascreen® BRAF RGQ PCR Kit detects four BRAF (V600) mutations and is used for melanomas.

The Therascreen® real-time PCR assays are designed on an identical principle which is each mutation-specific reaction mix uses a mutation-specific ARMS (see Box 4.2) primer to selectively amplify and enrich the mutated DNA, and then uses a labelled Scorpions® primer to detect the mutation during a second amplification reaction.

The Scorpions ® probe is FAM-labelled. The assay also contains an internal control labelled with HEX™ fluorophore targeting a region of the gene not known to contain SNP or mutation. Therefore, the internal control should always amplify. If it fails to do so then this may point to the presence of inhibitors in the DNA that may lead to false negative results.

The final choice of which of the CE-IVD-marked KRAS/EGFR/BRAF mutation assays to use is largely dependent on availability of appropriate laboratory equipment, experience, and cost of the kits.

4.2.1.3 Qualitative and quantitative PCR analyses

A melting curve analysis is an assessment of the dissociation characteristics of double-stranded DNA during heating; therefore, each PCR product present will be represented by a peak as they will have a different sequence content influencing the dissociation by heating of the two strands (see Figure 4.3).

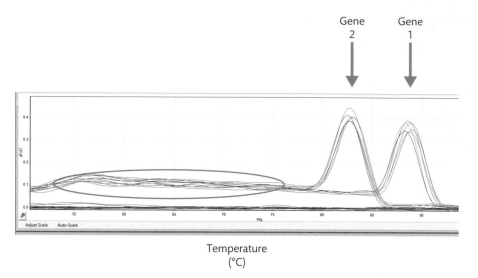

Temperature
(°C)

FIGURE 4.3

Example of a melting curve analysis performed after a RT-qPCR run with amplification of gene of interest (gene 1) and a housekeeping gene (gene 2) amplification. Each of the arrows corresponds to a specific gene amplification. As the sequence of the two amplified sequences is different, this results in two distinct peaks. The encircled region shows the absence of primer-dimer. If primer-dimrswere formed, there will be some small peaks appearing here. The x-axis represents the temperature and the y-axis is the detected fluorescence.

Like many protein assays, an unknown sample is compared with a standard curve, in order to extrapolate the starting concentration. A dilution series of known template concentrations, for example, a plasmid containing the target sequence, can be used to establish a standard curve for determining the initial starting amount of the target template in the test samples or for assessing the reaction efficiency. The log of each known concentration in the dilution series or the number of gene copies (x-axis) is plotted against the Ct value for that concentration (y-axis). The Ct (cycle threshold) is the numerical value, which correspond to a number of cycles that it takes for each reaction to reach an arbitrary amount of fluorescence during the exponential phase of the amplification. From this standard curve, information about the performance of the reaction, as well as various reaction parameters, such as the slope of the amplification curve, can be derived.

For mutation detection with real time PCR, such as the method used for the Therascreen®, (Qiagen, Venio, Netherlands) assay for EGFR, KRAS, and BRAF mutation screening from Qiagen, the Ct value is determined for the ARMS designed to target a specific mutation and compared with the internal control. The sample value, therefore, represents the calculated difference between the mut Ct and the control Ct, and is called the delta Ct (ΔCT = mut Ct − control Ct). A sample is deemed positive for the mutation if the ΔCT is less than a cut-off ΔCt value for a particular assay design. Each Therascreen® kit comes with a handbook that explains data interpretation and gives the cut-off values for each mutation detected (see EGFR detection with Therascreen®; see Figure 4.4).

The assay includes the amplification of an internal control, which corresponds to a region in the genome not known to harbour any hot-spot mutations or SNPs. This will assure the operator that no PCR inhibitors are present in the DNA sample being tested and that the quality of the template DNA is satisfactory. Therefore, if no amplification is obtained it means that the mutation is not present and the result is not a false negative.

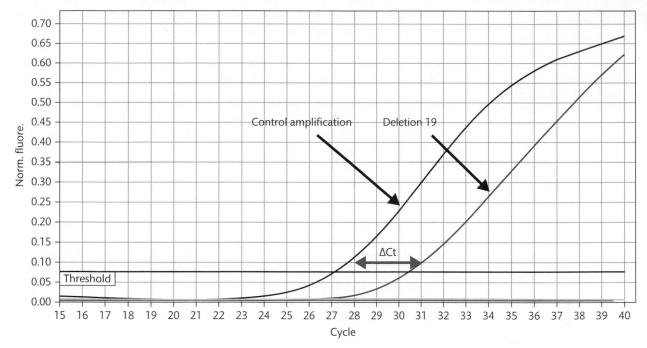

FIGURE 4.4

Example of detection of deletion 19 in EGFR using Therascreen® kit from Qiagen. Mutation assay cut-offs (ΔCt) for the mutation detected by the EGFR Therascreen assay. The x-axis represents the number of cycles and the y-axis is the fluorescence levels.

4.2.1.4 Microarrays-based techniques

Microarrays are high-throughput technologies (high density arrays) that emerged in the late 1990s, and have revolutionized the analysis of whole systems and are often referred to as 'OMICs' analyses. DNA microarrays are also referred to as DNA arrays, DNA chips, biochips, and GeneChips, and can simultaneously analyse thousands of features deposited on a support such as glass or silicon and do this in parallel for several samples.

For DNA microarrays, there are three main type of analysis that can be performed. These are genotyping with single nucleotide polymorphism (SNP) microarrays, molecular cytogenetics with array comparative genome hybridization (aCGH), and epigenetics with methylation arrays.

Multiple microarray platforms with different technological approaches exist and are produced by several companies, such as Affymetrix® (now part of the ThermoFischer Scientific group), Agilent® (Santa Clara, CA), and Illumina® (Illumina, San Diego, CA).

Affymetrix® technology is based on *in-situ* single-stranded DNA oligonucleotides (25-mers), synthesized and deposited on 5-inch square quartz wafers using photolithography and combinatorial chemistry. The DNA oligonucleotides correspond to thousands of sequences across the whole genome or transcriptome of an organism. Affymetrix offers a wide variety of microarrays, such as genome-wide and targeted genotyping microarrays for human disease research, e.g., SNP Array 6.0, mapping 10K, 100K and 500K Array sets.

Molecular cytogenetics arrays, such as CytoScan™ HD Array (ThermoFisher, Santa Clara, CA) and OncoScan® (ThermoFisher, Santa Clara, CA) microarrays, are used to screen for genome-wide copy number variations through the addition or deletion of specific regions of DNA. They may also be used for the investigation of **loss of heterozygosity (LOH)** and to profile solid tumour samples from degraded FFPE samples. Affymetrix is also well known for their Genechip® (ThermoFisher, Santa Clara, CA) microarrays for gene expression analysis, but this will be developed in the RNA section (see Section 4.2.2.2).

The Agilent® technology/platform is based on glass DNA microarrays that support 60-mer oligonucleotides probes, synthesized *in situ*, using Agilent's inkjet SurePrint technology. Examples of products for diagnostic testing are CGH & CGH + SNP microarrays [GenetiSure Postnatal Research CGH+SNP Array; GenetiSure Research CGH+SNP Array $(2 \times 400K)$] or SurePrint CGH & CNV microarrays. Agilent also offers custom-made microarrays using the same SurePrint technology.

The Illumina® platform, in contrast to Affymetrix and Agilent, uses BeadArray and Infinium assay technologies. BeadArray technology is based on 3-µm silica beads that self-assemble in microwells on either of two substrates—fibre optic bundles or planar silica slides. Each bead is covered with hundreds of thousands of copies of a specific oligonucleotide that act as the capture sequences in a given assay. Similar to other companies, Illumina has developed different microarrays to interrogate gDNA at both whole genome (Infinium Omni2.5-8) or exome (Infinium Omni2.5Exome-8) and targeted genome level. The latter range from relatively low coverage (1,756,820 fixed markers/SNPs with possibility of adding custom ones up to 245,000) to higher coverage (4,559,465 fixed markers/SNPs). Similarly, cytogenetic variants can be screened either at whole genome such as HumanCytoSNP-12 (12 samples) or targeted such as CytoSNP-850K.

All these platforms and applications are based on hybridization between two single-stranded DNA sequences. The different DNA fragments on the support are arranged in rows and columns, such that the identity of each fragment is known through its location on the array. The sample to be analysed is usually labelled with fluorescence-emitting dyes (Cy3/Cy5) or biotin. In the latter situation visualization is obtained using a fluorescent molecule, streptavidin-phycoerythrin, which binds to biotin (see Figure 4.5).

The conditions are optimized for an optimum temperature of hybridization and stringency of post-hybridization washes to keep the hybridization optimal for detection and quantification. Fluorescence signals are emitted after excitation by a laser beam (scanner) at the appropriate wavelength. The signals detected are then converted to numerical values that are analysed by sophisticated algorithms as discussed in Chapter 6, Section 6.2.1.1.

There are many applications of microarray technologies (see review by Heller, 2002). Early versions of DNA microarrays were designed to interrogate the level of expression of genes. Microarrays are now used for interrogating loss or gain of genetic material, to screen for SNPs, or to determine miRNA expression levels. An entire organism can be interrogated and several samples processed simultaneously. They can be used in various fields, including gene discovery, disease diagnosis, drug discovery, and toxicological research. Microarrays are used in research and industrial discovery settings, but more rarely used for diagnosis as their applications are currently for research use only. However, some of them are currently being included in clinical trials pointing to their potential use as an *in vitro* diagnostic. To apply microarrays to clinical settings, physicians need tools approved by regulatory agencies, such as the US Food and Drug Administration (FDA). Ongoing efforts towards the establishment of benchmark standards, assay optimization for clinical conditions, and demonstration of assay reproducibility are required to expand the clinical utility of microarrays.

Microarray Protocols (glass slide versus Affymetrix (Inc.) Genechip).

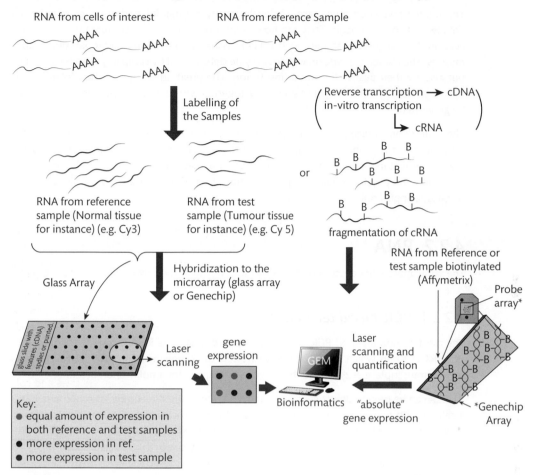

FIGURE 4.5

Microarray protocols: glass slide and Affymetrix Genechip®.

Key Point

A microarray is made of a supporting material (a glass slide, a silicon surface, or a bead) onto which thousands of DNA sequences (oligonucleotides or cDNA) are attached. These are either spotted on the surface or directly synthesized on the surface, and may represent the entire set of genes or DNA sequences of an organism. The DNA sequences are arranged in defined patterns and, consequently, allow hybrids formed with the samples to be precisely identified.

4.4 SELF-CHECK QUESTION

What are the different types of microarrays and what applications can they be used for?

4.2.1.4.1 Considerations to be taken for optimum results

The microarrays cited above are all developed by companies and are accompanied by detailed reliable protocols. They are now being optimized for small quantities of nucleic acids and for degraded samples, such as FFPE-tissue samples from cellular pathology laboratories. The choice of platform will not only be determined by sample type to be analysed, but also by their availability in the institution. The great advantage of these platforms is their high density, sensitivity, and specificity, together with the speed with which results can be generated.

The main disadvantage of microarrays is that they require some previous knowledge of the sequences, as the probes are synthesized on the support. This does not allow for full discovery of new 'features' in contrast to next-generation sequencing. The ongoing advances in this technology may well overtake the potential transition of microarrays from research use to diagnostic application.

4.2.2 RNA

The same approaches described previously for DNA can also be applied for the study of RNA.

4.2.2.1 PCR-based techniques

The most commonly used PCR-based technique for analysing RNA is reverse transcription-PCR (RT-PCR), in which mRNA is first converted to complementary DNA (cDNA) using a reverse transcriptase enzyme and then used as a template for amplification by PCR.

Cross reference

See Chapters 9, 10, and 13 for examples in haematological malignancies and sarcomas.

RT-PCR can be used to detect gene fusions that are the result of recurrent chromosomal translocations with many examples in haematological malignancies and sarcomas—There are also applications in infectious diseases. Another frequently used application of real-time RT-PCR is for detecting the expression of the panel of genes associated with a particular disease. This is the case for cancer, and particularly, for the diagnostic and prognostic testing of breast cancer (see Chapter 11).

The design of the primers is particularly important for RT-PCR in order to exclude amplification from the presence of gDNA 'contaminants'. To be able to discriminate amplification from gDNA, the forward primer and the reverse primer are usually designed to anneal in different exons, thus using size exclusion to prevent amplification of gDNA. Another precaution to take to avoid amplification from gDNA is to treat the RNA prior to use with DNAse I. Most of the extraction kits offer an optional DNAse on-column treatment.

RT-PCR can be done in a conventional way with analysis of the end-products on agarose gel, for example, when detecting the presence of a gene-fusion. However, real-time RT-PCR is more frequently used to detect the presence of gene fusions and to quantify genomic RNA from viral pathogens.

RT-qPCR is a reliable technique. However, there are some obstacles impeding a more extensive adoption of RT-qPCR assays for clinical use, such as assay quality assessment and standardization, which both affect reproducibility.

4.2.2.2 Qualitative and quantitative RT-PCR analyses

As for standard PCR analyses as described in Section 4.2.1.3, the same criteria apply for a standard RT-PCR analysis. If RT-PCR is used to identify the presence of a translocation, a single band at the expected size against a reference ladder and a positive control (known sample with the translocation in question) will be visualized in the electrophoretic gel.

When RT-qPCR is employed, several analytical approaches exist to assess the level of gene expression and, thus, quantify mRNA levels. In addition to the gene in question, a housekeeping gene is used as a reference to ensure that a similar amount of RNA has been converted into cDNA for each of the samples that will be used to assess the level of expression. What is important to consider is the use of a gene for which the expression will stay the same regardless of the cell type or tissue of origin, and that is not affected by any experimental treatment or conditions of cell culture. Sometimes two or three housekeeping genes are used. For more insight into the choice of the best housekeeping gene to use in analyses read Radonić et al. (2004), and Kozera and Rapacz (2013). As already described in the DNA section, an absolute Ct value will be recorded for each sample, and this will be used to quantify RNA against an experimental reference, using the housekeeping gene expression to normalize the data obtained. Of note, generally, a Ct value above 38 or 40 represents a gene for which no expression is detectable.

The following formulas will be used to determine the expression of the gene in question. Usually, the expression of a gene is determined compared with a 'reference' that can vary depending on the experiment. It could be the untreated cells or a normal tissue counterpart.

- *First step*: Normalize the expression level of gene in question to the expression level of the housekeeping gene (ΔCT) for each sample to be analysed (sample 1 is, for example, a cancer cell and sample 2 is a normal cell, but could also be treated and untreated) following the formulas:

$$\Delta CT\ sample1 = Ct\ (test\ gene) - Ct\ (Housekeeping\ gene)$$

$$\Delta CT\ sample2 = Ct\ (test\ gene) - Ct\ (Housekeeping\ gene)$$

- *Second step*: calculate the ΔΔCT that would be the relative expression of gene in question in sample 1 relative to sample 2 following the formulae:

$$\Delta\Delta CT = \Delta CT\ sample1 - \Delta CT\ sample2.$$

- This ΔΔCT will, therefore, correspond to a fold change in expression of the gene of interest in sample 1 (test sample) relative to sample 2 (which could be considered as the control sample, being either untreated cells or normal cells).

4.2.2.3 Microarrays-based techniques

Microarrays have been extensively used for analysing gene expression profiles and particularly in the field of oncology. In fact, the first DNA microarrays were developed with this application in mind. Other applications, such as genotyping, molecular cytogenetics, and methylation came later with the successes of gene expression microarrays (GEM).

GEM allow researchers to determine which genes are being expressed in a given cell type at a particular time and under particular conditions. They can be used to compare the gene expression in two different cell types or tissue samples; for example, healthy versus diseased tissue, or to examine changes in gene expression at different stages in the cell cycle or after treatment with a drug.

Companies, including those already mentioned for the analysis of DNA, have developed several platforms. In addition, in-house platforms were developed in the early 2000s, particularly in universities like Stanford in California, USA, using cDNA oligonucleotides spotted on a glass slide with a robot arrayer and hybridized with amplified single-stranded RNA labelled with different fluorochromes; for example, normal RNA in green with Cy3 and tumour RNA in red with Cy5.

Genechip® from Affymetrix® (ThermoFisher, Santa Clara, CA) have been one of the pioneers in the production of commercial platforms. Affymetrix's most popular Genechip®, the Human Genome U133 set of two microarrays (HG-U133 plus 2) contains over 1 million different oligonucleotides (25-mers), representing more than 33,000 of the best-characterized human genes. Each gene is represented by several probe sets with each probe set being represented by 11 oligonucleotides overlapping the sequence. The genechips are single-use only with one sample that is hybridized using a Genechip® microdevice. Complex bioinformatics is used afterwards to compare the scanning of different Genechips processed with different samples to determine the relative expression of the genes in the samples.

Cross reference

Gene expression microarrays analysis is covered in Chapter 6, Section 6.2.2.1.

One of the latest developments in the field of GEM has been the design of Genechip®, which can capture splice variants (Genechip Human Exon 1.0 ST Array). Alternative splicing of mRNA is a key molecular mechanism for increasing the functional diversity of the eukaryotic proteomes. A large body of experimental data implicates aberrant splicing in various human diseases. Moreover, in addition to coding RNA and alternative splicing, other types of RNA molecules such as non-coding RNA miRNAs can also be analysed using microarrays (Gene Chip miRNA 1.0 microarrays from Affymetrix and SurePrint Human miRNA Microarrays).

Genechip® has been used extensively in cancer research for tumour classification and identification of gene expression signatures to help predict prognosis and disease progression, particularly with identification of metastatic signatures (Perou et al., 2000; van't Veer et al., 2002, 2003). Although these studies have made great advancement on the molecular characterization of tumour subtypes in research settings, they are not commonly used in routine molecular diagnostics.

One of the main limitations for the use of GEM in clinical settings is that an individual tumour cannot be classified independently. It needs to be compared with other samples or 'standards', whose classification is known, and that is analysed under the same conditions as the tumour sample. This type of analysis also requires powerful bioinformatics, which presently limits its use in clinical settings. The issues of standardization and inter-laboratory discrepancies need also to be addressed before it can be widely implemented in clinical routine.

Cross reference

See Chapter 10, Section 10.7 for details about gene expression profiling of breast cancer.

Nonetheless, gene panels identified by translational research studies using these platforms have been used to provide diagnostic and prognostic tools. Examples of FDA approved tests include MammaPrint® test (Agendia, Irvine, CA), Oncotype DX and Prosigna® (NanoString, Ardmore, PA) for breast cancer. In particular, Oncotype Dx can be used to influence patient treatment and to predict cancer recurrence.

Key Point

mRNA expression can be assessed after conversion to gDNA using RT-PCR and RT-qPCR. Applications include cancer diagnosis and the identification of infectious agents. GEMs can also be used and have been extensively used in defining new cancer subtypes, for diagnosis and classification in the field of both haematological and solid tumours.

4.3 Protein analysis

The large-scale study of proteins is known as 'Proteomics'. The terms 'Proteomics' and 'Proteome' were first coined by a PhD student in the 1990s (Wasinger et al., 1995). They represent the combination of the words 'protein' and 'genome'. The key facet of any proteomics workflow is the ability to identify and quantify proteins of interest. Over the years, many approaches have been devised that range from single analyte to multi-analyte measurements. These rely on the use of the most sophisticated and sensitive instrumentation, although there are key components that need to be considered to achieve an optimal protocol for each study.

4.3.1 Protein identification and quantification

4.3.1.1 Western blotting

Historically, Western blotting has been used for the identification of proteins since the late 1970s (Towbin et al., 1979). This method relies on a two-step process, separation of proteins using gel electrophoresis, together with a **molecular weight of proteins** ladder, and then electrophoretic blotting, to appropriate membranes to be able to probe with specific antibodies to confirm the identity of the target protein (Figure 4.6).

This makes Western blotting an extremely powerful technique, despite its overall simplicity. Additional and unexpected bands can give further clues to the protein, e.g. several bands could indicate protein degradation; higher or lower bands could indicate specific protein post-translational modification (PTM), such as phosphorylation or glycosylation (see Sections 4.3.3.1 and 4.3.3.2). Alternate splicing may also cause unexpected size variations, as may the particular combination of charged amino acids found in the protein. Therefore, this method is still considered to be a valuable tool in scientific laboratories worldwide and will continue to be so. However, one of the major disadvantages of Western blotting is that it is a semi-quantitative tool and its success is dependent on the use of antibodies with confirmed specificity and affinity. Furthermore, a Western blot is a single analyte technique with the target of interest predetermined before evaluation.

> **Molecular weight of proteins**
> This is usually expressed in kilodaltons (kDa). For example, a molecule of a protein with molar mass of 35,000 g mol^{-1} has a mass of 35 kDa.

Key Point

The key to a good Western blot analysis is the use of specific and high affinity primary antibodies to detect proteins of interest.

4.5 SELF-CHECK QUESTION

What are the key advantages of using Western blotting?

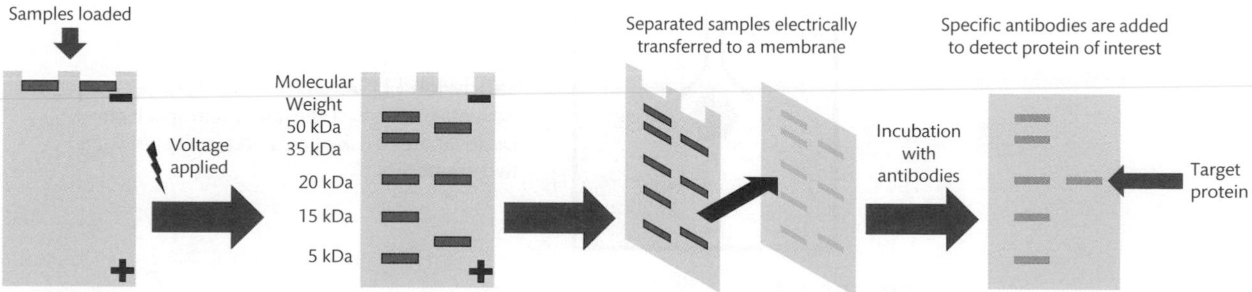

FIGURE 4.6
A schematic representation of the Western blotting process.
Note: The molecular weight markers (left) are run together with the test sample (right). These allow the determination of the molecular weight of the test sample band/s to be estimated.

4.3.1.2 Enzyme-linked immunosorbent assay (ELISA)

When a quantitative approach is required for the detection of proteins or peptides, the plate-based enzyme-linked immunosorbent assay (ELISA) technique is frequently used. There are numerous variations of ELISA available, such as direct, indirect, and sandwich (Figure 4.7).

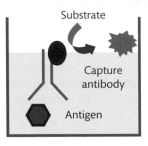

Direct ELISA
Advantages: quicker, reduced steps
Disadvantages: higher background signal

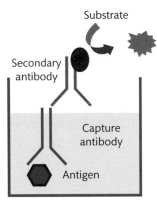

Indirect ELISA
Advantages: higher sensitivity
Disadvantages: risk of cross-reactivity between two antibodies

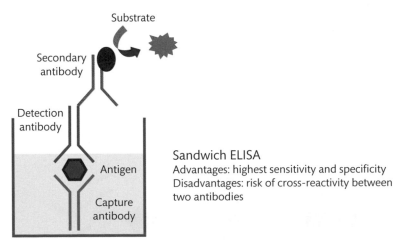

Sandwich ELISA
Advantages: highest sensitivity and specificity
Disadvantages: risk of cross-reactivity between two antibodies

FIGURE 4.7
An illustration of the most popular versions of ELISA that are utilized in the laboratory.

The assay works using a combination of antigen (immobilized on plate) and specific antibodies and measured using an enzyme reaction. The most crucial element of the assay is the addition of known quantities of the analyte that are used to produce a calibration or standard curve, allowing the calculation of the concentration of the protein in the sample being analysed.

ELISA measures a single analyte. However, the Luminex xMAP® (Luminex, Austin, TX) and Meso Scale Discovery (MSD; Rockville, MD) utilize a multiplex approach. The xMAP® allows simultaneous analysis of up to 500 targets from a small sample volume, by reading biological tests on the surface of microscopic polystyrene or magnetic beads called microspheres. The detection is performed in liquid array bioassay with small lasers, digital signal processors, photo detectors, and charge-coupled device imaging.

More recently, this technology has been further developed to an ultra-sensitive technique, such as single-molecule arrays (SIMOA) that can isolate and detect single enzyme molecules. Generally, ELISAs rely on a colour change, while the SIMOA is termed a digital ELISA and provides a binary output. Conventional ELISAs are limited to the picomolar (pg/mL) range, while the SIMOA is >1000-fold more sensitive and detects in the range of femtomolar (fg/mL). Additionally, traditional ELISAs require larger volumes of samples, while the SIMOA requires almost 50% less volume for comparative targets (Zetterberg et al., 2013).

Key Point

All plate-based assays are considered quantitative, because all targets are measured against a calibrator or standard curve.

4.6 SELF-CHECK QUESTION

What are the key differences between ELISA and multi-analyte assays?

4.3.1.3 Two-dimensional gel electrophoresis

For large scale single sample analysis two-dimensional electrophoresis (2-DGE) has been synonymous with proteomics. It still remains the best method for resolving highly complex protein mixtures. The technique is a combination of two different types of separation. In the first, the proteins are resolved on the basis of their isoelectric point by **isoelectric focusing** (IEF). In the second, focused proteins are further separated by electrophoresis on a polyacrylamide gel (Figure 4.8).

Thus, 2-DGE has the ability to resolve proteins from any type of protein mixture. In addition, due to the resolving power, 2-DGE can be used to visualize PTMs that are associated with the life cycle of a protein.

Recent proteomic approaches have seen an influx of technologies that can perform measurement of numerous proteins at the same time from single individual samples. These have ranged from gel-based two-dimensional differential gel electrophoresis—2-DIGE (Rabilloud and Lelong, 2011), plate-based multianalyte [Luminex xMAP®; Houser (2012) and MSD; Leng et al. (2008)] and array-based SomaLogic (Kraemer et al., 2011). Two-dimensional electrophoresis inherently has variability difficulties when comparing large data sets. What separates the original 2-DGE from 2D-DIGE is the use of fluorescent protein labels that allows the co-electrophoresis of multiple samples on a single gel.

Isoelectric focusing

A process of separating proteins or peptides based on their isoelectric point. All proteins carry either a positive or negative charge, and the isoelectric point is when the charge overall is neutral.

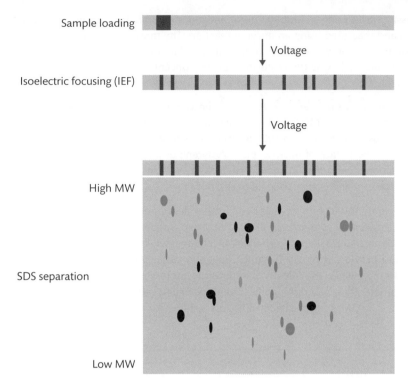

FIGURE 4.8
A schematic representation of two-dimensional electrophoresis (2-DGE).
Key: SDS, Sodium dodecyl sulfate; MW, molecular weight.

Key Point

Two-dimensional gel electrophoresis (2-DGE) is based on two properties of a protein—isoelectric point and molecular weight.

4.7 SELF-CHECK QUESTION

How does 2-DGE differ from 2D-DIGE?

4.3.1.4 Quantitation and identification of analytes

A key aspect of any quantitative approach is the accurate calculation of the concentration of the analyte in immunoassays or the identification of the spots in a 2-DGE gel. Many software packages are available to assist in this process. All ELISA and multi-analyte assays require 'curve fitting' software. Briefly, the better fit of the standard concentrations to a curve, the better the prediction would of the unknowns. Most software is of a simple linear model fitting type (line of 'best' fit). However, almost all modern immunoassays require a sigmoidal fitting. For this, a polynomial (non-linear) fitting is required, which essentially attempts to draw a dynamic line through all points on the graph.

As for the identification of the spots on a 2-DGE image, a number of steps is required that help in aligning, superimposing, and quantifying. In the first steps, the image is digitized and uploaded to the software (e.g. Progenesis; Non-linear dynamics). A reference image is assigned and all other images are then aligned by software superimposing until all the spots are aligned. Subsequently, the spots are then quantified and pixel intensity is obtained and statistically tested between different groups of gels.

4.3.2 Mass spectrometry

A mass spectrometer can help determine molecular masses by the measurement of the mass/charge (m/z) ratio of ions generated in a protein sample. A mass spectrum is a plot of the m/z acquired and this allows for the molecular composition of a certain sample or analyte to be accurately reported. Combined with a chemical label (discussed below) a relative abundance can also be determined. In proteomics, the analyte is usually a collection of peptides arising from a protein that has been enzymatically digested from clinical tissues, biofluids or cell culture.

In general, mass spectrometers have three main components—an ion source, mass analyser, and a mass detector. At the ion source, the analyte is converted into gaseous phase ions (ionization), which are then accelerated and separated according to their m/z within the mass analyser. The detector records the impact of the individual ions and their relative abundance in that sample. This information is needed to determine identification and quantification.

Ionization is generally achieved by two methods—matrix-assisted laser desorption/ionization (MALDI) and electrospray ionization (ESI).

In MALDI, the protein sample is combined with a 'matrix compound' that can absorb energy from a laser. This mixture is dissolved in an organic solvent, which then evaporates leaving just matrix crystals that contain the analyte(s) of interest. These crystals are targeted for excitation within the mass spectrometer chamber under high voltage. The energy from the laser is absorbed by the matrix crystals resulting in rapid **sublimation**. Emitted as heat, the gaseous ions accelerate towards the mass analyser detector.

ESI generally follows liquid chromatography (LC), where an analyte is diluted, separated and released from a bedded column with an increasing gradient of organic solvent (Figure 4.9).

LC is often utilized in proteomics to ensure a material is less complex when entering the mass spectrometer. Once approaching the mass spectrometer, the liquid solvent containing the analyte is forced through a narrow needle and released towards the mass spectrometer at high voltage. As the droplet (seen as a fine spray) enters the mass spectrometer, an inert gas (usually hydrogen) evaporates the organic solvent making the droplet progressively more positively charged. Once the charge of the droplets exceeds the '**Rayleigh's limit**' the droplets dissociate, resulting in the gas phase ions of the analyte being accelerated towards the detector. As ESI produces gaseous ions from a solution, it is more compatible with upstream protein fractionation (1-DGE/2-DGE) and LC (LC-ESI-MS). Therefore, ESI is the method of choice when examining complex samples, whereas MALDI is suitable for simple peptide mixtures.

Cross reference
Proteomic methods are discussed in Chapter 1, Section 1.4.1.

Sublimation
The phase transition of a substance directly from the solid to the gaseous phase without passing through a liquid phase.

Rayleigh's limit
Theoretical estimate of the maximum amount of charge a liquid droplet could carry.

Key Point

MALDI and ESI are methods used to ionize peptides for analysis by mass spectroscopy. MALDI excites ions with a laser to create a gaseous phase directly from solid material, whereas ESI uses the evaporation of a droplet to create a gaseous phase from a liquid phase. ESI is typically chosen for complex matrices such as cell lysates or biological fluids.

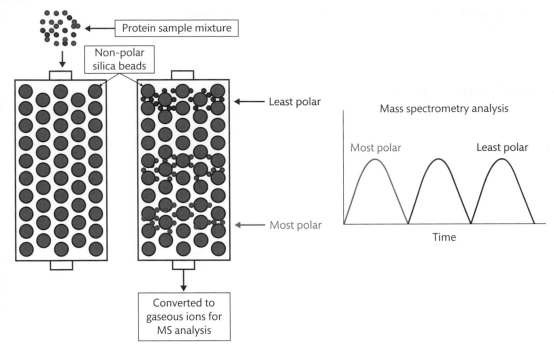

FIGURE 4.9
A schematic representation of liquid chromagraphic separation of a complex mixture prior to mass spectrometry analysis.

4.3.2.1 Mass analysers

Four main types of mass analysers are utilized in proteomics; triple quadrupole (TQ), time of flight (TOF), ion trap, and Fourier transform (FT).

TQ instruments utilize four electronically charged parallel metal rods to guide or selectively eliminate ions of any m/z in reaching the detector. A TQ has three quadrupoles and, therefore, it can be utilized for not only single MS (MS1) scans, but also multiple separation (MS/MS or MS2) scanning. The first quadrupole will be utilized for a typical MS1 scan and only intact peptide ions of a certain m/z will be selected to enter the second quadrupole. Here, the ions are fragmented by the collision with an inert gas (collision-induced dissociation; CID). These ion fragments are then focused and scanned by the third quadrupole, which generate a CID spectrum of one peptide. Once this is completed the first quadrupole will select intact peptides of a differing m/z.

Unlike TQ, TOF analysers do not require an electric field and simply uses the principle that ions of differing mass will occur within a given sample. In principal, smaller ions will arrive at a detector quicker than heavier ions if allowed to resolve in a flight tube (Glish and Vachet, 2003; El-Aneed et al., 2009).

More recently, TOF analysers have been coupled to a TQ analyser to create a highly sensitive hybrid analyser to report CID spectra. Ion traps separate ions in time, rather than by space. Ions of a certain m/z can be rejected from the ion trap chamber. An MS spectrum of intact peptides can be determined by gradually increasing the voltage and accelerating ions towards the detector. Alternatively, within the ion trap, ions can be fragmented by the insertion of helium gas; the increasing voltage then will send ion fragments and not intact ions towards the detector to create CID spectra.

The FT analysers are the latest model of analyser and the most complex. However, the resolution and mass accuracy far outperform other models. The principle of FT analysers is that ions within a magnetic field will orbit at the frequency that is in line with their m/z and, therefore, ions with a similar m/z will orbit around a magnetic field at the same frequency. Each group of ions with a similar m/z are then excited by an applied radio frequency. The change in ion velocity and individual excitation between electrodes is recorded and transformed (FT) into a series of frequencies and amplitudes that are further converted to m/z values to form a mass spectrum (Eliuk and Makarov, 2015).

4.3.2.2 Peptide mass finger printing

Peptide mass fingerprinting (PMF) is the term given to the process of identifying proteins using the information from intact peptide masses. Each protein can be uniquely identified by the masses of its corresponding peptides. This previously manual process has been automated with several widely available software packages performing protein-matching based upon peptide data. After the masses of the analyte have been determined by mass spectrometry, the identified spectra are matched against a sequence database (e.g. SWISS-PROT https://www.ebi.ac.uk/uniprot) within a program (e.g. Mascot, matrix science, or Sequest). The database carries out a theoretical protein digest employing the cleavage sites of the enzyme used in the experiment, typically trypsin. The estimated peptide masses of the derived peptides via a virtual digest are then correlated to the one derived by the real experiment. The proteins are then ranked in order of the best correlation with the virtual data, with a significance threshold for each match. In the same manner, CID fragment ions can be compared with a virtual digest and the degree of overlap between observed and predicted ions determines the protein and the probability of it being an accurate match.

The interpretation of CID data is more complex due to an added level of fragmentation and, therefore, increased number of 'series ions', N-terminal 'b series ions', and C-terminal 'y series ions' can aid this interpretation and the ordering of peptide sequence. These fragment ions can also be useful in the complete interpretation of CID spectra by *de novo* sequencing, which pieces all the evidence from the series ions together to identify a fragment peptide and, therefore, a protein match. Manual *de novo* sequencing can be extremely difficult and time-consuming. However, database algorithms have begun to perform automated *de novo* sequencing (e.g. PEAKS software). Conventional database probability matching fails to recognize novel peptides (arising from gene modifications), since it can only be matched to existing sequences in a known database from 'healthy individuals', although these databases are beginning to include data from disease phenotypes. The advantage of *de novo* sequencing is that it not reliant upon a database and a peptide can be investigated in relation to disease, even if its protein assignment is yet to be determined.

4.3.2.3 Isobaric labelling for mass spectrometry

In order to measure multiple samples and improve detection in proteomic experiments chemical labels such as tandem mass tags (TMT, Thermo Scientific, Santa Clara, CA) and isobaric tags for relative and absolute quantitation (iTRAQ, SCIEX) are utilized. For cell culture experiments, SILAC (Thermo Scientific) is often utilized. In all cases, this process involves a chemical reaction to help bind the tags of different molecular weights to the protein or peptide.

In principle, proteins or peptides are extracted from either cells or tissue, and chemically labelled with isobaric stable isotope tags, which will be unique for an individual patient or preparation. All labelled samples or preparations with unique tags can then be mixed together

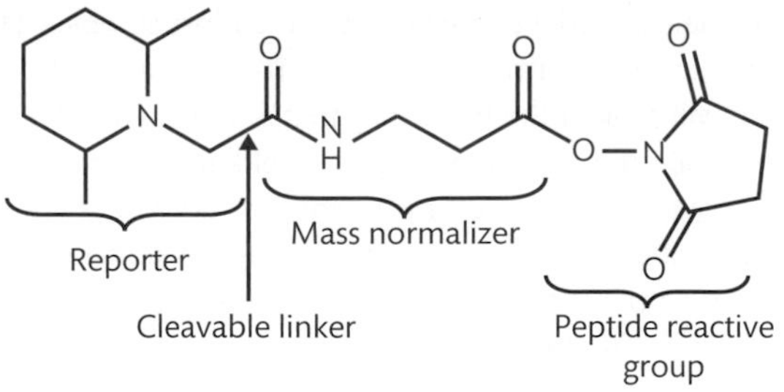

FIGURE 4.10
Functional regions of a tandem mass tag (TMT) structure. During MS2 fractionation by higher energy collision dissociation (HCD) the cleavable linker is broken to produce a unique Mass Reporter.
Redrawn from Thermo Scientific.

in a multiplexed manner, such as duplex, 6-plex or 10-plex from TMT, while iTRAQ is available in 4- and 8-plex. Next, TMT labelling will be considered as an example.

TMT chemical tags contain four regions—a mass reporter region, a cleavable linker, a mass normalizer, and a protein reactive group (Figure 4.10).

Although the full structure is chemically identical, each TMT tag has isotope replacements at differing locations. This ensures that the mass reporter has differing molecular masses in each individual tag. As the full structure and molecular weight of each TMT tag is identical, the differing TMT labels cannot be distinguished during protein fractionation, chromatography, or the first stage of fragmentation by mass spectrometry (MS1). The secondary stage of fragmentation (MS2) is achieved by high energy collision dissociation (HCD) and this cleaves off the peptide backbone and cleavable linker giving rise to normal fragment ions, and also ions regarding the relative amount of the peptide in the samples by TMT ions.

Key Point

Isobaric chemical labelling can be used to combine multiple samples, together or multiple preparations for direct comparison. This is advantageous as it reduces the cost and throughput of an experiment, but also allows the direct comparison of two or more groups (e.g. disease versus non-disease) in a single experiment. Each isobaric label is identical in mass, however, when fragmented by HCD, the chemical tag is cleaved and reveals the unique chemical label.

4.8 SELF-CHECK QUESTION

All complete TMT tags are identical in mass and structure. Using Figure 4.10, highlight which section(s) of a TMT tag is/are unique in mass.

4.3.3 Protein modifications

Almost all proteins undergo PTMs. These occur by a chemical alteration of specific amino acid residues or by the cleavage of the polypeptide backbone. There are several hundred reported chemical modifications, which vastly increase the diversity of protein that can be coded for by the standard complement of 20 amino acids. PTMs can be either permanent or reversible, and are shown to have a major influence on protein structure, protein–protein interaction, as well as biochemical activity. The addition of a phosphate molecule (phosphorylation) is the most common form of PTM and is critical for a number of cellular processes (see Section 4.3.3.1). Irregular PTM actions are often associated with disease, with PTM variants being used as disease biomarkers or targets for therapeutic intervention. It is clear, therefore, that PTMs have a central role in biological functions that are crucial to normal homeostasis (Liddy et al., 2013).

The investigation of PTMs at the proteomic level is becoming increasingly important as their roles in disease, particularly diseases associated with ageing (Cloos and Christgau, 2004), are being elucidated. Furthermore, genome studies cannot accurately predict protein modifications, even if a genetic signature does exist. The sheer volume and complexity of PTMs, plus the lack of suitable techniques for their measurement have been a limiting factor in their investigation in terms of normal function and disease. However, many modern proteomic approaches are now sensitive enough to measure and detect many modifications. For example, techniques previously discussed can separate modified proteins by size or PI, whereas 2-DGE can be stained with reagents that can distinguish certain modifications. Typically, the isolated PTM sample is enzyme digested, separated by LC and measured by mass spectrometry. There are some immunoassays that are sensitive in specifically capturing modified proteins (e.g. Luminex xMAP® for phosphorylated tau). In some cases, the stoichiometry of some PTMs is very low and, therefore, an affinity-based enrichment is needed to aid the detection by mass spectrometry. Database searching of the mass spectra will identify the peptide with a certain modification and will determine in which amino acid residue that alteration occurred. One must be wary of artificial modifications introduced into the sample by the process of certain separation, digestion, and chemical labelling techniques.

Key Point

PTMs are amino acid-level dynamic or permanent changes that happen to almost all proteins. The addition of a phosphate molecule (phosphorylation) is the most common form of PTM.

4.3.3.1 Phosphoproteomics

Phosphorylation is a ubiquitously expressed PTM and is the principal modification in biological regulation in humans. Phosphorylation is well recognized as a key regulator of enzyme activity and plays major roles in gene expression, cell division, and signal transduction. In disease, faulty phosphorylation is most commonly associated with cancer and neurodegenerative diseases such as Alzheimer's disease (AD). The enzymes that phosphorylate proteins are termed kinases and those that dephosphorylate are termed phosphatases, their targets are known as phosphoproteins. It has been estimated that there are up to 500 kinase and 100 phosphatase encoding genes in the human genome, which act upon ~100,000 potential phosphosites in the human proteome (Cohen, 2001).

All phosphoprotein analysis begins with a sample that contains both phosphorylated and non-phosphorylated proteins, and the initial step would typically be an in-solution enzyme digestion followed by peptide fractionation. The main aim is to identify and isolate the phosphoprotein, the site of action, and if possible, the abundance. As mentioned previously many PTMs, including phosphorylation, are in low abundance in complex samples and, therefore, enrichment is generally required. The affinity purification method involves antibodies that bind to phosphorylated proteins, which are held within a column, while non-phosphorylated peptides are allowed to flow through the column. The held phosphorylated peptides or proteins are removed from the column using an elution buffer. Antibodies that target specific phosphorylation sites (tyrosine, serine, and threonine) are also commercially available. IMAC (see Figure 4.11) is another approach that is taken to enrich a sample for phosphoproteins and occurs only after protein digestion, and is more widely employed due to its compatibility to downstream analysis by mass spectrometry.

The analysis of phosphoproteins has been revolutionized by mass spectrometry. Previously, phosphoproteins were radiolabelled with ^{32}P, separated by 2D peptide mapping, identified by autoradiography, and sequenced by Edman degradation. In its simplest form, mass spectrometry will recognize the mass shift of 79.983 daltons of a single phosphate modification on an amino acid residue. Furthermore, phosphopeptides have other characteristic fragment ions such as $H_2PO_4^-$, PO_3^- and PO_2^-, which have masses of 96.987, 79.971, and 62.972 daltons, respectively.

4.9 SELF-CHECK QUESTION

Why is sample enrichment needed for phosphoproteomics and what are the most common amino acid sites where phosphorylation occurs?

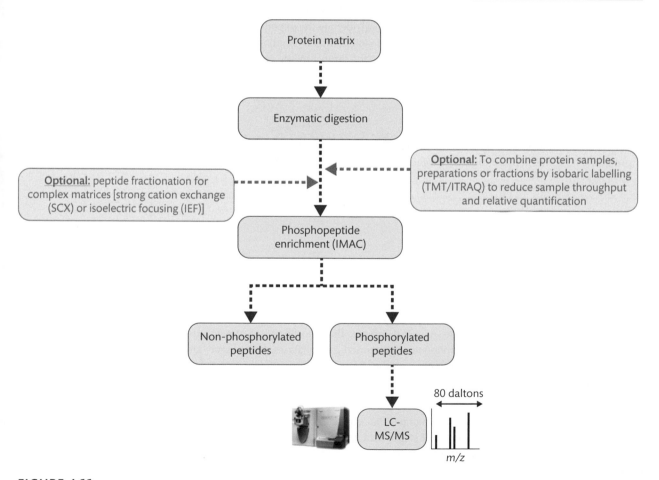

FIGURE 4.11

Immobilized metal-affinity chromatography (IMAC) pipeline for phosphoproteomics using mass spectrometry.

4.3.3 2 Glycoproteomics

Glycosylation is the addition of a short-chain carbohydrate residue to a protein during or after synthesis. It is a common event with almost 50% of proteins being glycosylated and three major types occur; N-linked glycosylation (unique to eukaryotes), O-linked, and glycosylphosphatidylinositol (GPI) anchors. A known major role is in controlling protein–protein interaction and, therefore, has a critical role in cell signalling and cell recognition in development of immune response. The alterations and/or deficits to the glycan composition of glycoproteins have been attributed to a number of diseases (e.g. hepatic cancer and rheumatoid arthritis).

Like phosphoproteomics, successful glycoprotein detection requires a separation technique and enrichment. Furthermore, the glycan(s) need to be removed from the protein to confirm the type. This can be achieved by peptide-N-glycosidase F (PNGase F). However, this is only applicable for N-linked glycans. SYPRO and Pro-Q emerald are glycoprotein dyes and coupled to 2-DGE have been useful in the enrichment of glycoproteins using gel-based techniques. However, 2-DGE has been shown to isolate glycoforms prior to mass spectrometry analysis. Lectin-affinity columns are also utilized for glycoproteins. Lectins are proteins that bind carbohydrates, with different ligands being specific for certain glycan residues. Lectin-affinity chromatography after 2-DGE protein spot detection can also be applied to glycopeptides post enzymatic digestion.

At the whole proteome level, glycoprotein analysis (both protein and glycan detection) is at a higher level of complexity compared with phosphoproteomics, unless the glycan chain is relatively simple to deduce. Therefore, this discipline of proteomics still requires much further development so that high throughput techniques and extensive examination of glycoproteins can be achieved.

4.3.4 Proteomics applications

The application of the proteomic technologies discussed in this chapter can be used to solve and investigate ongoing disease or predict disease conversion. In a medical context, high-throughput proteomic technologies are generally used to investigate biological markers that are indicators of a physiological or pathological condition (disease biomarkers) or how an individual is responding to certain treatment (toxicity biomarkers).

Disease state biomarker discovery is arguably the most attractive use for proteomics. In the early 1990s 2-DGE and mass spectrometry were successfully utilized to identify biological markers for ovarian and prostate cancers (CA125 and prostate-specific antigen, respectively). This approach has been routinely used to research cardiovascular, infectious, and in more recent years neurodegenerative diseases.

Neurodegenerative diseases, such as Alzheimer's disease, have an added challenge in that the diseases organ is not accessible during life and is protected from the peripheral system by the blood–brain barrier. Studies in *post mortem* brains identified the key pathological protein signatures of the disease (amyloid-beta and hyperphosphorylated tau). These proteomic discoveries in the mid-1980s have been critical in understanding the pathogenesis of Alzheimer's disease and has led to *in vivo* neuroimaging of these proteins to highlight individuals as a risk marker for Alzheimer's disease in living subjects (Klunk et al., 2004). Further to this, the advancement of assay-based technologies has enabled the measurement of amyloid-beta and tau in cerebrospinal fluid (CSF; Blennow et al., 2001) and, more recently, in blood plasma (Mattsson et al., 2016). However, measurement of amyloid-beta and tau alone cannot diagnose Alzheimer's disease with accuracy in biofluids. Therefore, proteomic investigation for Alzheimer's disease, using the latest assay-based and mass spectrometry-based technologies, continues to be performed, to discover novel CSF (Ollson et al., 2016) and/or blood signatures

(Hye et al., 2014; Ashton et al., 2015) of preclinical prediction of Alzheimer's. A biofluid proteomic signature of Alzheimer's is of critical importance to the field as it would greatly facilitate the diagnosis, monitoring, and successful participant selection for Alzheimer's drug trials, which to date have been largely unsuccessful.

4.4 **Future directions**

The PCR methods presented in this chapter for the analysis of nucleic acids are widely used in diagnostic laboratories, but they are now being challenged by the advent of high throughput technologies such as NGS. For 'field work' and 'point of care' testing PCR-based technologies will probably continue to be used, but in the laboratory setting NGS may take over as the preferred method for nucleic acid analysis. Before it does so the management of big data sets, the streamlining of bioinformatics pipelines together with standardized interpretation have all to be addressed. These challenges are further discussed in Chapter 6, Section 6.2.1.3.

Proteomic analysis has evolved in much the same way as for DNA and RNA. High-throughput protein identification and quantification, such as large-scale mass spectrometry and multi-analyte immunoassays, now dominate the focus of biochemistry laboratories investigating disease trends. The former, in particular, has created a huge bioinformatic challenge, as the significant advancements in proteomic technologies now yield a multitude of information, which contain many answers to disease pathogenesis.

A recently described and potentially diagnostically important development has seen the identification of heterogeneity in frozen sections of colonic tumours using mass spectroscopy (Inglese et al., 2017). Could this methodology be applied for prognostic and predictive purposes as the analysis of nucleic acids and proteins in tissue sections and homogenates thereof is currently practised? Time will tell!

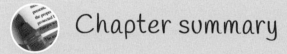 Chapter summary

In this chapter we have:

- Introduced a range of techniques that can be used for the identification of nucleic acids and proteins/peptides in homogenate sample preparations.

- Described the different molecular techniques developed around PCR.

- Described microarrays technologies for the high-throughput investigation of whole genomes or transcriptomes.

- Provided descriptions of Western blot, ELISA, and 2-DGE methods for the identification of proteins.

- Outlined the principles of mass spectrometry and described the various protocols that can be employed.

- Discussed methodological options for the investigation of post-translational modifications.

- Considered ways in which nucleic acids and proteins can be quantified using these methods.

- Provided examples of the application of the technologies in the context of molecular diagnosis and the understanding of disease processes.

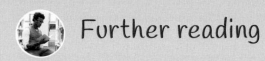

Further reading

- Abersold R. Mass spectrometry-based proteomics. Nature 2003; 422: 198–207.

- Perez-Diez A, Morgun A, and Shulzhenko N. *Microarrays for Cancer Diagnosis and Classification in Madame Curie Bioscience Database.* Austin (TX), LandEs Bioscience, 2000–2013.

- Li PCH, Abootaleb Sedighi, Lin Wang. *Microarray Technology: Methods and Applications.* New York, Springer, 2015.

- Leng SX, McElhaney JE, Walston JD, et al. ELISA and multiplex technologies for cytokine measurement in inflammation and aging research. Journal of Gerontology A Biological Sciences and Medical Sciences 2008; 63: 879–884.

- Qian WJ. High-throughput proteomics using Fourier transform ion cyclotron resonance mass spectrometry. Expert Review Protoeomics 2004; 1: 87–95.

References

- Amary MF, Pauwels P, Meulemans E, et al. Detection of β-catenin mutations in paraffin-embedded sporadic desmoid-type fibromatosis by mutation-specific restriction enzyme digestion (MSRED): an ancillary diagnostic tool. American Journal of Surgical Pathology 2007; 31: 1299–309.

- Ashton, NJ, Kiddle, SJ, Graf J, et al. Blood protein predictors of brain amyloid for enrichment in clinical trials? Alzheimers & Dementia (Amsterdam) 2015; 1: 48–60.

- Blennow K, Vanmechelen E, Hampel H. CSF total tau, Abeta42 and phosphorylated tau protein as biomarkers for Alzheimer's disease. Molecular Neurobiology 2001; 24: 87–97.

- Chamberlain JS, Chamberlain JR, Fenwick, RG, et al. Diagnosis of Duchenne and Becker muscular dystrophies by polymerase chain reaction. A multicenter study. Journal of the American Medical Association 1992; 267: 2609–2615.

- Cloos PA, Christgau S. Post-translational modifications of proteins: implications for aging, antigen recognition, and autoimmunity. Biogerontology 2004; 5: 139–158.

- Cohen P. The role of protein phosphorylation in human health and disease. The Sir Hans Krebs Medal Lecture. European Journal of Biochemistry 2001; 268: 5001–5010.

- Dausset J, Cann H, Cohen D, Lathrop M, and Centre d'Etude du Polymorphisme Humain (CEPH): Collaborative Genetic Mapping of the Human Genome. Genomics 1990; 6: 575–7.

- Didenko, VV (Ed.) *Fluorescent Energy Transfer Nucleic Acid Probes.* New York, NY, Humana Press, 2006.

- El-Aneed A, Cohen A, Banoub J. Mass spectrometry, review of the basics: electrospray, MALDI, and commonly used mass analyzers. Applied Spectroscopy Reviews 2009; 44: 210–230.

- Eliuk S, Makarov A. Evolution of Orbitrap Mass Spectrometry Instrumentation. Annual Review of Analytical Chemistry 2015; 8: 61–80.

- Fodde R, Losekoot M. Mutation detection by denaturing gradient gel electrophoresis (DGGE) Human Mutation 1994; 3: 83–94.

- Glish GL, Vachet RW. The basics of mass spectrometry in the twenty-first century. Nature Reviews: Drug Discoveries 2003; 2: 140–150.

- Heller MJ. DNA Microarray Technology: Devices, Systems and Applications. Annual Review of Biomedical Engineering 2002; 4: 129–153.

- Houser B. Bio-Rad's Bio-Plex® suspension array system, xMAP technology overview. Archives of Physiology and Biochemistry 2012; 118: 192–196.

- Hye A, Riddoch-Contreras J, Baird AL, et al. Plasma proteins predict conversion to dementia from prodromal disease. Alzheimers & Dementia 2014; 10: 799–807.

- Inglese P, McKenzie JS, Mroz A, et al. Deep learning and 3D-DESI imaging reveal the hidden metabolic heterogeneity of cancer. Chemical Sciences 2017; 8: 3500–3511.

- Kakavas VK, Plageras P, Vlachos TA, Papaioannou A, Noulas VA. PCR-SSCP: a method for the molecular analysis of genetic diseases. Molecular Biotechnology 2008; 38: 155–163.

- Klunk WE, Engler H, Nordberg A, et al. Imaging brain amyloid in Alzheimer's disease with Pittsburgh Compound-B. Annals of Neurology 2004; 55: 306–319.

- Koressaar T, Remm M. Enhancements and modifications of primer design program Primer3. Bioinformatics 2007; 23: 1289–1291.

- Kozera B, Rapacz M. Reference genes in real-time PCR. Journal of Applied Genetics 2013; 54: 391–406.

- Kraemer S, Vaught JD, Bock C, et al. From SOMAmer-based biomarker discovery to diagnostic and clinical applications: a SOMAmer-based, streamlined multiplex proteomic assay. PLoS One 2011; 6: e26332.

- Leng SX, McElhaney JE, Walston JD, et al. Elisa and multiplex technologies for cytokine measurement in inflammation and aging research. Journal of Gerontology A Biological Sciences and Medical Sciences 2008; 63: 879–884.

- Liddy KA, White MY, Cordwell SJ. Functional decorations: post-translational modifications and heart disease delineated by targeted proteomics. Genome Medicine 2013; 5: 20.

- Mattsson N, Zetterberg H, Janelidze S, et al. Plasma tau in Alzheimer's disease. Neurology 2016; 87: 1827–1835.

- Milbury CA, Li J, Liu P, Makrigiorgos GM. COLD-PCR: improving the sensitivity of molecular diagnostics assays. Expert Reviews in Molecular Diagnosis 2011; 11: 159–169.

- Moutou C, Garde N, Viville S. Multiplex PCR combining deltaF508 mutation and intragenic microsatellites of the CFTR gene for pre-implantation genetic diagnosis (PGD) of cystic fibrosis. European Journal of Human Genetics 2002; 10: 231–238.

- Mullis KF, Faloona F, Scharf S, Saiki R, Horn G, et al. Specific enzymatic amplification of DNA in vitro: The polymerase chain reaction. Cold Spring Harbor Symposium on Quantitative Biology 1986; 51: 263–273.

- Navarro E, Serrano-Heras G, Castaño MJ, Solera J. Real-time PCR detection chemistry. Clinica Chimica Acta 2015; 439: 231–250.

- Notomi T, Okayama H, Masubuchi H, et al. Loop-mediated isothermal amplification of DNA. Nucleic Acids Research 2000; 28: e-63.

- Ollson B, Lautner R, Andreasson U, et al. CSF and blood biomarkers for the diagnosis of Alzheimer's disease: a systematic review and meta-analysis. Lancet: Neurology 2016; 15: 673–684.

- Perou CM, Sørlie T, Eisen MB, et al. Molecular portraits of human breast tumours. Nature 2000; 406: 747–752.

- Rabilloud T, Lelong C. Two-dimensional gel electrophoresis in proteomics: a tutorial. Journal of Proteomics 2011; 74: 1829–1841.

- Radonić A, Thulke S, Mackay IM, et al. Guideline to reference gene selection for quantitative real-time PCR. Biochemistry & Biophysics Research Communications 2004; 313: 856–862.

- Tao Z-Y, Zhou H-Y, Xia H, et al. Adaptation of a visualized loop-mediated isothermal amplification technique for field detection of Plasmodium vivax infection. Parasite Vectors 2011; 4: 115.

- Towbin H, Staehelin T, Gordon J. Electrophoretic transfer of proteins from polyacrylamide gels to nitrocellulose sheets: procedure and some applications. Proceedings of the National Academy of Sciences USA 1979; 76: 4350–4354.

- Untergasser A, Cutcutache I, Koressaar T, et al. Primer3—new capabilities and interfaces. Nucleic Acids Research 2012; 40: e115.

- van't Veer LJ, Dai H, van de Vijver MJ, et al. Gene expression profiling predicts clinical outcome of breast cancer. Nature 2002; 415: 530–536.

- van't Veer LJ, Dai H, van de Vijver MJ, et al. Expression profiling predicts outcome in breast cancer. Breast Cancer Res 2003; 5: 57–58.

- Wang Y-H, Barker P, Griffith J. Visualization of diagnostic heteroduplex DNAs from cystic fibrosis deletion heterozygotes provides an estimate of the kinking of DNA by bulged bases. Journal of Biological Chemistry 1992; 267: 4911–4916.

- Wasinger VC, Cordwell SJ, Cerpa-Poljak A, et al. Progress with gene-product mapping of the Mollicutes: *Mycoplasma genitalium*. Electrophoresis 1995; 16: 1090–1094.

- Zetterberg H, Wilson D, Andreasson U, et al. Plasma tau levels in Alzheimer's disease. Alzheimers Research & Therapy 2013; 5: 9.

Useful websites

- LAMP isothermal PCR: https://www.neb.com/applications/dna-amplification-pcr-and-qpcr/isothermal-amplification.
- SNP checking when designing PCR primers: Genome browser such as UCSC https://genome.ucsc.edu.
- SNPCheck v3: https://secure.ngrl.org.uk/SNPCheck/snpcheck.htm.
- Swiss-Prot: https://www.ebi.ac.uk/uniprot.

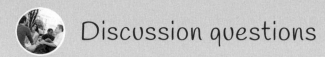 Discussion questions

4.1 What are the different variants of PCR and how do they differ from 'standard' protocol?

4.2 What are the principles and distinguishing features of Western blotting and ELISA methods as used for the identification of proteins?

Answers to the self-check questions and tips for responding to the discussion questions are provided on the book's accompanying website:

⊙ Visit: www.oup.com/uk/warford

5

Sequencing Technologies

Nadège Presneau and Rifat Hamoudi

Learning objectives

After studying this chapter, you should be able to:

- Describe the principles for first-generation sequencing, i.e. Sanger sequencing and pyrosequencing methods.

- Understand the principles of sequencing by synthesis (SBS) together with the major technologies developed for second-generation sequencing (Illumina and Ion Torrent).

- Discuss the principles of third-generation sequencing that allows sequencing of single molecules in real time.

- Appreciate the concept of fourth-generation sequencing, which combines *in situ* cellular localization with gene expression sequencing.

- Describe some applications of sequencing technologies.

- Understand the advantages and challenges of each technology.

5.1 **Introduction**

This chapter presents an overview of sequencing methods. Nucleic acid sequencing has undergone rapid and impressive development since the 1970s. The 'Sanger' sequencing method developed at that time was used to produce the first draft of the human genome (International Human Genome Sequencing Consortium, 2004) and this has been a fundamental springboard to the understanding of genomics in health and disease. The multinational effort to produce this draft took several years and this was due, in part, to the rate-limiting nature of the underlying 'first-generation sequencing' technology.

Pyrosequencing was introduced in 2005 by 454 Life Sciences (now Roche) and was used to sequence the genome of James Watson, the co-discoverer of the structure of DNA (Wheeler et al., 2008). This represents an adaptation of the 'Sanger' method, but does not require the use of fluorescently labelled dyes.

While both methodologies are still used, considerable improvements have been made in throughput through the introduction of massively paralleled sequencing. This 'next or second generation sequencing' has reduced sample size and assay cost, combined with increased speed and accuracy. The Illumina platform (known at the time as Solexa, a company based in Cambridge, UK) was the first of this type to be commercialized. A few years later Life Technologies commercialized their platform called the Ion Torrent Personal Genome Machine (PGM), which is now owned by ThermoFisher Scientific. Like first-generation technologies this technology uses PCR amplification for the library preparation of nucleic acid samples that are subsequently sequenced.

'Third-generation' technologies removed the need for PCR library preparation and have allowed the sequencing of single molecules in real time. The possibility of combining in situ cellular localization with gene expression sequencing provides a further tantalizing advance that has been termed 'fourth-generation sequencing'.

Today, a variety of sequencing methods mean that DNA and RNA together with epigenetic changes can be investigated rapidly and thoroughly. Each platform produces different types of raw data and, therefore, requires distinctive bioinformatic pipelines to analyse and extract meaningful information. These are considered in detail in Chapter 6, but are covered briefly here. Accordingly, the focus of this chapter is to describe the principles of the main sequencing technologies, to indicate advantages and constraints, and to provide some examples of research and diagnostic applications.

5.2 **First-generation sequencing**

Two sequencing techniques were developed independently in the 1970s:

- the chain termination method by Sanger and Coulson in the UK (Sanger and Coulson, 1975; Sanger et al., 1977);
- the chemical degradation method by Maxam and Gilbert in the USA (Maxam and Gilbert, 1977).

The 'Sanger' method became the technique of choice and is still extensively used.

5.2.1 Sanger sequencing

The method developed by Sanger uses chemically altered deoxynucleotide triphosphates 'dideoxy' bases to terminate (chain termination) newly synthesized DNA fragments at specific bases of either A, C, T, or G. These fragments are then size-separated by electrophoresis, and the DNA sequence can be read and reconstructed.

5.2.1.1 *Principles*

1. Anneal a short oligonucleotide primer onto the DNA fragment to be sequenced; this is usually a PCR product. The primer anneals to a specific known region on the template DNA and will act as a starting point for complementary strand synthesis by the DNA polymerase.

2. In the presence of DNA polymerase, catalytic polymerization of deoxynucleotide triphosphates (dNTP) onto the DNA occurs. The polymerization is extended until the enzyme incorporates a fluorescently labelled dideoxynucleotide triphosphate (ddNTP) 'terminator' into the growing chain. The terminator will block further strand synthesis because the ddNTP lacks the hydroxyl group at the 3' position of the sugar component; see Figure 5.1.

Chemical structure of deoxynucleotide and dideoxynucleotide

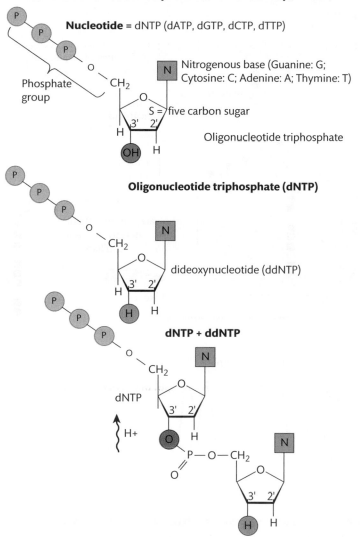

FIGURE 5.1
Chemical structure of a deoxynucleotide (dNTP) and dideoxynucletide (ddNTP).

This group is needed for the next nucleotide to be attached; chain termination therefore occurs whenever the enzyme incorporates a ddNTP.

3. Separate the fragments of different sizes that have been generated. Although this was achieved previously by using very thin polyacrylamide gels, it is now done in very small capillary tubes containing polyacrylamide gel. The polyacrylamide gel contains urea, which denatures the DNA so the newly synthesized strands dissociate from their templates. Electrophoresis is carried out at high voltage, causing the gel to heat up to 60°C and above, so the strands cannot re-anneal. The final products of the sequencing reaction consist of a set of fragments of different lengths each, fluorescently labelled at their 3′ ends.

As each of the four ddNTPs has a different fluorescent dye attached they can be detected by a CCD camera and translated into a chromatogram (Figure 5.2). As the fluorescently labelled extension products from a sequencing reaction are electrophoresed past the laser detection area of a gel, each discrete base can be 'called'. Over the course of several hours up to 600 bases can be accurately called, depending on the initial quality of the template.

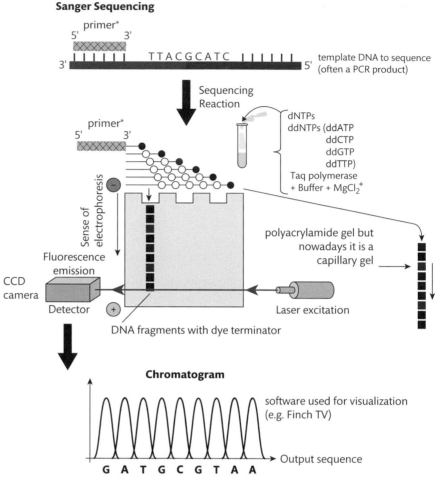

Sanger Sequencing

FIGURE 5.2
Schematic workflow of Sanger sequencing.

*primer can be the same that generated the PCR products.

4. Analysis: the biggest issue for Sanger sequencing is base-calling accuracy. Many bioinformatic algorithms were introduced to measure the quality of the base called in sequencing data. The best algorithm that was used for most mutational screening and the Human Genome Project was the Phred base. This was based on calculating several parameters related to peak shape and height at each base in the chromatogram and using those parameters to look up corresponding quality scores in tables. Quality scores range from 4 to about 60, with higher values corresponding to higher quality. The quality scores are logarithmically linked to error probabilities, as shown in Table 5.1. It has been shown that Phred's error probabilities are very accurate (Ewing and Green, 1998). This high accuracy has also been observed for sequences generated at different laboratories, each using a different combination of sequencing enzymes, fluorescent dyes, and gel run conditions.

5.2.1.2 Applications

Sanger sequencing, and its subsequent modifications, was used for the human genome project from 1990 to 2003 (International Human Genome Sequencing Consortium, 2004). In addition, Sanger sequencing has been used to make significant discoveries that have had an impact on medical treatment. Examples include the identification of *BRCA1* (Friedman et al.,

TABLE 5.1 Phred quality scores for sequencing

Phred quality score	Probability that the base is called wrong	Accuracy of the base call
10	1 in 10	90%
20	1 in 100	99%
30	1 in 1000	99.9%
40	1 in 10,000	99.99%
50	1 in 100,000	99.999%

1994) and *BRCA2* (Wooster et al., 1995) in hereditary breast and/or ovarian cancer. In 2002, this technology was also used to identify a common *BRAF* V600E mutation as a prevalent mutation present in many cancers (Davies et al., 2002).

Key Point

Sanger sequencing, currently referred to as first-generation sequencing, is based on sequencing by synthesis (SBS), which means the incorporation and, subsequently, the detection of fluorescently labelled dideoxyribonucleotides, each representing either an A, C, G, or T base.

5.1 SELF-CHECK QUESTION

What is the difference between a deoxyribonucleotide and a dideoxyribonucleotide, and what is the function of each in the Sanger sequencing reaction?

5.2.2 Pyrosequencing

Pyrosequencing was developed at the Royal Institute of Technology (KTH) in Stockholm (Ronaghi et al., 1996, 1998) and is an alternative method of Sanger DNA sequencing, based on the 'sequencing by synthesis (SBS)' principle and the detection of released pyrophosphate (PPi) during DNA synthesis. Pyrosequencing can sequence more rapidly than using the Sanger method. Typical run times are 10 minutes for 96 samples for DNA fragments of up to 100 bp in a semi-high throughput manner, which makes it ideal for molecular disease diagnosis. As opposed to Sanger sequencing no gels or dye labels are needed.

5.2.2.1 *Principle*

Pyrosequencing employs four enzymes working in a cooperative fashion. These are: DNA polymerase, ATP sulfurylase, luciferase, and apyrase. The substrates are adenosine 5′ phosulfate (APS) and luciferin. The method is summarized below and illustrated in Figure 5.3.

1. A sequencing primer is hybridized to a single-stranded DNA template, and incubated with the four enzymes and the substrates adenosine 5′ phosphosulfate (APS) and luciferin.

2. The first of four dNTPs is added to the reaction. The DNA polymerase catalyses the incorporation of the dNTP into the DNA strand, when it is complementary to the base in the

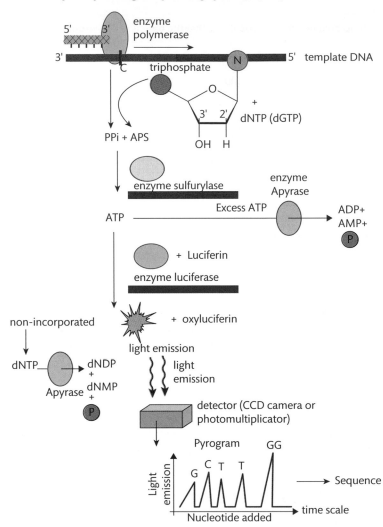

FIGURE 5.3
Pyrosequencing method.

template strand. Each incorporation event is accompanied by release of PPi in a quantity equimolar to the amount of incorporated nucleotide.

3. The released PPi is quantitatively converted by the ATP sulfurylase to ATP in the presence of APS. This ATP drives the luciferase-mediated conversion of luciferin to oxyluciferin, which generates visible light in amounts that are proportional to the amount of ATP produced. The light is detected by a charge-coupled device (CCD) camera and seen as a peak in a pyrogram. Each light signal is proportional to the number of nucleotides incorporated.

4. Apyrase, a nucleotide degrading enzyme, continuously degrades unincorporated dNTPs, and the excess ATP. When the degradation is complete, another dNTP is added with the reaction beginning again at step 2.

As mentioned previously, the addition of dNTPs is performed one at a time. It should be noted that deoxyadenosine alpha-thio triphosphate (dATPαS) is used as a substitute for the natural deoxyadenosine triphosphate (dATP) since while the DNA polymerase efficiently uses it, it is not recognized by the luciferase.

As the process continues, the complementary DNA strand is built up and the nucleotide sequence is determined from the signal peak in the pyrogram (Figure 5.3).

5.2.2.2 Benefits and applications

The benefits of pyrosequencing are read-length, speed, accuracy, flexibility, parallel processing, throughput, and cost, together with the automation of sample handling and preparation. Furthermore, the technique avoids the need for labelled primers, labelled nucleotides, and gel electrophoresis as required for Sanger sequencing. A limitation inherent to the pyrosequencing approach is the detection of long homopolymers (repeated nucleotides), where sequencing errors can occur.

Pyrosequencing has been successfully used for confirming sequencing and also for *de novo* sequencing. Currently, the method is being used for many applications such as SNP genotyping, identification of bacteria fungal and viral typing (Ahmadian et al., 2000, 2006). Moreover, the method has demonstrated the ability to determine difficult secondary structures (Ronaghi et al., 1998) and perform mutation detection (Ahmadian et al., 2000; Garcia et al., 2000), DNA methylation analysis (Uhlmann et al., 2002), multiplex sequencing (Gharizadeh et al., 2003, 2006), tag sequencing of cDNA library (NordstrÖm et al., 2001) and clone checking (Nourizad et al., 2003). Whole genome sequencing has also been undertaken using this sequencing method (Margulies et al., 2005).

Key Point

Pyrosequencing uses four different enzymes that convert PPi pyrophosphates to a fluorescent signal, which indicates the incorporation and identification of a dNTP into the template DNA being sequenced.

5.2 SELF-CHECK QUESTION

What are the features that differentiate pyrosequencing from Sanger sequencing?

5.3 Second-or next-generation sequencing

Illumina and Ion Torrent are the two main commercial leaders for second-generation sequencing platforms. The two platforms are based on SBS technology, but differ in the chemistry they use (see Sections 5.3.1.1 and 5.3.2.1). Illumina technology is based on fluorescent dideoxynuclotides, like those used for Sanger sequencing, although in 2018, Illumina started moving towards semiconductor-based technology by releasing the iSeq DNA sequencing instrument (https://emea.illumina.com/systems/sequencing-platforms/iseq.html).

5.3.1 Illumina

5.3.1.1 Principle

Illumina has developed several instruments that differ from each other only by the sequencing capacity (Table 5.2). All the instruments follow the same principle, which is similar to capillary Sanger sequencing in which a DNA polymerase catalyses the incorporation of fluorescently labelled dNTPs into a single DNA strand template during sequential cycles of DNA synthesis. Incorporation of nucleotides is identified by fluorophore excitation during each cycle at the

TABLE 5.2 Illumina platforms

	Typical run time	Maximum output	Maximum read per run	Maximum read length (bp)	Main applications
Benchtop sequencers					
MiniSeq	4–24 h	7.5 Gb	25 millions	2 × 150 bp	Small whole genome (bacteria, virus); targeted sequencing (DNA amplicons; transcripts (RNA), microRNAs)
MiSeq	4–55 h	15 Gb	25 millions	2 × 300 bp	Small whole genome (bacteria, virus); targeted sequencing (DNA amplicons; transcripts (RNA), microRNAs)
NextSeq	12–30 h	120 Gb	400 millions	2 × 150 bp	Small whole genome (bacteria, virus); targeted sequencing (DNA amplicons; transcripts (RNA), microRNAs); methylation sequencing
Production scale sequencers					
HiSeq (2500/3000/4000)	1–3.5 days (HiSeq 3000/4000) 7 h to 6 days (HiSeq 2500)	1500 Gb	5 billions	2 × 150 bp	Same as benchtop sequencers, but also large whole genome (human, animal), whole transcriptome; gene expression profiling with mRNA-Seq
HiSeq X	Less than 3 days	1800 Gb	6 billions	2 × 150 bp	Large whole genome sequencing
NovaSeq	19–40 h	6000 Gb	20 billions	2 × 150 bp	Same as benchtop sequencers but also large whole genome (human, animal), whole transcriptome; gene expression profiling with mRNA-Seq

Information taken from Illumina website.

point of incorporation. The main difference between capillary Sanger sequencing and Illumina sequencing is that, instead of sequencing only a single DNA fragment at one time, millions of fragments are sequenced in a massively parallel fashion. The Illumina workflow includes three basic steps. A detailed animation of SBS sequencing is available at www.illumina.com/SBSvideo.

5.3.1.1.1 Library preparation

DNA or cDNA is randomly fragmented to lengths around 200 bp and adapters (specific to the Illumina platform) are ligated on both 5′ and 3′ ends. Newly developed protocols have combined the fragmentation and adapter ligation into a single step making the process more efficient. Those fragments containing the adapters are then amplified by PCR and purified to produce a library. The library is loaded into a flow cell where each of the fragments is captured on the support surface (glass slide with lanes), which contains oligonucleotides that are complementary to the two adapters. Each of the fragments is then PCR amplified directly on the surface into distinct, clonal clusters (same exact molecule) using 'bridge amplification'. These monoclonal clusters are then sequenced.

5.3.1.1.2 Sequencing

The Illumina platform uses a proprietary reversible terminator-based method (similar in principle to the Sanger sequencing terminator chain reaction) that detects the incorporation of the fluorescently labelled ddNTPs incorporated into the single strand DNA template. As part of this process, 3' OH groups are blocked to ensure the incorporation of one nucleotide at a time. After each cycle, excitation and reading of the incorporated fluorescence, a complex chemistry reverts the ddNTPs into conventional dNTP to allow the incorporation of the next ddNTPs following the DNA template during the next cycle. This process is repeated for a determined number of cycles. Cycle number will depend on the platform used, but varies from 36 cycles to several hundred cycles when the HiSeq method is used. Along with the forward reads, reverse reads are also generated thereby creating paired end **contiguous sequences**. This allows the two strands to be sequenced, thereby improving the reading accuracy and detection of potential variants.

Contiguous sequences
This is a set of overlapping DNA segments that together represent a consensus region of DNA.

5.3.1.1.3 Data analysis

Illumina's raw data consist of image files that capture the fluorescence. The image is processed using image processing techniques to generate .bcl files that consist of base calls with quality scores. The data is de-multiplexed and FASTQ files generated, followed by sequence filtering and quality control. The newly identified sequence reads (paired-ends) presented in FASTQ files are analysed by aligning the generated sequence to a reference genome sequence, to generate **Binary Alignment Map (BAM)** files. The BAM files are then interrogated for misalignment of sequences to identify variants and generate what is referred to as variant call format (VCF) files, which can be visualized using other bioinformatics tools such as the integrated genome viewer (IGV). The VCF files containing the genomic data can be securely transferred, stored, analysed, and shared on a baseSpace sequence Hub that is available from Illumina (www.illumina.com/informatics/research/sequencing-data-analysis-management/basespace.html). The process is summarized in Figure 5.4.

Binary alignment map (BAM)
Binary version of a SAM file. A SAM file (.sam) is a tab-delimited text file that contains sequence alignment data.

While Illumina have developed their own bioinformatics tools, it is also possible to use free software to construct the bioinformatics pipeline. This can be carried out using a suite of tools such as Galaxy (https://galaxyproject.org/admin/get-galaxy/). Galaxy consists of all the main software needed for sequence quality control, alignment, and some variant calling. The bioinformatician can build the pipeline by utilizing individual software packages from within the Galaxy suite to produce a custom workflow (Hillman-Jackson et al., 2012; Blankenberg and Hillman-Jackson, 2014).

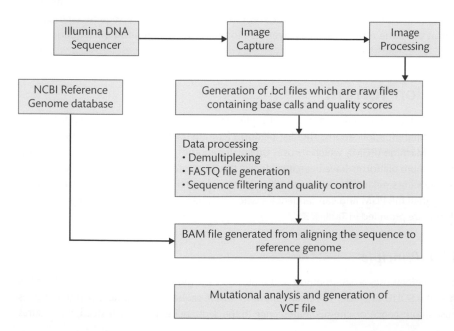

FIGURE 5.4
Workflow of DNA sequencing using Illumina platform. The image is processed using image processing techniques to generate .bcl files, which consist of base calls with quality scores. The data is demultiplexed and FASTQ files generated followed by sequence filtering and quality control. The newly identified sequence reads (paired-ends) presented in FASTQ files are analysed by aligning the generated sequence to a reference genome sequence, to generate binary alignment map (BAM) files. The BAM files are then interrogated for misalignment of sequences to identify variants and generate what is referred to as variant call format (VCF) files.

5.3.1.2 Key applications

The applications of NGS with the Illumina platforms are various and linked to the type of platform used. These range from whole genome sequencing applied, for example, to human, bacteria and viruses, whole exome sequencing (largely applied to *Homo sapiens*) and targeted sequencing (amplicons overlapping specific regions of the DNA). RNA sequencing is generally used to identify novel RNA variants and splice sites, or to quantify mRNAs for gene expression profiling. Other applications, such as genome-wide methylation or DNA–protein interactions (ChipSeq), can also be performed with NGS.

5.3.1.3 Alternative platforms

The Illumina platforms vary from benchtop sequencers such as the miniSeq systems, the MiSeq or the NextSeq series to production-scale sequencers (NovaSeq and HiSeq series). For platform comparisons consult the Illumina.com website (https://www.illumina.com/systems/sequencing-platforms.html) and Table 5.2. Benchtop sequencers are most frequently used in diagnostic laboratories. These machines have the advantages that they take up a minimum of space, and allow quick and efficient small whole-genome sequencing (bacteria, virus), targeted gene sequencing (amplicons—gene panel such as cancer gene panel), or targeted expression profiling with mRNA-Seq and gene expression profiling, which are relevant for molecular pathology diagnostics. Novaseq and HiSeq X series are machines that allow large whole-genome sequencing and are usually bought by sequencing centres, such as the Wellcome Trust Sanger Institute in Cambridge in UK or the Max Planck Institute for Molecular Genetics in Berlin in Germany.

> ## Key Point
>
> **Illumina sequencing is based on sequencing by synthesis. It uses fluorescent dideoxy-nucleotides (similar to Sanger sequencing) that are incorporated in a cyclic manner depending on the target DNA being sequenced. The detection is, therefore, based on laser excitation that will record the fluorescence emitted at each cycle corresponding to the incorporation of a ddNTP in all the different clusters generated during the library preparation. As millions of different molecules of DNA are sequenced in parallel at the same time, which has given rise to the term 'next generation sequencing'.**

5.3.2 Ion Torrent

Ion Torrent has three main platforms with different capacities of sequencing input that are used for different applications. The first platform (commercialized in 2010) was the Personal Genome Machine (PGM), which focuses on small genomes and targeted sequencing. Since then two more platforms have been introduced; the Ion Proton (2012) that allows higher output sequencing similar to NextSeq and HiSeq (Illumina) and the S5, in 2014, which has a higher capacity than the PGM and can sequence a whole exome in 2 hours. Examples of Ion Torrent platforms are provided in Table 5.3.

5.3.2.1 Principle

This technology uses a SBS method and emulsion PCR (emPCR) similar to other platforms such as the SOLiD and 454. The main difference in the sequencing technology is that it does not use fluorescence or chemiluminescence in the sequencing process. Instead, it measures

TABLE 5.3 Ion Torrent platforms

	Sequencing chips	Typical run time (hours)	Maximum output	Maximum read per run	Maximum read length (bp)	Main applications
Ion PGM	Ion 314™ Ion 316™ Ion 318™	2–4 3–5 4–7	30–100 Mb 300–1 Gb 600–2 Gb	400–550 thousands 2–3 M 4–5.5 M	200–400	Low-throughput gene panel (targeted DNA) and viral/microbial sequencing; targeted RNA
Ion S5	Ion 510 Ion 520 Ion 530 Ion 540	2.5–4	300–500 Mb 600–1 Gb 3–4 Gb 10–15 Gb	2–3 M 4–6 M 15–20 M 60–80 M	200–400 (600 bp only with Ion chips 520 and 530)	Gene panel (targeted DNA), viral and microbial genome, exome sequencing, targeted RNA, RNA-seq
IonS5 XL	Ion 510 Ion 520 Ion 530 Ion 540	2.5–4	300–500 Mb 600–1Gb 3–4 Gb 10–15 Gb	2–3 M 4–6 M 15–20 M 60–80M	200–400 (600 bp only with Ion chips 520 and 530)	Gene panel (targeted DNA), viral and microbial genome, exome sequencing, targeted RNA, RNA-seq
Proton	Ion PI chip	2–4	Up to 10 Gb	60–80 M	200–400	Human-scale whole-genome sequencing ChIP sequencing, whole transcriptome sequencing, exome sequencing, methylation analysis, gene expression by sequencing, small genome sequencing, *de novo* sequencing, small RNA sequencing

Information taken from ThermoFisher Scientific website.

the hydrogen ions released during base incorporation into newly synthesized DNA via a semiconductor technology. As the technology does not rely on optical detection, this results in higher speed, lower running and maintenance cost, and a smaller instrument size. Sequence analysis is also independent of sample degradation, which is inherent when using FFPE material, which can auto-fluoresce and thus interfere with Illumina's fluorescent-based technology. In addition, the Illumina SBS method requires a minimum fragment length of around 150–200 bp, but the semiconductor-based method is agnostic to fragment length and, hence, works with any degraded sample. A video showing ion-sequencing technology in action is available for viewing at https://www.youtube.com/watch?v=ZL7DXFPz8rU.

The workflow follows similar steps as described for the Illumina platforms and consists of four major steps: library construction, template preparation, sequencing, and analysis.

5.3.2.1.1 Library construction

This is a common process, and involves taking DNA or cDNA, fragmenting it to a uniform size (around 200–400 b) followed by adding sequencing adapters specific to the Ion Torrent technology. Multiplex PCR can also been used to generate small DNA fragments of regions of interest to generate small fragments of Ion Torrent has specific kits for a variety of applications, including DNA fragments for small genomes, total RNA-seq, and targeted sequencing panels (cancer panels).

5.3.2.1.2 Template preparation/amplification

The fragments generated during the library preparation are chemically attached to beads and amplified using emulsion PCR. Beads coated with complementary primers are mixed with a dilute aqueous solution containing the fragments to be sequenced along with the necessary

PCR reagents. This solution is then mixed with oil to form an emulsion of microdroplets. The concentration of beads and DNA fragments is adjusted such that each microdroplet contains only one of each. Clonal amplification of each DNA fragment is then performed within the microdroplets. Each microdroplet constitutes a mini-PCR reaction. Following amplification, the emulsion is partitioned using organic extraction and centrifugation, with the amplified beads being enriched in a glycerol gradient and unamplified beads pelleting at the bottom. While the emulsion PCR process is effective, it is slow and complicated. Since the first launch of the PGM Ion Torrent have improved the emPCR and introduced new equipment such as the Ion Touch 2, the Avalanche, and Ion Chef to reduce the hands-on time by automating the process.

5.3.2.1.3 Sequencing

What really differentiates Ion Torrent's systems is the sequencing technology. Unlike other platforms, the Ion Torrent system measures the direct release of hydrogen ions from the reaction. The lack of optics also means that slow image scans and storage of large image files is avoided, promoting faster data generation. Parallel 200 bp reads take about 2 hours to generate depending on the type of library prep kit used as well as the type of chip used –see Table 5.3. Finally, the lack of fluorescence or chemiluminescence means that the system can use unmodified nucleotides, which are cheaper to purchase and better tolerated by DNA polymerase, making them ideal for processing intact and degraded nucleic acid samples.

5.3.2.1.4 Data analysis

As the Ion Torrent systems generate standard output files such as FASTQ, data analysis can be carried out using third party software. In addition to a variety of available third party analysis solutions, Ion Torrent offers the 'Torrent Browser' software, which acts as the primary interface for a number of basic functions, such as sequence alignment and quality control. Similar to Illumina, a cloud-based solution called 'Ion Reporter' serves as a front end for a variety of open source analysis solutions.

Ion Torrent sequencer generates .dat files of the electrical signals' raw traces. The signal processing step generates well files by converting the raw traces into a single number per flow per well. The BaseCaller converts the well files information into a sequence of bases and writes the sequence into an unaligned BAM file. The BAM file is aligned using Torrent Mapping Alignment Program (TMAP). The signal processing step also marks several types of low-quality reads, but does not remove them. Examples include polyclonal reads, reads with high signal processing residual that indicate an ambiguous signal value, and reads that do not contain a valid nucleic acid library sequence tag.

The first step in analysing the Ion Torrent sequence data is base calling, which is summarized in Figure 5.5. These steps are generally analysed automatically using the Torrent Suite Software analysis pipeline after the run is generated (https://github.com/iontorrent/TMAP).

Further analysis for DNA sequencing can be carried out using the Ion Reporter software that compares any variants with mutation databases such as SNPdb, a database of SNPs, and multiple small-scale variations that include insertions/deletions, microsatellites, and non-polymorphic variants (https://www.ncbi.nlm.nih.gov/snp/). ClinVar, a freely accessible public database that archives and aggregates information about relationships among variation and human health (https://www.ncbi.nlm.nih.gov/clinvar/), is also available to identify clinically important mutations. A further database cataloguing mutations in cancer may also be interrogated (see https://cancer.sanger.ac.uk/cosmic). RNA-seq is analysed in a similar way as described previously, but starting from BAM files.

Both Ion Torrent and Illumina BAM files can be viewed using IGV software (Robinson et al., 2011) where the sequencing tracks are aligned to a reference sequence. An example of sequencing visualization is provided in Figure 5.6, showing next-generation sequence of KRAS exon 2.

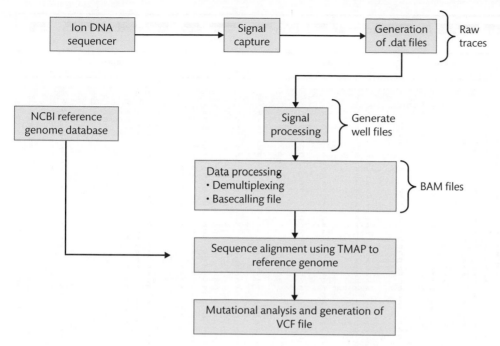

FIGURE 5.5

Workflow of DNA sequencing using Ion Torrent technology. The raw data is captured as digital signal, which is processed to generate raw traces in .dat files. The signals are processed to generate well files. The well files undergo demultiplexing, where each sequence track is linked to the sample it was generated from. This is followed by base calling where the sequences are aligned to each other to generate binary alignment map (BAM) file. The sequence in the BAM file is aligned to reference genome sequence to generate aligned sequence, which is used to call variants in variant call file (VCF) template.

5.3.2.2 Key applications

Similar to the Illumina platform, the key applications vary from sequencing small whole genomes, such as bacteria and viruses, and amplicons and gene panels with some medical applications, such as cancer panels with the PGM/ionS5 platform (Hsiao et al., 2016; Indugu et.al., 2016) to human exome sequencing (Damiati et al., 2016) and whole human genome sequencing with the IOnS5™ XL (Bertolini et al., 2015) and entire mitochondrial genome on the ion S5. Ion Torrent technology is currently used more for the diagnosis and prognosis of human diseases using targeted genetic panels.

Key Point

Ion Torrent employs a SBS, but it relies on the detection of variation in pH every time a nucleotide is incorporated through the release of a H^+ ion. Accordingly, each 'reaction chip' represents a miniature pH meter. With this technology, there is no need for fluorescent labels and no optical analysis is required.

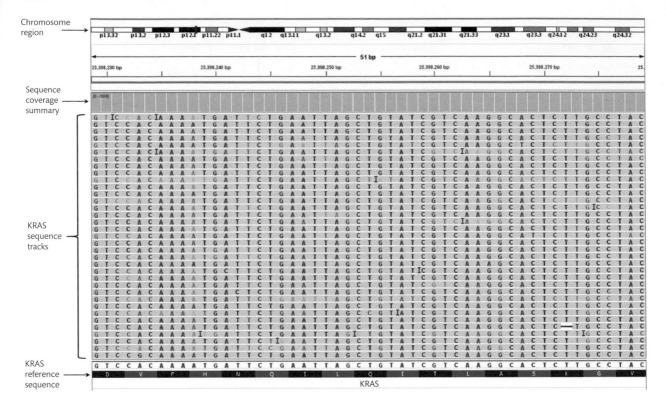

FIGURE 5.6

NGS sequence of KRAS exon 2 obtained using Ion Torrent PGM sequencer. BAM file containing the sequence information for exon 2 of KRAS obtained with Ion Torrent platform have been visualized here with IGV (Integrative Genome Viewer) software available from the Broad Institute (http://software.broadinstitute.org/software/igv).

5.3.3 Ligation-based NGS

In 2008, Applied Biosystems introduced the ABI SOLiD® (Sequencing by Oligo Ligation and Detection; ThermoFisher, Santa Clara, CA) platform that uses a sequencing chemistry, based upon DNA catalysed ligation. This platform is now commercialized by ThermoFisher. The chemistry is different from Ion Torrent and Illumina as the sequencing is by ligation, rather than by synthesis.

The specific process couples oligonucleotide adapter-linked libraries of DNA fragments (gDNA) formed by nebulization, sonication, or enzymatic digestion that are linked to two universal adapters (P1 and P2) with magnetic beads. Each bead–DNA complex is then amplified by emPCR to generate polymerase colonies that are called 'polonies'.

After amplification, the beads are selected to keep the ones that contain a molecule of DNA amplified on them using polystyrene beads coated with P2 (one of the two universal adapters linked to the DNA fragments). After enrichment, the beads are covalently attached to the surface of a specially treated glass slide that is placed into a fluidics cassette within the SOLiD sequencer. The ligation-based sequencing process starts with the annealing of a universal sequencing primer that is complementary to the SOLiD-specific adapters on the library fragments. The addition of a limited set of semi-degenerate fluorescently dyed octamer oligonucleotides and DNA ligase is automated by the instrument. DNA ligation occurs when a matching octamer hybridizes to the DNA fragment sequence adjacent to the universal primer 3' end. In these octamers, the dinucleotide positions on the template sequence that are currently interrogated are separated by 5-bp from the dinucleotide position interrogated in the next ligation cycle allowing sequence determination following fluorescence emission. The fluorescent dye is cleaved in each cycle to allow the next probe to attach. A video of the SOLiD sequencing can be viewed at: https://www.dnatube.com/video/27908/Solid-DNA-sequencing-animation.

While this technology provides high-throughput and parallel sequencing, and is cost effective, it does not handle palindromic sequences well. These sequences are made up of double-stranded DNA and/or RNA that are the same when read from 5' to 3' on one strand and 5' to 3' on the other, complementary, strand.

5.4 **Third-generation sequencing**

As opposed to second-generation sequencing described above that requires PCR for the sample preparation, third-generation sequencing does not require this procedure to be used. The principle of third-generation sequencing is to sequence in a 'live' manner. Sequencing of longer fragments—ultralong reads of 104–106 bases are possible—is combined with a higher accuracy. At the time of writing, two companies have developed such approaches—Oxford Nanopore Technologies (Oxford, UK) and Pacific Biosciences (PacBio, California, USA).

5.4.1 Oxford Nanopore Technologies (ONT)

ONT licensed core nanopore sequencing patents in 2007 and began strand sequencing efforts in 2010. The technology is based on the research on nanopores conducted in the early 1990s by the academic laboratories led by David Deamer, at the University of California Santa Cruz, and by George Church and Daniel Branton at Harvard University (Deamer et al., 2016).

The ONT method is based on electronic single molecule-sensing technology requiring no amplification, labelling, or optical instrumentation. The concept is based on the generation of an ionic current burst every time a nucleotide traverses through a nanopore, see video at https://www.youtube.com/watch?v=GUb1TZvMWsw.

A nanopore, as its name indicates, is a pore of a nanometer size. It is a complex of proteins set in an electrically-resistant polymer membrane. Hence, the nanopore is used as a single detector. The sequencing devices are, therefore, arrays of nanopores set in an electrically resistant artificial membrane similar to a cell membrane. A current is passed through each nanopore, and when molecules such as proteins or DNA pass through or near the nanopores, they create characteristic disruptions in current. Measurement of that current makes it possible to identify the molecule in question.

Two main devices have been developed by the company: the MinION, a single-use DNA sequencing pocket-sized USB-compatible molecular sensing device, which could see its application in a variety of settings, such as analysis of clinical pathogens, species identification in the field (POC testing) and medical applications, including precision medicine, such as identifying **structural variations** and breakpoints associated with various diseases from ultralong read sequences. There is also a larger-throughput product called PromethION, which will allow users to process larger samples or greater numbers of smaller ones.

The MinION is a flow cell bearing up to 2048 individually accessible nanopores that can be controlled in groups of 512 by an application-specific integrated circuit (ASIC). Prior to sequencing,

Structural variation
Generally defined as a region of DNA approximately 1 kb and larger in size, and can include inversions and balanced translocations or genomic imbalances (insertions and deletions).

adapters are ligated to both ends of gDNA or cDNA fragments. These adapters facilitate strand capture and loading of a processive enzyme at the 5′- end of one strand. The enzyme is required to ensure uni-directional single-nucleotide displacement along the strand on a mil-lisecond time scale. The adapters also concentrate DNA substrates at the membrane surface proximal to the nanopore, boosting the DNA capture rate by several thousand-fold. In addition, a hairpin adapter permits contiguous sequencing of both strands of a duplex molecule by cova-lently attaching one strand to the other. Upon capture of a DNA molecule in the nanopore, the enzyme processes along one strand (the 'template read'). After the enzyme passes through the hairpin, this process repeats for the complementary strand (the 'complement read').

5.4.2 Single-molecule real-time (SMRT) sequencing, Pacific BioSciences (PacBio)

Their first instrument, the PacBio RS, appeared in 2010. The principle of SMRT PacBio sequenc-ing is to capture sequence information during the replication process of the target DNA mol-ecule. The template, called a SMRTbell, is a closed, single-stranded circular DNA that is created by ligating hairpin adapters to both ends of a target double-stranded DNA molecule. When a sample of SMRTbell is loaded onto a chip called a SMRT cell (see video https://www.youtube.com/watch?time_continue=4&v=WMZmG00uhwU) the SMRT cell diffuses into a sequencing unit called a zero-mode waveguide (ZMW). In each ZMW, a single polymerase is immobilized at the bottom, which can bind to either hairpin adapter of the SMRT cell and start the replica-tion. Four fluorescent-labelled nucleotides, which generate distinct emission spectra, are added to the SMRT cell. As a base is held by the polymerase, a light pulse is produced that identifies the base. The replication processes in all ZMWs of a SMRT cell are recorded as a 'movie' of light pulses, and the pulses corresponding to each ZMW can be interpreted to be a sequence of bases (called a continuous long-read, CLR). This approach can then deliver reads of over 10 kb.

The typical throughput of the PacBio RS II system is 0.5–1 billion bases per SMRT cell. The entire workflow compared with second-generation sequencing is fast, as from template preparation to primary base call analysis takes less than a day. The technology is particularly suited in building high quality *de novo* assemblies of very complex genomes. Re-sequencing of even highly curated genomes, such as the human genome, can help identify more structural variants and also closes gaps in some regions of the genome. It also simplifies sequence assembly by spanning repeat regions and enhances the detection of copy number variations. Because no DNA amplification is required, the system can reduce certain artefacts and biases in genome coverage.

PacBio does not yet provide the high throughput offered by SBS techniques, such as Illumina HiSeq 2500. A major challenge of PacBio sequencing is that the error rate of continuous long-read (CLR) is relatively high at around 15% (http://www.pacb.com/wp-content/uploads/2015/09/Perspective_UnderstandingAccuracySMRTSequencing1.pdf). However, because the errors are randomly distributed in CLRs, the error rate can be reduced by generating reads with sufficient sequencing passes. Thus, a coverage of 10 passes yields >95% accuracy (Eid et al., 2009).

Key Point

The advantages of third-generation sequencing is that no prior PCR amplification of the DNA is needed for sequencing, allowing rapid sequencing and reducing potential errors introduced during the DNA amplification. It is worth noting that Ion Torrent and Illumina have engineered the enzymes used for their library prep to minimize these errors.

5.4.3 Other platforms

At the time of writing this chapter several companies are developing their own NGS platforms with the goal of decreasing the time of library preparation, augmenting the read length, enhancing accuracy together with shorter turnaround times. Some of these platforms are also trying to find their own niche away from the market leaders.

Of the most notable is Genia Technologies (a company founded in 2009 and acquired in 2014 by Roche), which is a nanopore-based sequencing company. The technology is very similar to that used by Oxford Nanopore Technology. What makes it distinctive is that Genia's technology uses a NanoTag sequencing approach, developed in collaboration with Columbia and Harvard Universities, USA (Fuller et al., 2016). Each nucleotide is engineered to carry a different tag. DNA sequences are not identified by detecting the nucleotides themselves as with the ONT method, but by measuring the current changes caused by the passage of each of four different tags that are released from the incorporated nucleotide during the polymerase reaction. The technology uses a DNA replication enzyme to sequence a template strand as base-specific engineered tags cleaved by the enzyme are captured by the nanopore. As the cleaved tags travel through the nanopore, they attenuate the current flow across the membrane in a sequence-dependent manner. DNA sequences are identified from the residual currents of the cleaved tags that flow through the nanopore/DNA complex. Because of the sensitivity of Genia's analog circuitry underneath each sensor, the platform allows sequencing of single DNA molecules that enables the detection of DNA mutation at very low frequency that other technologies might miss due to the sequencing of multiple DNA strands and pooling of the results.

GnuBIO is a droplet-based sequencing technology developed by the Biorad company and employs technology developed by the team of David Weitz at Harvard University, USA (Abate et al., 2013). In the GnuBIO system, each droplet works as a unique reaction vesicle. The droplet acts as microfluidics to perform the biochemical reactions for sequencing inside of tiny picolitre-sized aqueous drops. Each droplet acts as a discrete reaction vessel, like a miniature test tube, where the sequencing assay is performed. Hence, compared with Illumina or Ion Torrent platforms the entire workflow (target selection/enrichment, DNA amplification, DNA sequencing, and analysis) is done within one single instrument.

Lastly, NabSys is a technology that uses electronic detection based on semiconductor measurement. The platform works by adding sequence-specific tags to long DNA molecules. These tagged DNA molecules are then driven through the nanodetectors at high velocity (over 1 million bases per second) using a combination of electrophoretic and hydrodynamic control. The detector reports the locations of the sequence-specific tags on the molecules. The NabSys suite of software tools then analyses the information. Detailed information about the nanodetector and the process of DNA tagging are currently not published by NabSys. The technology can be used for structural analysis of sequence data.

5.4.4 Applications of third-generation sequencing

Third-generation sequencing technologies have been particularly used to produce highly accurate *de novo* assemblies of hundreds of microbial genomes (Koren et al., 2013; Loman et al., 2015) and highly contiguous sets of overlapping DNA segments that together represent consensus regions of DNA for many dozens of plant and animal genomes thus enabling new insights into evolution and sequence diversity (Chen et al., 2014; Berlin et al., 2015; Gordon et al., 2016). In particular, the PacBio platform has been shown to have clinical applications, such as HLA typing for transplantation (Chang et al., 2014), and determining the strain of the hepatitis B virus in infected patients (Li et al., 2017). This technology has also been successfully applied to the study of microbiomes (Surana et al., 2017). Transcriptome research for the identification of new fusion transcripts or splicing isoforms together with new exons and transcripts, the characterization of structural variations, such as translocations, provide examples of further applications (Weirather et al., 2015). RNA sequencing, using Iso-Seq also developed by PacBio, allows splicing mutations and levels of alternative spliced isoforms to be identified. These have been associated with a variety of diseases and particularly with cancer. In addition, third-generation sequencing is playing a key role in advancing the field of epigenetics, with novel DNA modifications being detected (Rhoads and Au, 2015).

5.5 **Future directions**

Next-generation sequencing is now starting to be more affordable (particularly in respect of reagents), but nonetheless certain platforms still need, in the first instance, a substantial investment for equipment purchase. As discussed, third-generation sequencing holds great promise, but some technical improvements both in methodology and bioinformatics analysis need to be made, and costs need to go down further to allow it to be more widely adopted in clinical practice and particularly in diagnostic laboratories.

Cross reference

In situ hybridization is covered in Chapter 3, Section 3.5.

The sequencing methods described in this chapter are all based on the use of homogenate samples and as such they provide pooled outputs without any cellular localization. While this is perfectly acceptable for many applications they cannot provide information on gene expression localized in individual cells. *In situ* hybridization methods can provide this, but only, at best, for a handful of transcripts at a time. Single-cell transcriptome sequencing (RNA-seq) can also be used, but there is no subcellular localization. 'Fourth-generation sequencing' seeks to bring sequencing of gene expression together with individual *in situ* cellular localization (Mignardi and Nilsson, 2014).

By combining the technologies of sequencing and *in situ* hybridization, fluorescent *in situ* RNA sequencing (FISSEQ) reports both the expression and localization of thousands of RNA transcripts simultaneously (Lee et al., 2014, 2015). FISSEQ relies on rolling circle amplification to create a three-dimensional grid of 'DNA nanoballs,' each representing a single RNA molecule, locked in its original subcellular location. Ligation-based sequencing then reads 27–30 bases from each nanoball, and the fluorescent output is read in three dimensions using confocal microscopy. The method identifies several thousand transcripts per experiment showing relatively good concordance with both RNA-seq and DNA microarray datasets for moderately expressed genes. The potential application of this technology is to identify biomarkers for use in directing targeted therapy.

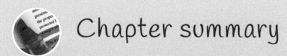 Chapter summary

This chapter has:

■ Provided an overview of sequencing technologies.

■ Defined and outlined the concepts of the different sequencing methods.

■ Presented some major applications for which those technologies can be used.

■ Discussed the advantages and limitations of each sequencing method.

Further reading

- Alekseyev YO, Fazeli R, Yang S, et al. A next-generation sequencing primer—how does it work and what can it do? Academic Pathology 2018; 5: 1–11.

- Cummings CA, Peters E, Lacroix L, Andre F, Lackner MR. The role of next-generation sequencing in enabling personalized oncology therapy. Clinical Translation Sciences 2016; 9: 283–292.

- Goodwin S, McPherson JD, McCombie WR. Coming of age: ten years of next-generation sequencing technologies. Nature Reviews: Genetics 2016; 17: 333–351.

- Ilyas M. Next-generation sequencing in diagnostic pathology. Pathobiology 2017; 84: 292–305.

- Levy SE, Myers RM. Advancements in next-generation sequencing. Annual Review of Genomics and Human Genetics 2016; 17: 95–115.

- Loman NJ, Misra RV, Dallman TJ, et al. Performance comparison of benchtop high-throughput sequencing platforms. Nature Biotechnology 2012; 30: 434–439.

- Merriman B, Rothberg JM. Progress in Ion Torrent semiconductor chip based sequencing. Electrophoresis 2012; 33: 3397–3417.

- Wu W, Choudry H (Eds). *Next Generation Sequencing in Cancer Research, Volume 1: Decoding the Cancer Genome*. Berlin, Springer, 2013.

 # References

- Abate AR, Hung T, Sperling RA, et al. DNA sequence analysis with droplet-based microfluidics. Lab on a Chip 2013; 13: 4864–4869.

- Ahmadian A, Gharizadeh B, Gustafsson AC, et al. Single-nucleotide polymorphism analysis by pyrosequencing. Analytical Biochemistry 2000; 280: 103–110.

- Ahmadian A, Ehn M, Hober S. Pyrosequencing: history, biochemistry and future. Clinica Chimica Acta 2006; 363: 83–94.

- Berlin K, Koren S, Chin CS, et al. Assembling large genomes with single-molecule sequencing and locality-sensitive hashing. Nature Biotechnology 2015; 33: 623–630.

- Bertolini F, Scimone C, Geraci C, et al. Next generation semiconductor based sequencing of the donkey (*Equus asinus*) genome provided comparative sequence data against the horse genome and a few millions of single nucleotide polymorphisms. PLoS One 2015; 10: e0131925.

- Blankenberg D, Hillman-Jackson J. Analysis of next-generation sequencing data using Galaxy. Methods in Molecular Biology 2014; 1150: 21–43.

- Chang CJ, Chen PL, Yang WS, Chao KM. A fault-tolerant method for HLA typing with PacBio data. BMC Bioinformatics 2014; 15: 296.

- Chen X, Bracht JR, Goldman AD, et al. The architecture of a scrambled genome reveals massive levels of genomic rearrangement during development. Cell 2014; 158: 1187–1198.

- Damiati E, Borsani G, Giacopuzzi E. Amplicon-based semiconductor sequencing of human exomes: performance evaluation and optimization strategies. Human Genetics 2016; 135: 499–511.

- Davies H, Bignell GR, Cox C, et al. Mutations of the BRAF gene in human cancer. Nature 2002; 417: 949–954.

- Deamer D, Akeson M, Branton D. Three decades of nanopore sequencing. Nature: Biotechnology 2016; 34: 518–524.

- Eid J, Fehr A, Gray J, et al. Real-time DNA sequencing from single polymerase molecules. Science 2009; 323: 133–138.

- Ewing B, Green P. Base-calling of automated sequencer traces using Phred. II. Error probabilities. Genome Research 1998; 8: 1860–194. Available at: https://genome.cshlp.org/content/8/3/186.full (accessed 14 October 2018).

- Friedman LS, Ostermeyer EA, Szabo CI, et al. Confirmation of BRCA1 by analysis of germline mutations linked to breast and ovarian cancer in ten families. Nature: Genetics 1994; 8: 399–404.

- Fuller CW, Kumar S, Porel M, et al. Real-time single-molecule electronic DNA sequencing by synthesis using polymer-tagged nucleotides on a nanopore array. Proceedings of the National Academy of Sciences, USA 2016; 113: 5233–5238.

- Garcia CA, Ahmadian A, Gharizadeh B, Lundeberg J, Ronaghi M, Nyrén P. Mutation detection by pyrosequencing: sequencing of exons 5-8 of the p53 tumor suppressor gene. Gene 2000; 253: 249–257.

- Gharizadeh B, Ghaderi M, Donnelly D, et al. Multiple-primer DNA sequencing method. Electrophoresis 2003; 24(7–8): 1145–1151.

- Gharizadeh B, Akhras M, Nourizad N, et al. Methodological improvements of pyrosequencing technology. Journal of Biotechnology 2006; 124: 504–511.

- Gordon D, Huddleston J, Chaisson MJ, et al. Long-read sequence assembly of the gorilla genome. Science 2016; 352: aae0344.

- Hillman-Jackson J, Clements D, Blankenberg D, et al. Using Galaxy to perform large-scale interactive data analyses. Current Protocols in Bioinformatics 2012; 38(1): 10.5.1–10.5.47.

- Hsiao YP, Lu CT, Chang-Chien J, Chao WR, Yang JJ. Advances and applications of Ion Torrent personal genome machine in cutaneous squamous cell carcinoma reveal novel gene mutations. Materials 2016; 9: E464.

- Indugu N, Bittinger K, Kumar S, Vecchiarelli B, Pitta D. A comparison of rumen microbial profiles in dairy cows as retrieved by 454 Roche and Ion Torrent (PGM) sequencing platforms. Peer Journal 2016; 4: e1599.

- International Human Genome Sequencing Consortium. Finishing the euchromatic sequence of the human genome. Nature 2004; 431: 931–945.

- Koren S, Harhay GP, Smith TP, et al. Reducing assembly complexity of microbial genomes with single-molecule sequencing. Genome Biology 2013; 14: R101.

- Lee JH, Daugharthy ER, Scheiman J, et al. Highly multiplexed subcellular RNA sequencing in situ. Science 2014; 343: 1360–1363.

● Lee JH, Daugharthy ER, Scheiman J, et al. Fluorescent in situ sequencing (FISSEQ) of RNA for gene expression profiling in intact cells and tissues. Nature Protocols 2015; 10: 442–458.

● Li J, Wang M, Yu D, et al. A comparative study on the characterization of hepatitis B virus quasispecies by clone-based sequencing and third-generation sequencing. Emergency Microbes Infection 2017; 6: e100.

● Loman NJ, Quick J, Simpson JT. A complete bacterial genome assembled de novo using only nanopore sequencing data. Nature: Methods 2015; 12: 733–735.

● Margulies M, Egholm M, Altman WE, et al. Genome sequencing in microfabricated high-density picolitre reactors. Nature 2005; 437: 376–380.

● Maxam AM, Gilbert W. A new method for sequencing DNA. Proceedings of the National Academy of Sciences, USA 1977; 74: 560–564.

● Mignardi M, Nilsson M. Fourth-generation sequencing in the cell and the clinic. Genome Medicine 2014; 6: 31.

● Nordström T, Gharizadeh B, Pourmand N, Nyren P, Ronaghi M. Method enabling fast partial sequencing of cDNA clones. Analytical Biochemistry 2001; 292(2): 266–271.

● Nourizad N, Gharizadeh B, Nyrén P. Method for clone checking. Electrophoresis 2003; 24: 1712–1715.

● Robinson JT, Thorvaldsdóttir H, Winckler W, et al. Integrative genomics viewer. Nature: Biotechnology 2011; 29: 24–26.

● Rhoads A, Au KF. PacBio sequencing and its applications. Genomics Proteomics & Bioinformatics 2015; 13: 278–289.

● Ronaghi M, Karamohamed S, Pettersson B, Uhlén M, Nyrén P. Real-time DNA sequencing using detection of pyrophosphate release. Analytical Biochemistry 1996; 242: 84–89.

● Ronaghi M, Uhlén M, Nyrén P. A sequencing method based on real-time pyrophosphate. Science 1998; 281: 363.

● Sanger F, Coulson AR. A rapid method for determining sequences in DNA by primed synthesis with DNA polymerase. Journal of Molecular Biology 1975; 94: 441–448.

● Sanger F, Nicklen S, Coulson AR. DNA sequencing with chain-terminating inhibitors. Proceedings of the National Academy of Sciences, USA 1977; 74: 5463–5467.

● Surana NK, Kasper DL. Moving beyond microbiome-wide associations to causal microbe identification. Nature 2017; 552: 244–247.

● Uhlmann K, Brinckmann A, Toliat MR, Ritter H, Nürnberg P. Evaluation of a potential epigenetic biomarker by quantitative methyl-single nucleotide polymorphism analysis. Electrophoresis 2002; 23: 4072–4079.

● Weirather JL, Afshar PT, Clark TA, et al. Characterization of fusion genes and the significantly expressed fusion isoforms in breast cancer by hybrid sequencing. Nucleic Acids Research 2015; 43: e116.

● Wheeler DA, Srinivasan M, Egholm M, et al. The complete genome of an individual by massively parallel DNA sequencing. Nature 2008; 452: 872–876.

- Wooster R, Bignell G, Lancaster J, et al. Identification of the breast cancer susceptibility gene BRCA2. Nature 1995; 378: 789–792.

 # Useful websites

Second-generation sequencing

Illumina

- Portal for videos: www.illumina.com/SBSvideo.

- Principle of technology: www.youtube.com/watch?annotation_id=annotation_1533942809&feature=iv&src_vid=HMyCqWhwB8E&v=fCd6B5HRaZ8.

- Bioinformatics: www.illumina.com/informatics/research/sequencing-data-analysis-management/basespace.html.

- Bioinformatics: https://galaxyproject.org/admin/get-galaxy/.

- Instrumentation: https://www.illumina.com/systems/sequencing-platforms.html.

Ion Torrent

- Principle: https://www.youtube.com/watch?v=ZL7DXFPz8rU.

- Bioinformatics: https://github.com/iontorrent/TMAP.

- Integrative Genome Viewer software available from the Broad Institute: http://software.broadinstitute.org/software/igv.

Genomic databases

- Bioinformatics for SNP analysis: https://www.ncbi.nlm.nih.gov/snp.

- Database for mutation analysis: https://www.ncbi.nlm.nih.gov/clinvar.

- Catalogue of somatic mutations in cancer: https://cancer.sanger.ac.uk/cosmic.

Ligation-based sequencing

- Solid sequencing platform: https://www.dnatube.com/video/27908/Solid-DNA-sequencing-animation.

Third-generation sequencing

Oxford Nanopore Technologies

- Principle: https://www.youtube.com/watch?v=GUb1TZvMWsw.

Pacific BioSciences

- Principle: https://www.youtube.com/watch?time_continue=4&v=WMZmG00uhwU.

- Sequencing accuracy: http://www.pacb.com/wp-content/uploads/2015/09/Perspective_
UnderstandingAccuracySMRTSequencing1.pdf.

 Discussion question

5.1 What are the factors and challenges of implementing next-generation sequencing in a molecular diagnostic laboratory?

Answers to the self-check questions and tips for responding to the discussion question are provided on the book's accompanying website:

Visit www.oup.com/uk/warford

6

Molecular Analysis and Interpreting Molecular Data

Rifat Hamoudi and Anthony Warford

Learning objectives

After studying this chapter, you should be able to:

- Understand the types of data generated from homogenate and intact cell and tissue preparations.

- Appreciate how the data can be analysed using qualitative, semi-quantitative, and quantitative methods.

- Give examples of where bioinformatics is being used to answer questions through the interrogation of DNA and RNA target sequences.

- Understand the principles of the use of bioinformatics to provide information on epigenetic changes and DNA–protein relationships.

- Give examples where qualitative and semi-quantitative methods can be applied to the analysis of intact samples for nucleic acids and proteins.

- Have an overview of the application of telepathology, and image capture and analysis for diagnosis without the need for microscopy.

6.1 Introduction

Cross reference
See Chapter 3, Section 3.4 for more details on flow cytometry.

All molecular techniques produce data, and this can be in many and various forms. As has been seen, data analysis is sometimes integral to the method. Examples include flow cytometry for intact cells, together with ELISA and mass spectrometry, for proteins and peptides, respectively, and PCR for nucleic acids as considered in Chapter 4,. The reader is referred to 'Chapters 3 and 4' for information about their integration into the work flow of these methods. This chapter will focus on the nature and types of data generated from homogenate, and intact cell and tissue preparations, and their analysis where this is clearly distinct from its initial generation.

The type of analysis is driven primarily by whether it is applied to homogenate or intact samples. For the former, this can be quantitative and uses the stepwise application of complex bioinformatic algorithms. The use of bioinformatics has become ever more important and especially so since the advent of next generation sequencing (NGS). It should be noted that it is still very much a work in progress. Indeed, the volume of data from different 'OMIC' modalities, genomic, **transcriptomic**, **epigenetic** and **proteomic**, continues to increase seemingly exponentially (Cook et al., 2016). The challenge is to capture this data in a form where it can be recovered and analysed for immediate clinical and research use, and to store it for potential future review in a form where it may well need to be read from different software platforms.

For intact samples, analysis is often qualitative or semi-quantitative, and applied to microscopy of samples in which specific nucleic acids or proteins have been identified using *in situ* hybridization (ISH) or immunocytochemical (ICC) methods, respectively. This analysis can be supplemented by the use of microscope slide scanning technology that allows diagnosis to be made at distant sites.

In this chapter data analysis will be considered under two main headings; those that apply to homogenate preparations for DNA and RNA analysis, and those that are used for intact sample analysis for nucleic acids and proteins.

Cross reference
See Chapter 5 for more details on sequencing technologies.

Transcriptomics
Study of RNA molecules in a cellular population.

Epigenetic
Changes arising from non-genetic influences on gene expression.

Proteomics
Study of protein expression in a cellular population.

Key Point

Many types of data are derived from molecular assays. Those derived from homogenate assays can be subject to quantitative analysis that may include the use of complex bioinformatic algorithms; see Table 6.1 for a selection as quoted in the text. Those from intact samples are often analysed using qualitative or semi-quantitative methods.

6.2 Data analysis of homogenate preparations

Homogenate samples obtained from fresh blood and snap-frozen tissue, for example, usually have intact nucleic acids and proteins. On the other hand, formalin-fixed paraffin embedded (FFPE) tissues are degraded due to cross-linking effects of fixation, and especially so with respect to nucleic acids. Other sample sources may also be degraded. For example, saliva naturally contains enzymes such as RNase (Palanisamy and Wong, 2010). Therefore, molecular analysis and data interpretation may differ depending on the sample source and its preparation.

Cross reference
See Chapter 1 Box 1.3 for more information on formalin fixation.

6.2.1 Analysis of DNA-based data

DNA-based analysis is important in diagnostic medicine and generally consists of the identification of structural changes, mutation detection and copy number variation within a region in the DNA.

Cross reference
The analysis of PCR assays is described in Chapter 4.

TABLE 6.1 A selection of bioinformatic programs for analysis of DNA and RNA

Software name	Application	Software description
Bowtie	Aligning short and long reads of DNA sequence	Extremely fast, general purpose short read aligner
BumpHunter	Identification of differentially methylated regions from methylome array data	Identifying differentially methylated regions in epigenetic epidemiology studies
BWA	Aligning short and long alignment to reference genome	Mapping low-divergent sequences against a large reference genome, such as the human genome
COMET	Segmentation of WGBS to identify differentially methylated regions from low coverage methylome data	Dynamically segments WGBS methylomes into blocks of COMETs from which lost information can be recovered in the form of differentially methylated COMETs (DMCs)
Cufflinks	Identification of fusion transcripts and expression values through RNA-Seq assembly and alignments into a parsimonious set of transcripts	Assembles transcripts, estimates their abundances, and tests for differential expression and regulation in RNA-Seq samples
DESeq	Identification of differentially expressed genes from RNA-seq data	Test for differential expression based variance estimates
DMRcate	Identification of differentially methylated regions from methylome array and using WGBS data	De novo identification and extraction of DMRs from the human genome using WGBS and Illumina Infinium Array (450K and EPIC) data
EdgeR	Identification of differentially expressed genes from RNA-seq data	Finds differentially expressed genes and transcripts using range of statistical methodologies
FunNorm	Normalization of methylation data from Illumina Infinium arrays	Functional normalization (FunNorm) is a between-array normalization method for the Illumina Infinium Human Methylation 450 platform. It removes unwanted variation by regressing out variability explained by the control probes present on the array
MiNiFi	Analysis and visualization of methylation array data	Analyse and visualize Illumina Infinium methylation arrays
PLINK	GWAS analysis	Analysis of genotype and phenotype data
SNPTEST	GWAS analysis	Analysis of single SNP association in genome-wide studies
STAR	Extracting gene expression counts from aligned sequences	Mapping of large sets of high-throughput RNA sequencing reads to a reference genome
TopHat	Discovery of fusion products from RNA-seq data	Aligns RNA-seq reads to the genome using Bowtie to discover splice sites

Key: WGBS, whole-genome bisulfite sequencing

6.2.1.1 DNA microarrays

Cross reference

Microarrays-based techniques are discussed in Chapter 4, Section 4.2.1.4.

DNA microarrays are used to physically map and characterize regions of the DNA. aCGH uses genome-mapped, and sequence-verified genomic plasmid clones, such as Bacterial Artificial Chromosomes (BAC), which consist of around 350,000 bp of DNA sequence arrayed on glass slides as hybridization targets for normal and diseased sample DNA. They allow high-throughput screening and have been successfully used in the identification of large chromosomal gains, rearrangement, or deletion (Watkins et al., 2010). Such arrays are amenable to degraded nucleic acid extracted from FFPE tissue (Johnson et al., 2006).

The raw data from aCGH is in the form of numerical values, representing the hybridization of the diseased and normal samples to the clones on the slides. The initial step is to convert this to $\log_2$ ratio of disease/normal numerical value along the sequence of the arrayed BAC plasmid clones (Jarmuz et al., 2006) as shown in Figure 6.1a. The second step in the data analysis is to ensure that the quality of the array is acceptable, and this is done by checking the aCGH ratio of positive and negative clones. Once these are verified, the third step is to plot the ratio along the consecutive sequence clones and identify any amplification, gains, and losses. Generally, a ratio of zero means there are no chromosomal abnormalities, a high ratio indicates chromosomal gain and a low ratio a loss as shown in Figure 6.1b. Because aCGH is amenable to analysing data from degraded FFPE samples, various smoothing algorithms such as Hidden Markov model and circular binary segmentation were implemented to identify abnormal chromosomal patterns embedded in noise (Hsu et al., 2011).

A key issue is to identify a control area within the DNA that is not co-amplified with the region of interest. In aCGH, this is done by searching the chromosomal regions for flat aCGH ratio (i.e. ratio equal to zero) and identifying genes that map to that region. For example, Luan et al.

(A)

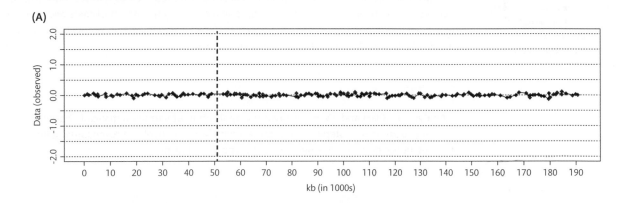

(B)

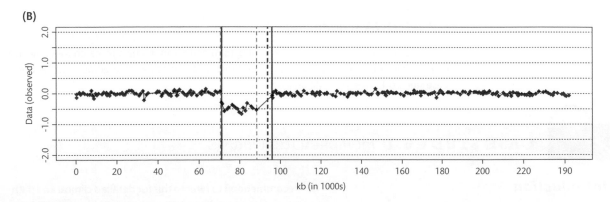

FIGURE 6.1

Array comparative genome hybridization. aCGH of chromosome 1 using BAC clones. Each dot represents a BAC clone. (a) The $\log^2$ ratio of disease/normal for normal sample. (b) The $\log^2$ ratio of disease/normal for disease state showing deletion from 70,000 to around 95,000 bp on chromosome 1.

(2010) used aCGH data to identify *MAGT2* gene as control for the detection of amplification of *SELPLG* and *CORO1C* genes in primary effusion lymphoma.

6.2.1.2 Genome-wide association studies

Genome-wide association studies (GWAS) provide a higher resolution DNA map of regions exhibiting sets of genetic variants based on a large number of SNPs present in different individuals to correlate their association with a disease. GWAS using SNP arrays have been used in many polygenic trait studies to delineate regions that are significantly altered between the case and the control groups. Each DNA sample is genotyped at around 400,000–1 million SNPs in cases and controls. The aim of a GWAS is to identify SNPs, where one allele is significantly more common in disease than in control cases (Liu et al., 2017). The data analysis is based on the **odds ratio** and is typically calculated using p-value derived from chi-squared test. There are many software packages that can be used for this process. The most common software are SNPTEST and PLINK (Pei et al., 2010).

GWAS identifies a large number of marker loci (e.g. SNPs or dinucleotide repeats) tested for association with the disease. Because of the large number of marker loci and genetic heterogeneity amongst individuals in a population, error or bias can occur and be detrimental to the GWAS analysis. Therefore, in terms of data analysis, the initial step is the removal of false-positive associations. Several quality control steps are required to remove individuals or markers with particularly high error rates. If, as advised, many thousands of cases and controls have been genotyped to maximize the power to detect association, the removal of a handful of individuals should have little effect on overall power.

Odds ratio

A measure of association between an exposure and an outcome. The OR represents the odds that an outcome will occur given a particular exposure, compared to the odds of the outcome occurring in the absence of that exposure. Odds ratios are most commonly used in case-control studies; however, they can also be used in cross-sectional and cohort study designs.

Key Point

Analysis of aCGH and GWAS involves the application of various statistically based software programs in a stepwise fashion to identify, respectively, chromosomal gains or losses, or the presence of single nucleotide polymorphisms.

6.1 SELF-CHECK

What are the similarities and distinctions for the genomic analysis pathways for aCGH and GWAS?

6.2.1.3 DNA sequencing

NGS, as described in Chapter 5, is beginning to migrate from being a research tool to assisting in making informed clinical choices, for example, in the selection of targeted drug therapy; see Case study 6.1, 'Homogenate analysis'.

CASE STUDY 6.1   Homogenate analysis

Introduction

The following *hypothetical* case study is based on an *actual* case report authored by Shen et al. (2018) and the reader is recommended to refer to this for detailed clinical and pathological content. Here, the reader might like to consider:

- What is the relevance of the summary of the patient's information to treatment with gefitinib?

- Why would the use of unfixed bone biopsy be preferred to use of decalcified and FFPE tissue sections for NGS?

- Why is a full, rather than a partial response to targeted drug therapy difficult to achieve?

This case highlights how with developments in targeted therapy, traditional treatment based on histopathological diagnosis has become limited. NGS, with the advantages of high sensitivity, high throughput, and low sample quantity, can play an important role in targeted therapy. Particularly in patients diagnosed via small biopsy samples, NGS can provide maximal tumour genomic assessment, determine potential therapeutic targets, and evaluate tumour heterogeneity at the gene level that can be used to stratify patient therapy.

Case study

A 58-year-old woman with no history of smoking and of Chinese ethnic grouping presented with a severe cough and haemoptysis was admitted to hospital. A core biopsy from the tumour mass, which was located centrally near the left main bronchus, was taken for histological examination. This was examined after FFPE by morphological and immunocytochemical analysis, and the tumour was diagnosed as non-small cell lung cancer of adenocarcinoma subtype.

Due to the proximity of the tumour to the bronchus surgery to remove the growth was not feasible and chemotherapy was initiated resulting in a partial response. Meanwhile, unstained sections from the tumour biopsy were homogenized and analysed by NGS for the presence

of EGFR mutations. This revealed an EGFR 19 exon deletion, an example of which is shown below.

This profile indicated that the patient could benefit from targeted therapy employing gefitinib (Iressa), a small molecule inhibitor of EGFR. The drug was administered and the partial response was prolonged for 6 months before symptoms indicated the return of aggressive disease and the presence of potential bone metastases.

A biopsy was taken from one of the potential bone metastatic sites and divided with half being processed for FFPE examination after decalcification, and the remainder being frozen without fixation then analysed by NGS. The results of histological examination were similar to the original lung biopsy sample. However, the NGS results indicated the presence of the EGFR 19 exon deletion *and* exon 20 T790M mutation in the metastatic tumour deposit. This showed that the molecular profile of the lung cancer had changed with extra EGFR mutations being acquired, possibly through intratumoural heterogeneity. This would explain the eventual failure of the gefitinib EGFR targeted therapy.

As a result of these findings the patient was prescribed osimertinib (Tagrisso) as a second-line targeted EGFR therapy. This small molecule drug binds irreversibly to EGFR with a T790M mutation and to EGFR with a L858R mutation, and with an exon 19 deletion. Again, a partial response, this time lasting 9 months, was achieved before the tumour once again reasserted itself and the patient succumbed to the disease.

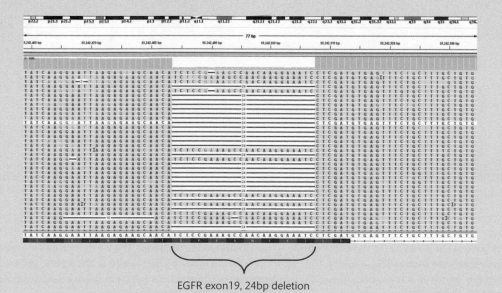

EGFR exon19, 24bp deletion

Cross reference
See Chapter 5 for more
information on sequencing
technologies.

Since the advent of NGS, data generation has increased exponentially and bioinformatics has become the real bottleneck in processing DNA-sequencing data. Many software programs were developed to using a modification of Phred for base-calling accuracy as was used for first-generation (Sanger) sequencing. In NGS, a sequence quality score of 20 and higher is considered to be acceptable. As NGS involves sequencing the same DNA area multiple times, displaying the results involves each sequencing track being mapped over the reference sequence of the region that is being sequenced.

The raw data generated by NGS instruments is in FASTQ format, see Figure 6.2. The first step in bioinformatic analysis is to check the quality of the FASTQ file. This is done by using summary statistics of the base-calling accuracy across the sequence base pairs to determine accuracy

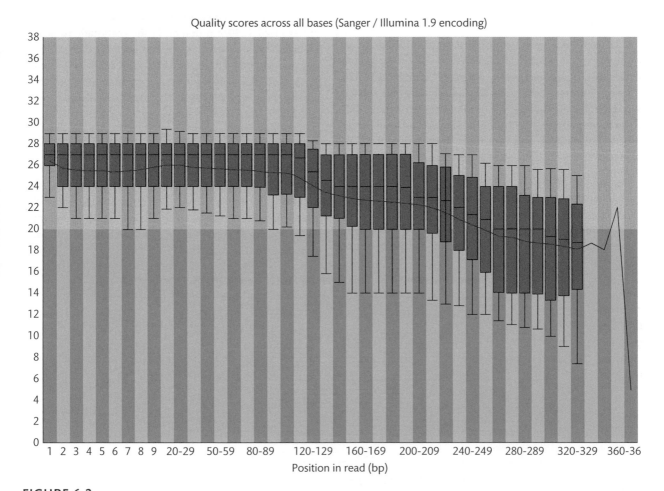

Quality scores across all bases (Sanger / Illumina 1.9 encoding)

Position in read (bp)

FIGURE 6.2

FASTQ quality control file. A FASTQ file shows the quality across all bases in the sequence. The x-axis indicates the read position and the y-axis indicates the base-calling accuracy. The colours indicate the following: red, which is <20 is poor; yellow, between 20 and 28 is acceptable; above 28 is very good. The whiskers indicate the variation in base quality at specific read positions. The figure shows that all bases called until base position 160 have quality score above 20, which is acceptable.

along the sequence. The FASTQ tool produces a basic text and a HTML output file that contain the results of the analysis, including:

- Basic statistics.
- Per base sequence quality.
- Per sequence quality scores.
- Per base sequence content.
- Per base GC content.
- Per sequence GC content.
- Per base N content, when the read at a particular base cannot be determined accurately.
- Sequence length distribution.
- Sequence duplication levels.
- Overrepresented sequences.
- Kmer content. This is a motif (or a small word) of length k observed more than once in a genomic or sequenced sequence. It is commonly used for searching repeat sequence across the genome.

If the sequence quality is satisfactory, the next step is to align the raw sequence data to a reference genome sequence. The most common alignment software are BWA and Bowtie (Giannoulatou et al., 2014). However, the algorithms used for these are optimal for shorter sequences; for longer sequences the alignment can be very time consuming. The resulting alignment produces a complex file format that describes the alignment details. Alignment output files can be generated in either SAM or its compressed format, BAM file format, with the latter being the more popular file format. BAM contains comprehensive raw data of genome sequencing, consisting of a compressed binary representation of the sequence alignment map. Following alignment, the **variant-calling** step is carried out by identifying misaligned base pair sequences. This generates a file such as VCF that describes the gene sequence variations across the entire genome sequence.

Variant calling
The process by which the variants are identified from sequence data.

Currently, the most common variant-calling software are VarScan and Mutect2, which are used for point mutation screening. For insertion and deletion mutations, the commonly used software is Pindel (Ye et al., 2009). Nevertheless, variant-calling algorithms are very much a work in progress. A paper using lung and melanoma NGS sequencing data distributed to various international genome centres showed discrepancies between the various bioinformatics pipelines used in each centre (Wang et al., 2013). Part of the problem in variant calling at the genomic level is that all current bioinformatics softwares apply the same parameters across the whole of the genome, not taking into account differences in GC content, gene density, and repeat regions from chromosome to chromosome.

In addition to bioinformatics problems, degraded DNA requires different parameters to be developed as highlighted in the recently published RING trial that showed that the DNA integrity is highly variable between different laboratories (Kapp et al., 2015). Thus, current variant-calling software, which identifies true mutants in blood and fresh-frozen tissue, may produce false positive and false negative calls on data generated from FFPE or saliva when the DNA is degraded.

Finally, most of the NGS bioinformatics pipelines start by aligning the sequence data to a reference sequence, which means that structural variants, such as genomic translocations and areas of the genome that undergo somatic hypermutation or other biological phenomena, will not be detected easily, leading to the development of *de novo* alignment algorithms.

> ## Key Point
>
> Next-generation sequencing, involving massively parallel sequencing, can produce sequence data for an entire genome. Accordingly, the complexity of analysis required for the latter is much greater than for first-generation sequencing and is still evolving.

6.2 SELF-CHECK

What does the FASTQ file serve as and what type of information can be found in this file?

6.2.2 Analysis of RNA-based data

RNA consists of several species, each with different function. The common RNA species of interest are messenger RNA (mRNA) that codes for protein assembly and smaller RNA species, such as microRNA (miRNA), which is non-coding RNA and is involved in RNA silencing and post-transcriptional regulation of gene expression (Bartel, 2004). Additional RNAs comprise a group of non-coding species that includes transfer RNA and ribosomal RNA. The most commonly investigated type of RNA is mRNA and the field associated with it is referred to as transcriptomics. This section will focus on mRNA and miRNA analysis in various samples with the exception of RT-PCR analysis that is covered in Chapter 4, Section 4.2.2.1.

6.2.2.1 Gene expression microarrays analysis

For a full description of the techniques please read Chapter 4, Section 4.2.2.3. During the late 1990s gene expression microarray studies became ubiquitous as the power analysis of the transcriptome to identify key signature profiles that can explain biological mechanisms in various species became apparent (Eisen et al., 1998).

In 2000, to characterize the gene expression signature in lymphoid malignancies, a seminal translational study was carried out on follicular lymphoma, chronic lymphocytic leukaemia, and the clinically heterogeneous diffuse large B-cell lymphoma (DLBCL). Based on gene expression profiling of DNA microarrays two new molecularly distinct sub-classes of a DLBCL were identified (Alizadeh et al., 2000). The first typed expressed genes characteristic of germinal centre B-cells and the second type expressed genes induced during *in vitro* activation of peripheral blood B-cells. A significant increase in survival rate was observed in patients with germinal centre B-like DLBLC, compared with those with activated B-like DLBCL. This was the first study showing the molecular classification of tumours based on gene expression profiles and had enabled the identification of a new subtype of lymphoid malignancies that otherwise looked morphologically the same and could, therefore, benefit from a different treatment modality.

Gene expression microarray output is an image of the scanned array (Figure 6.3). Image analysis is carried out to extract the intensity for each spot or probe feature on the array. Image analysis consists of gridding that aligns a grid to the spots on the array, segmentation that identifies the shape of each spot, and finally, intensity extraction, which measures the intensity for each spot against any surrounding background. Following image analysis, the data file corresponds to the raw gene expression microarray data, consisting of mapped probe positions and the gene names for each position, as well as the numerical value that is often transformed into a logarithmic value corresponding to the expression of each gene in each sample. Data analysis from the raw data consists of data pre-processing followed by filtering of non-variant probes

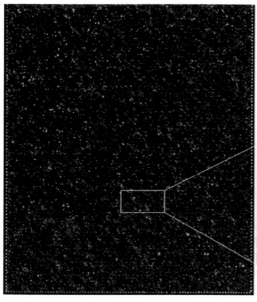
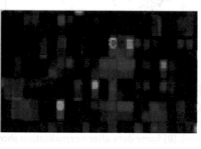

FIGURE 6.3
Gene expression array image.
Raw image from gene expression microarray file showing each pixel intensity determined by the expression of level of a gene in the specific sample hybridized on the array. The rectangle shows the zoomed in array pixels.

and, finally, identification and visualization of differentially expressed genes ($\log_2$ ratios) and possibly correlating them to clinicopathological data.

Data pre-processing consists of background correction, normalization, and summarization. Background correction involves the subtraction of background signal from the spot intensity to get a more accurate estimate of the biological signal from the spot. Normalization removes systematic variation in a microarray experiment, which affects the measure of specific gene expression levels. Normalization for a two-channel array (Cy5 and Cy3 dyes) is carried out within the array, to remove systematic differences due to intensity and location dependent dye biases. To visualize the quality of the normalization of the data **MA plots** are used (Dudoit et al., 2002).

There are two methods of normalization for two-channel arrays. *Global LOWESS* (LOcally WEighted Scatter plot Smoothing) algorithm, is a non-linear regression of $\log_2$ ratios against the average $\log_2$ intensity. It computes local linear regressions that are joined together to form a smooth curve. This normalization takes into account intensity artefacts. *Print-tip LOWESS* normalization, which is local linear regression computation, is limited to a single print-tip group. The effect of each can be seen in Figure 6.4.

For single-colour arrays, such as Affymetrix or Illumina, normalization is carried out between arrays and consists of various algorithms, which can be grouped into the following four sequences of events:

1. *Adjust the arrays* using some control or housekeeping genes that would be expected to have the same intensity level across all of the samples.

2. *Adjust using spike control.* Spike controls are nucleic acids with a known concentration and abundance that can be used to assess the quality of the array data obtained for the test samples.

3. *Multiply each array* by a constant to make the mean (median) intensity the same for each individual array (global normalization).

4. *Match the percentiles* of each array (quantile normalization).

MA plot
This is an application of a Bland–Altman plot for visual representation of genomic data. The plot visualizes the differences between measurements taken in two samples, by transforming the data onto M (log ratio) and A (mean average) scales, then plotting these values.

to divide each mapped read by the number of matched positions in the genome in question. Cufflinks uses fragments per kilobase of exon per million fragments mapped (FPKM) (Roberts et al., 2011). More sophisticated normalization algorithms are now available, such as quantile-based trimmed mean of median values (TMM) and Linear Models for Microarray Analysis (LIMMA; Ritchie et al., 2015).

The normalized RNA sequencing count data is then subjected to differential gene expression analysis. Many models have been developed for this. The most common software for carrying this out is DESeq (Anders and Huber, 2010; Conesa et al., 2016) followed by EdgeR (Robinson et al., 2010) and Cuffdiff (Trapnell et al., 2012).

These analyses can be carried out using freely available bioinformatics pipelines, for instance on Galaxy (https://galaxyproject.org/), that contain all the software needed for analysing NGS RNA sequencing data. RNA sequencing analysis is summarized in Figure 6.6.

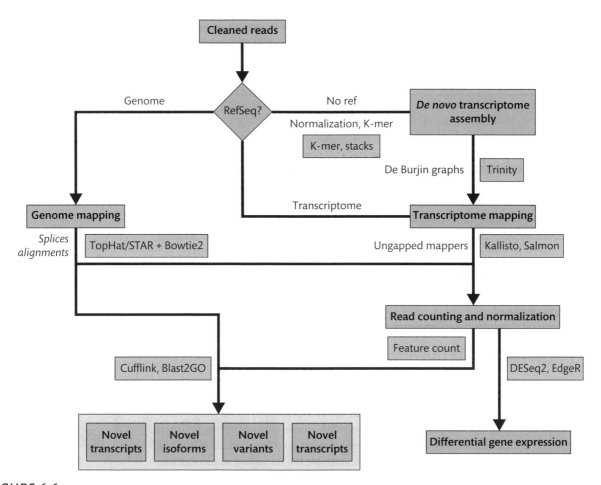

FIGURE 6.6

RNA sequencing analysis workflow. Generally, the steps involve cleaning the sequence followed by alignment to a reference human genome sequence to identify novel transcripts, isoforms, splice variants, and fusion transcripts or to carry out differential gene expression analysis using transcriptome mapping. If the RNA sequence does not map to a reference sequence, then *de novo* transcriptome assembly will be carried out to join together the RNA sequence fragments and use those to derive differentially expressed genes.

The grey boxes show the name of the currently commonly available software used at various steps of RNA sequencing analysis.

Key Point

RNAseq, in some instances, is replacing microarrays for assessing gene expression, as it does not imply prior knowledge of the target genome, it can identify new transcripts, such as alternative transcripts or translocation, e.g. in cancer. The downside of RNAseq is that the bioinformatics is not presently as well developed as for microarrays.

6.4 SELF-CHECK

Why is it better to use RNA sequencing to detect fusion transcript over gene expression microarrays?

6.2.2.2 ChIP sequencing

ChIP sequencing, also known as ChIP-seq, is a method used to analyse protein interactions with DNA (www.illumina.com/Documents/products/datasheets/datasheet_chip_sequence.pdf). ChIP-seq is used primarily to determine how transcription factors and other chromatin-associated proteins influence phenotype-affecting mechanisms (Johnson et al., 2007). A control sample is needed for ChIp-seq as open chromatin regions are fragmented more easily than closed regions. As the repetitive sequences (30–70 bp) tend to be enriched it is always important to compare the ChIP-seq peak with the same region in a matched control. Sequencing of the repetitive regions is followed by alignment and peak calling when comparing with control sample. Peak calling will identify enriched areas that, when integrated with other data, can lead to subpeak identification and motif analysis. With ChIP-seq, the depth of sequencing is very important as the more prominent peaks are identified with fewer reads, whereas weaker peaks require greater depth. For these, around 30–40 million reads per sample is usually sufficient. The bioinformatics pipeline for ChIP-seq is summarized in Figure 6.7.

6.2.3 Epigenetics

Epigenetics refers to changes in the chromosome that affect gene activity and expression. Unlike genomics, such changes do not involve change in the nucleotide sequence and occur through DNA methylation and histone modification, each of which alters how genes are expressed without altering the underlying DNA sequence. The term 'epigenome' refers to the global analyses of epigenetic changes across the entire genome.

DNA methylation involves the addition of a methyl group to the 5-carbon position of cytosine residues of the dinucleotide CpG (adjacent C and G bases linked by a phosphate bond). Enzymes that add a methyl group are called DNA methyltransferases. DNA methylation is implicated in the repression of transcriptional activity. Therefore, a common way to study epigenetics is to subject the DNA to bisulfite conversion. This involves treatment of DNA with bisulfite, which converts cytosine residues to uracil leaving the 5-methylcytosine (mC) residues unaffected; see Figure 6.8.

6.2.3.1 Targeted methylation

Targeted methylation is focused on the study of the DNA methylation patterns at each CpG site within the gene of interest. This can provide better base calling than whole methylome and

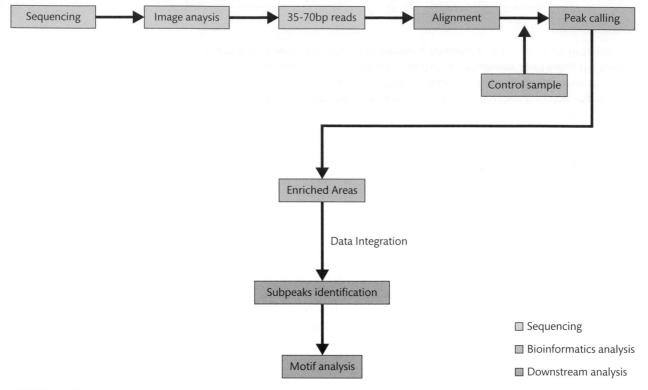

FIGURE 6.7

ChIP-seq bioinformatics workflow. ChIP-seq workflow showing the raw data captured as an image and processed to obtain 35–70 bp reads. The reads are aligned to control sample to generate peak calls for transcription factor binding site. Those enriched areas undergo further downstream analysis to identify sub-peak identification leading to motif analysis.

Cross reference

More information on pyrosequencing is provided in Chapter 5, Section 5.2.2.

is, therefore, useful when studying the methylation of a small number of genes. One method is to carry out PCR using primers that are specific to the bisulfite-converted DNA strand. This is usually called methylation-specific PCR. Another method is to detect the methylated bases using pyrosequencing). Pyrosequencing methylation is more sensitive than methylation-specific PCR.

FIGURE 6.8

DNA bisulfite conversion. A DNA sample is incubated with sodium bisulfite that converts unmethylated C into U (uracil), but leaves methylated C unmodified, which is why the first CG underlined in green remains unchanged. PCR is then used to change the UG to TG. This generates a DNA strand that is differentiable upon subsequent sequencing.

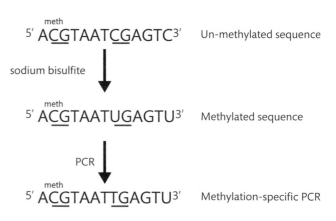

6.2.3.2 Epigenetic microarray

Since epigenetic changes tend to be global, epigenetic arrays have been used in numerous studies. The most common epigenetic arrays are the Illumina arrays and these are labelled according to the number of individual CpG sites spotted on them. The initial array from Illumina was the Infinium 27k array, which contained 27,578 individual CpG sites spread across 14,495 genes. However, this has now been superseded by the Infinium 850k or Methylation EPIC array that interrogates 850,000 methylation sites across the genome at single nucleotide resolution.

The raw data from an Illumina epigenetic array is an intensity data file (IDAT) that consists of the position of the mapped probes, as well as the numerical data corresponding to the methylation value. For each CpG, there are two measurements—a methylated intensity (M) and un-methylated intensity (U). These intensity values can be used to determine the proportion of methylation at each CpG locus. Methylation levels are commonly reported as beta values.

The bioinformatics workflow involved in analysing epigenetic data from epigenetic arrays can be summarized by the following steps:

- The initial step is *quality control*, which consist of calculating detection p-values (significance) for every CpG in every sample indicative of the quality of the signal. Tools such as MiNiFi calculate detection p-values by comparing the total signal (M + U) for each probe to the background signal level estimated from the negative control probes. Bad quality probes are removed by filtering the data with a p-value > 0.05 as being bad quality—see Figure 6.9.

- Following the removal of the bad quality probes, *data pre-processing* is carried out similar to that described for gene expression microarray in Section 6.2.2.2.

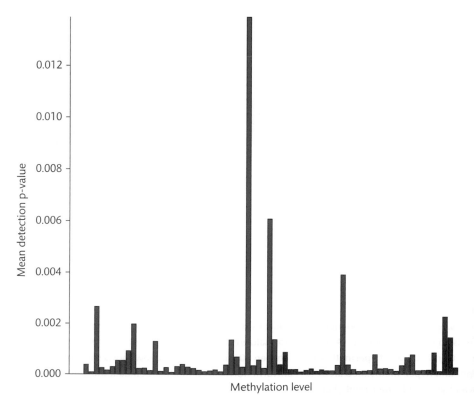

FIGURE 6.9

Mean detection of methylation levels. The methylation level detected in different samples based on mean detection p-value. The colour shows different pathological stages from the same disease. Each bar indicates a different patient.

In summary, the starting point for analysis is the same as DNA sequencing and RNAseq as FASTQ files (Bolger et al., 2014). A bisulfite-converted human genome reference is generated using Bowtie2, and the EpiGenome library sequence data in question are, therefore, aligned to the reference genome. Methylation information is extracted from the output SAM file, and genome tracks are extracted for visualization and reporting of downstream differential methylation calculations. Differences between lower methylation structures such as DMP, DMR, and higher methylation structures such as co-differentially methylated regions (COMET), which captures the key areas that are methylated across the whole methylome, can be seen in Figure 6.12.

Key Point

Methylation of DNA bases identifies changes in the promoter part of the gene and, thus, plays an important role in epigenetics. Targeted methylation provides high coverage of changes in CpG islands, which exist at promoter region of the genome, whereas whole genome bisulfite sequencing provides an important picture of methylation changes across the whole genome, which can, in turn, provide information on gene silencing in disease. Methylation analysis involve analysis at DMP. DMR involves grouping of DMPs according to biological information and using bioinformatics techniques such as clustering to capture the methylated regions. More recently, methods that capture the whole methylome such as COMET have shown to be important in identifying data across the methylome.

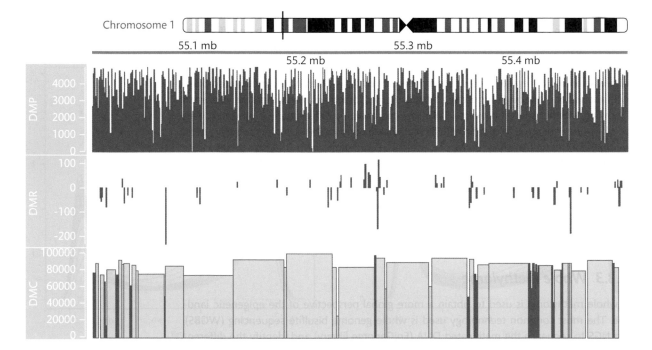

FIGURE 6.12

CpG methylation pattern of a region of chromosome 1. Methylation pattern from a region in chromosome 1 highlighted by the red line. The methylation is shown at the level of differential methylated point (DMP), differentially methylated region (DMR), and differential co-methylation using COMET algorithm (DMC). DMP shows variable levels of CpG methylation across the region. DMR delineates some of the methylated regions. However, DMC is the best at detecting the methylation level and variation across the region. Modified from Libertini et al. (2016).

6.3 Analysis of intact sample preparations

By contrast to the quantitative analysis that is essential for many homogenate assays, that for intact preparations is seldom quantitative, but usually reported using semi-quantitative methods. The main reason for this contrast is that there is usually no absolute reference standard for intact preparations. A secondary issue is the heterogeneity often present in the 'end product' of an intact sample procedure.

6.3.1 Analysis of genes in interphase nuclei

As described in Chapter 3, ISH assays are available to demonstrate genes and their rearrangement in interphase nuclei. As discussed in Chapters 9, 10, and 11, the demonstration of genes has become an important tool for precision medicine guiding specific therapeutic options.

The prime diagnostic example of this is the assessment of *HER2* gene copy number change for breast, gastric, and gastro-oesophageal cancers. In the ISH assay, one option is to undertake hybridization, on the same sample preparation, with the test probe for HER2 and a repetitive centromeric sequence (CEP17), as the control probe that is also present on chromosome 17. The centromeric probe hybridization not only verifies that hybridization to the chromosome is possible, but the signal it provides can be assessed alongside that for HER2 to determine the expression ratio. In an alternative ISH procedure, where only the HER2 probe is used, then counting the number of hybridized 'gene dots' in intact and non-overlapping nuclei is done to determine whether the criterion for over-expression of the gene has been met or exceeded. Whichever ISH method is adopted the results must be interpreted according to strict guidelines (Bartlett et al., 2011; Wolff et al., 2014; Koopman et al., 2015; Rakha et al., 2015; see Box 6.1) so that the patient sample can be assessed as suitable or otherwise for anti-HER2 monoclonal therapy such as that provided by trastuzumab (Herceptin). This serves as an example of an amalgam of qualitative analysis through the need to identify suitable nuclei, and quantitative analysis, due to the counting of ISH signals in these nuclei.

BOX 6.1 Interpretation of immunocytochemical and *in situ* hybridization for HER2 staining of breast, gastric, and gastro-oesophageal cancer samples

Immunocytochemistry

For a breast cancer section to be scored as positive for the receptor, intense and complete cell membrane staining must be observed in ≥10% tumour cells. This is referred to as a 3+ result and is a strong indicator for HER2 targeted therapy. A 2+ result is regarded as borderline. In this situation, although ≥10% of tumour cells are appropriately stained, only moderate staining intensity is observed. In this situation, ISH is undertaken to assess gene copy change. In situations where only weak staining of tumour cells is observed by ICC a score of 1+ is given and the case is regarded as negative for HER2 in respect of the use of targeted therapy. This analysis algorithm is also applied to gastric and gastro-oesophageal cancer samples with one significant modification. As the tumour cells often sit on a basement membrane then complete circumferential membrane staining for HER2 is not observed. Accordingly, staining of the lateral and apical membranes of the cells is analysed for intensity and percentage of stained cells.

In situ *hybridization*

For this assessment great care must be taken to identify the interphase nuclei to be counted. This can be particularly challenging when fluorescent ISH (FISH) techniques are used. Only non-overlapping and well-preserved tumour cell nuclei should be assessed, and a minimum of 20 of these need to be present. In situations where a dual probe ISH technique has been employed then a case where the HER2/CEP17 ratio is ≥2.0 with an average HER2 copy number ≥4.0 signals per cell will be reported as positive. When using only the HER2 probe the average copy number over the cells counted must be ≥6 for positive receptor status to be reported. Cases are reported as negative as an predictive indicator for HER2 therapy when the average HER2/CEP17 ratio is ≤2.0 with an average HWE2 copy number ≤4.0 signals per cell or, when using only the HER2 probe alone, the average copy number is ≤4 copies per cell. In equivocal cases then the signals in up to a further 40 nuclei, preferably from three distinct tumour regions, should be counted, and the same assessment criteria applied, but beyond this no further assessment of a case is recommended.

A further example qualitative and quantitative analysis is provided by the demonstration of the oncogenic gene fusion product EML4-ALK. This is found in a small percentage of non-small cell lung carcinomas. When present, it indicates suitability for targeted therapy using the kinase inhibitor Crizotinib or, in some countries, Alectinib. In the break-apart ISH procedure (Figure 6.13) the presence of the translocation/fusion gene product is recorded as the separation of two hybridized probes to reveal distinct colours in intact nuclei. When interpretation is undertaken, the qualitative element relies on the identification of 50 intact non-overlapping tumour nuclei followed by the counting of the number with break-apart signals in them. The threshold for a positive EML4-ALK sample is reached when 15 or more nuclei show the break-apart signal (McLeer-Florin et al., 2012; Choi et al., 2015; see Case study 6.2).

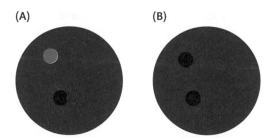

(A)   (B)

FIGURE 6.13
Break apart EML-4 ALK FISH. (A) Normal nucleus. (B) EML4-ALK fusion nucleus. The two probes, one carrying red and the other green fluorescent labels, used in the ISH assay hybridize to different regions near and on the ALK gene present on chromosome 2 (2p23). In (A) normal signals for the probes may be close and, therefore, retain their separate colours or be superimposed and show as a single yellow dot. (B) When fusion occurs with the EML-4 gene, present on the same chromosome then the ALK gene is interrupted and the probes are now separated to provide a break-apart signal. In this Figure only the most simple and distinctive FISH results are illustrated. In pathological situations interpretation can be complex and, therefore, qualitative segregation between normal and EML4 -ALK fusion cells must precede the counting of nuclei.

CASE STUDY 6.2 Intact sample analysis

Eligibility of a case of lung cancer for targeted therapy

A 65-year-old man presented with shortness of breath, persistent cough, and haemoptysis. He was referred by his GP for a chest X-ray that revealed a centrally placed mass in the right lower lobe of his lung. Due to its position surgery to remove the mass was not possible. Instead, to establish a diagnosis, a biopsy was taken and a FFPE block was prepared. Sections of this were stained using haematoxylin and eosin, and these revealed a pleomorphic population of large vacuolated cells together with a mixture of inflammatory cells. To confirm the morphological diagnosis of non-small cell lung cancer (NSCLC) a panel of antibodies were applied to additional sections using ICC. The majority of the vacuolated cells were TTF-1 and Cytokeratin 5/6 and 7 positive, but negative for calretinin, CD56 and synaptophysin. Accordingly, the tumour was confirmed as NSCLC.

To determine whether the patient was suitable for targeted therapy molecular analysis was undertaken for EGFR mutation status, PD-L1, and for EML4-ALK fusion using appropriate technologies. The results for EGFR and PD-L1 did not suggest that the patient would benefit from targeted therapy against these receptors. However, the ISH break-apart assay demonstrated the following image collated staining patterns in 50 non-overlapping intact cancer cell nuclei.

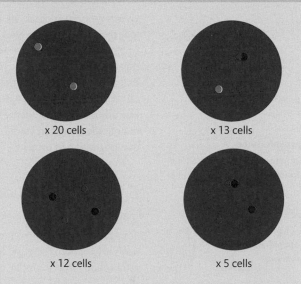

x 20 cells   x 13 cells

x 12 cells   x 5 cells

The top row of cells were classified as negative with respect to EML4-ALK fusion, but the bottom row demonstrated patterns of staining indicative of the presence of the fusion. As the total number of cells demonstrating EML4-ALK fusion patterns was 17 this exceeded the minimum requirement for 15/50 cells to demonstrate evidence of fusion. Accordingly, targeted therapy using Crizotinib or Alectinib was recommended to treat this patient with inoperable NSCLC.

Key Point

Analysis of gene detection in intact nuclei typically involves the counting of *in situ* hybridization signals in a defined number of intact nuclei from which a diagnostic assessment of significance can be made.

6.6 SELF-CHECK

In a break-apart assay what type of *in situ* hybridization signal would indicate the presence of gene fusion in an intact nucleus?

6.3.2 Analysis of RNA expression

As discussed in Chapter 4, and in the present chapter, the expression levels of mRNA and microRNA can be determined using a variety of methods on homogenate samples. The output can be quantitative using qRT-PCR or using NGS. This data is very valuable in determining change of RNA expression and its relationship to disease. However, in tissue samples it cannot provide precise information on the cell type of cells harbouring individual RNA species. Here the use of ISH methods is essential.

Using methods such as the rolling circle amplification and branched DNA techniques, very low copy numbers of RNA can be detected (Larsson et al., 2010; Wang et al., 2012). Indeed, single molecule RNA detection has been claimed (Krzywkowski et al., 2017) and this would represent the scaling of the highest resolution of quantified analysis. However, for the present, this must be the exception, rather than the rule. This is especially so when routine FFPE tissues, rather than cells are used due to the deleterious effects of formalin and subsequent paraffin wax embedding on RNA preservation. Microscopic interpretation is either qualitative or semi-quantitative, and can use methods that have evolved for the analysis of immunocytochemical demonstration of proteins (Table 6.2).

Cross reference

See Chapter 9 for more details on lymphoma.

Cross reference

See Chapter 11 for more details on epithelial tumours and melanoma.

6.3.3 Analysis of protein expression

Immunocytochemistry (ICC) is widely used for diagnostic purposes. Panels of antibodies are often employed to aid the differential diagnosis of a cancer. For example, this is undertaken for the diagnosis of lymphoma where B- and T-cell lymphomas are distinguished and their sub-classification established. Other examples of the use of panels include those used for epithelial tumours, In all of

TABLE 6.2 Semi-quantitative scoring methods for assessment of protein and RNA staining by immunocytochemistry and *in situ* hybridization

Scoring system	Principle	Reference
Allred	An estimation of the percentage of the relevant stained cells across the whole of the preparation is made. The percentage of relevant cells that are stained is estimated as either 1 = <1%; 2 = 5–10%; 3 = 11–33%; 4 = 34–66%; 5 = ≥67%* and average intensity as either 1, for weak, 2, for moderate, or 3, for strong staining across the sample. The score is arrived at by adding the percentage score and intensity score together giving a final score between 0–8.	Allred (2010)
H score	Staining intensities are recorded for 100 relevant cells. For weak staining each cell is counted as 1, for moderate staining, 2, and for strong staining, 3. A cumulative score, between 1 and 300, is then produced. This method is slow to undertake, but precise.	McCarty et al. (1986)
Quickscore	Similar to the Allred scoring system, but the final score is a multiplication of the percentage and intensity of staining. The percentage of relevant cells that are stained is estimated as either 1 = 0–4%; 2 = 5–19%; 3 = 20–39%; 4 = 40–59%; 5 = 60–79%; 6 = 80–100%* and average intensity as either 1, for weak, 2, for moderate, or 3, for strong staining across the sample. The score is produced by multiplying the percentage score and intensity score together giving a final score between 0 and 18.	Detre et al. (1995)

*Original categories for the scoring of oestrogen and progesterone expression in breast cancer. These can be changed for the scoring of other RNA or protein molecules demonstrated by ISH and ICC, respectively.

Adapted from Warford and Jasani, 2016.

these situations, the pathologist analyses the stained slides in a purely qualitative manner, observing the presence or absence of immunostaining and correlating the histological patterns of expression with those known to be associated with the pathology.

It is perhaps surprising to note in the field of precision medicine that the analysis of some ICC results can appear to be as simple as to ask and then answer the question, 'Is the tumour positive for the predictive biomarker?' This would seem, on the surface, to be true for PD-L1 ICC, where the presence of just 1% positive tumour cells is enough to indicate that the patient is a candidate for anti PD-1 monoclonal antibody therapy using nivolumab. However, the reality is somewhat more complex. In this example ICC must be undertaken using FDA approved antibody kits on dedicated automated platforms and a range of positive and negative controls must be included (Phillips et al., 2015). Once the latter have been confirmed as suitable, then the test section/s of tissue must be precisely analysed according to set guidelines to identify positive tumour cells and to exclude normal immune cells that may stain, together with any non-specific reactions. Indeed, such is the importance of these analyses to patient management that special training programmes are run for pathologists who will interpret the ICC-stained slides.

As discussed in Chapter 10, ICC is used in invasive breast cancer as a predictive indicator for targeted therapy, using the anti-oestrogen drug tamoxifen (Jordan, 2006) and/or endogenous oestrogen depleting aromatase inhibitors (Geisler, 2008). Interpretation of oestrogen and progesterone (ER/PR) ICC is strictly controlled by guidelines (Hammond et al., 2010) and involves the use of semi-quantitative microscopic interpretation employing either Allred, H Score, or Quickscore methodologies—see Table 6.2. For a tumour to be classified as positive for ER/PR ≥1% of tumour cell nuclei must be immunoreactive. However, it has been suggested that a score of ≥10% positive tumour cells may be a more reliable indicator of suitability for hormonal therapeutic intervention (Nofech-Mozes et al., 2012).

ICC is also used as the primary tool for determining the HER2 status of invasive breast, gastric, and gastro-oesophageal cancer. Its use is preferred over ISH as it is quicker and cheaper to perform, and interpretation is easier. As with the determination of ER/PR, status guidelines for interpretation are in place and these involve the use of a distinctive semi-quantitative scoring system (Wolff et al., 2014; Koopman et al., 2015; Rakha et al., 2015). The guidance for interpretation is summarized in Box 6.1.

Key Point

The analysis of RNA gene expression and protein expression in intact sample preparations is usually undertaken using qualitative or semi-quantitative methods. The latter are based on grading the intensity and percentage or number of cells that are stained.

6.7 SELF-CHECK

What would be the H score (see Table 6.2), for a population of 100 cancer cells exhibiting the following staining for protein X?

Fifty cells were unstained, 20 were stained at weak intensity, 20 were stained at moderate intensity, and 10 were stained at strong intensity.

6.4 **Image capture and analysis**

The microscopic image capture of 'regions of interest' of cell and tissue preparations is, of course, relevant for publications and presentations. 35-mm film has been replaced by digital imaging and, with this, has come a challenge that the latter should not be manipulated, but always be a true reflection of the original. This is an absolute requirement when imaging is used in safety testing undertaken under **Good Laboratory Practice** regulations. It is also strongly supported for all microscopic image capture that is incorporated into publications, so that readers may be assured that what is viewed is what was originally seen (Deutsch et al., 2008).

Automated whole-slide scanning of stained slides has emerged recently, and can be undertaken using a variety of instruments that produce high resolution scans in a few minutes in brightfield mode (Al-Janabi et al., 2012; Figure 6.14). The automated scanning of fluorescently stained preparations can also be undertaken, but at slower speed as focusing is more difficult. Although whole-slide image files are very large, this no longer represents an issue for storage. However, the use of 'platform-independent' storage media is important to allow their viewing using diverse imaging software. Whole-slide scanning has made a significant contribution to the construction of atlases that record gene expression by ICC and ISH (www.proteinatlas.org, www.brain-map.org). For diagnostic

Good Laboratory Practice
An internationally recognized standard applied for the preclinical safety testing of substances.

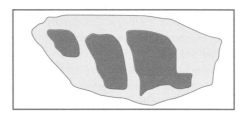

Stained section (blue) containing diseased areas (yellow) is scanned

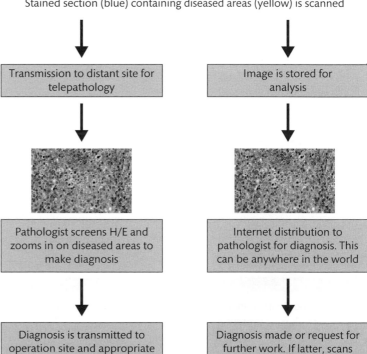

Transmission to distant site for telepathology	Image is stored for analysis

Pathologist screens H/E and zooms in on diseased areas to make diagnosis	Internet distribution to pathologist for diagnosis. This can be anywhere in the world

Diagnosis is transmitted to operation site and appropriate action taken	Diagnosis made or request for further work. If latter, scans are sent to make firm diagnosis

FIGURE 6.14
Applications for whole-slide scanning.
Note: The use of automated image analysis to distinguish and quantitate tumour cells is still developmental and is not presently recommended for determination of suitability of patient samples for targeted therapy.

purposes the automated analysis of whoe-slide images could provide for quicker and more accurate scoring of ISH and ICC-stained slides linked to predictive tests such as HER2. While algorithms have been developed none, as yet, have been recommended for use (Rahka et al., 2015).

Telepathology has emerged as a means of diagnostic reporting, using either selected microscopic images or whole-slide scans of stained slide preparations, produced at distant locations. The principle is that the stained preparation is imaged at the place of preparation, transmitted electronically and then interpreted at another location by a pathologist. The use of whole-slide imaging is advantageous, as it provides an unbiased platform for interpretation. Telepathology has been applied to situations where a rapid diagnosis is required when a pathologist cannot be on site. For example, during the intraoperative assessment of breast cancer telepathology can be used to determine if a sentinel lymph node sample contains metastatic cells and, accordingly, whether the operation should be concluded with a simple mastectomy or more radical surgery (Têtu et al., 2012). Diagnostic accuracy using telepathology has been shown to be equivalent to on-site microscopy (Chorneyko et al., 2002; Jukić et al., 2011). It has been used in Canada where the distance between the sites may be several hundreds of miles, and accessibility is an issue (Têtu et al., 2012). However, in a subtle variation, the time taken for a pathologist to travel across a large city to be present to diagnose a lesion during intraoperative assessment can also be removed by using telepathology.

Key Point

Image capture of intact preparations allows transmission for diagnosis at distant sites (telepathology) or for subsequent analysis. The latter is still developmental in terms of diagnostic application.

6.8 SELF-CHECK

In what circumstances would automated image analysis be of use in the diagnosis of cancer?

6.5 **Future directions**

Homogenate samples produce large amounts of data from various 'OMIC' technologies. There is urgent need to develop more sophisticated bioinformatic algorithms to integrate the data generated from the various sources and correlate them with clinical pathology data in order to provide meaningful reports that will aid, ultimately, in the diagnosis of disease and management of patients. These algorithms need to be fast and able to extract real data from noise. Most current bioinformatics algorithms, such as those used for sequence alignment, are modifications of the original algorithms developed in the 1980s. In addition, novel mathematical algorithms and software need to be developed to integrate the 'OMICs' data from multiple sources into mathematical models that are able to correlate and mine them at multiple levels. Additionally, as data of similar types can be generated from different instruments, there is an urgent need for standardized analysis and reporting to allow common data sets to be produced.

Intact samples are useful for the visualization of the location of biomarkers within their cellular architecture. However, the biggest challenges are to automatically extract data from stained preparations and to quantify this so as to provide precise information of disease perturbations. As emphasized in other chapters in this book, this is particularly important when these changes are linked to precision medicine and targeted therapy. Automated image capture and analysis algorithms need to be validated to the point where they can reliably take over from present manual semi-quantitative scoring methods such as H-score and Quickscore.

Chapter summary

This chapter has sought to:

■ Distinguish between qualitative, semi-quantitative, and quantitative analysis of samples.

■ Provide examples and explain the application of bioinformatic algorithms to data obtained from homogenate preparations for DNA and RNA analysis.

■ Describe qualitative and semi-quantitative analysis methods for intact cell and tissue samples for nucleic acids and proteins.

■ Discuss the application of telepathology and image capture and analysis for diagnosis without the need for microscopy.

Further reading

● Costa-Silva J, Domingues D, Lopes FM. RNA-Seq differential expression analysis: an extended review and a software tool. PLoS One 2017; 12: e0190152.

● Kim H, Kim J, Selby H, et al. A short survey of computational analysis methods in analysing ChIP-seq data. Human Genomics 2011; 5, 117–123.

● Pevsner J (Ed.) *Bioinformatics and Functional Genomics*, 3rd edn. Oxford, Wiley Blackwell, 2015.

● Robinson MD, Pelizzola M. Computational epigenomics: challenges and opportunities. Frontiers in Genetics 2015; 6: 88.

References

● Adil Butt M, Pye H, Haidry RJ, et al. Upregulation of mucin glycoprotein MUC1 in the progression to esophageal adenocarcinoma and therapeutic potential with a targeted photoactive antibody-drug conjugate. Oncotarget 2017; 8: 25080–25096.

● Alizadeh AA, Eisen MB, Davis RE, et al. Distinct types of diffuse large B-cell lymphoma identified by gene expression profiling. Nature 2000; 403: 503–511.

● Al-Janabi S, Huisman A, Van Diest PJ. Digital pathology: current status and future perspectives. Histopathology 2012; 61: 1–9.

● Allred DC. Issues and updates: evaluating estrogen receptor-a, progesterone receptor, and HER2 in breast cancer. Modern Pathology 2010; 23: S52–59.

● Anders S, Huber W. Differential expression analysis for sequence count data. Genome Biology 2010; 11: R106.

● Bartel DP. MicroRNAs: genomics, biogenesis, mechanism, and function. Cell 2004; 116: 281–297.

● Bartlett JM, Starczynski J, Atkey N, et al. HER2 testing in the UK: recommendations for breast and gastric *in-situ* hybridisation methods. Journal of Clinical Pathology 2011; 64: 649–653.

- Bolger AM, Lohse M, Usadel B. Trimmomatic: a flexible trimmer for Illumina sequence data. Bioinformatics 2014; 30: 2114–2120.

- Choi IH, Kim DW, Ha SY, Choi YL, Lee HJ, Han J. Analysis of histologic features suspecting anaplastic lymphoma kinase (ALK) expressing pulmonary adenocarcinoma. Journal of Pathology and Translational Medicine 2015; 49: 310–317.

- Chorneyko K, Giesler R, Sabatino D, et al. Telepathology for routine light microscopic and frozen section diagnosis. American Journal of Clinical Pathology 2002; 117: 783–790.

- Conesa A, Madrigal P, Tarazona S, et al. A survey of best practices for RNA-seq data analysis. Genome Biology 2016; 17: 13.

- Cook CE, Bergman MT, Finn RD, et al. The European Bioinformatics Institute in 2016: data growth and integration. Nucleic Acids Research 2016; 44: D20–26.

- da Fonseca RR, Albrechtsen A, Themudo GE, et al. Next-generation biology: sequencing and data analysis approaches for non-model organisms. Marine Genomics 2016; 30: 3–13.

- Detre S, Saccani Jotti G, Dowsett M. A 'quickscore' method for immunohistochemical semiquantitation: validation for oestrogen receptor in breast carcinomas. Journal of Clinical Pathology 1995; 48: 876–878.

- Deutsch EW, Ball CA, Berman JJ, et al. Minimum information specification for in situ hybridization and immunohistochemistry experiments (MISFISHIE). Nature: Biotechnology 2008; 26: 305–312.

- Dudoit S, Yang YH, Callow MJ, Speed TP. Statistical methods for identifying genes with differential expression in replicated cDNA micro array experiments. Statistica Sinica 2002; 12: 111–139.

- Eisen MB, Spellman PT, Brown PO, Botstein D. Cluster analysis and display of genome-wide expression patterns. Proceedings of the National Academy of Sciences, USA 1998; 95: 14863–14868.

- Fortin JP, Triche TJ, Jr, Hansen KD. Preprocessing, normalisation and integration of the Illumina HumanMethylationEPIC array with minfi. Bioinformatics 2017; 33: 558–560.

- Geisler J. Aromatase inhibitors: from bench to bedside and back. Breast Cancer 2008; 15: 17–26.

- Giannoulatou E, Park SH, Humphreys DT, Ho JW. Verification and validation of bioinformatics software without a gold standard: a case study of BWA and Bowtie. BMC Bioinformatics 2014; 15(Suppl 16): S15.

- Hammond ME, Hayes DF, Dowsett M, et al. American Society of Clinical Oncology/College of American Pathologists guideline recommendations for immunohistochemical testing of estrogen and progesterone receptors in breast cancer (unabridged version). Archives of Pathology & Laboratory Medicine 2010; 134: 48–72.

- Hamoudi RA, Appert A, Ye H, et al. Differential expression of NF-kappaB target genes in MALT lymphoma with and without chromosome translocation: insights into molecular mechanism. Leukemia 2010; 24: 1487–1497.

- Hsu FH, Chen HI, Tsai MH, et al. A model-based circular binary segmentation algorithm for the analysis of array CGH data. BMC Research Notes 2011; 4: 394.

- Irizarry RA, Bolstad BM, Collin F, Cope LM, Hobbs B, Speed TP. Summaries of Affymetrix GeneChip probe level data. Nucleic Acids Research 2003; 31: e15.

- Jarmuz M, Ballif BC, Kashork CD, Theisen AP, Bejjani BA, Shaffer LG. Comparative genomic hybridization by microarray for the detection of cytogenetic imbalance. Methods in Molecular Medicine 2006; 128: 23–31.

- Johnson NA, Hamoudi RA, Ichimura K, et al. Application of array CGH on archival formalin-fixed paraffin-embedded tissues including small numbers of microdissected cells. Laboratory Investigations 2006; 86: 968–978.

- Johnson DS, Mortazavi A, Myers RM, Wold B. Genome-wide mapping of *in vivo* protein-DNA interactions. Science 2007; 316: 1497–1502.

- Jordan VC. Tamoxifen (ICI46,474) as a targeted therapy to treat and prevent breast cancer. British Journal of Pharmacology 2006; 147(Suppl 1): S269–276.

- Jukić DM, Drogowski LM, Martina J, Parwani AV. Clinical examination and validation of primary diagnosis in anatomic pathology using whole slide digital images. Archives of Pathology & Laboratory Medicine 2011; 135: 372–378.

- Kapp JR, Diss T, Spicer J, et al. Variation in pre-PCR processing of FFPE samples leads to discrepancies in BRAF and EGFR mutation detection: a diagnostic RING trial. Journal of Clinical Pathology 2015; 68: 111–118.

- Koopman T, Louwen M, Hage M, Smits MM, Imholz AL. Pathologic diagnostics of HER2 positivity in gastroesophageal adenocarcinoma. American Journal of Clinical Pathology 2015; 143: 257–264.

- Krueger F, Andrews SR. Bismark: a flexible aligner and methylation caller for Bisulfite-Seq applications. Bioinformatics 2011; 27: 1571–1572.

- Krzywkowski T, Hauling T, Nilsson M. *In situ* single-molecule RNA genotyping using padlock probes and rolling circle amplification. Methods in Molecular Biology 2017; 1492: 59–76.

- Larsson C, Grundberg I, Söderberg O, Nilsson M. *In situ* detection and genotyping of individual mRNA molecules. Nature: Methods 2010; 7: 395–397.

- Libertini E, Heath SC, Hamoudi RA, et al. Information recovery from low coverage whole-genome bisulfite sequencing. Nature Communications 2016; 7: 1–10.

- Liu JZ, Erlich Y, Pickrell JK. Case-control association mapping by proxy using family history of disease. Nature: Genetics 2017; 49: 325–331.

- Luan SL, Boulanger E, Ye H, et al. Primary effusion lymphoma: genomic profiling revealed amplification of SELPLG and CORO1C encoding for proteins important for cell migration. Journal of Pathology 2010; 222: 166–179.

- Maksimovic J, Phipson B, Oshlack A. A cross-package bioconductor workflow for analysing methylation array data. F1000 Research 2016; 5: 1281.

- McCarty KS Jr, Szabo E, Flowers JL, et al. Use of a monoclonal anti-estrogen receptor antibody in the immunohistochemical evaluation of human tumors. Cancer Research 1986; 46(Suppl): 4244s–4248s.

- McLeer-Florin A, Moro-Sibilot D, Melis A, et al. Dual IHC and FISH testing for ALK gene rearrangement in lung adenocarcinomas in a routine practice: a French study. Journal of Thoracic Oncology 2012; 7: 348–54.

- Nofech-Mozes S, Vella ET, Dhesy-Thind S, Hanna WM Cancer care Ontario guideline recommendations for hormone receptor testing in breast cancer. Clinical Oncology 2012; 24: 684–696.

- Palanisamy V, Wong DT. Transcriptomic analyses of saliva. Methods in Molecular Biology 2010; 666: 43–51.

- Pei YF, Zhang L, Li J, Deng HW. Analyses and comparison of imputation-based association methods. PloS One 2010; 5: e10827.

- Phillips T, Simmons P, Inzunza HD, et al. Development of an automated PD-L1 immunohistochemistry (IHC) assay for non-small cell lung cancer. Applied Immunohistochemistry and Molecular Morphology 2015; 23: 541–549.

- Rakha EA, Pinder SE, Bartlett JM, et al. Updated UK Recommendations for HER2 assessment in breast cancer. Journal of Clinical Pathology 2015; 68: 93–99.

- Ritchie ME, Phipson B, Wu D, et al. Limma powers differential expression analyses for RNA-sequencing and microarray studies. Nucleic Acids Research 2015; 43: e47.

- Roberts A, Pimentel H, Trapnell C, et al. Identification of novel transcripts in annotated genomes using RNA-Seq. Bioinformatics 2011; 27: 2325–2329.

- Robinson, MD, McCarthy, DJ, Smyth, GK. EdgeR: a Bioconductor package for differential expression analysis of digital gene expression data. Bioinformatics 2010; 26: 139–140.

- Selvaraj S, Natarajan J. Microarray data analysis and mining tools. Bioinformation 2011; 6: 95–99.

- Shen X, Shen J, Zhang H, et al. Detection and monitoring of driver mutations by next generation sequencing in squamous cell lung cancer patient and possible predictive biomarker of third generation EGFR tyrosine kinase inhibitors. Thoracic Cancer 2018; 9: 181–184.

- Subramanian A, Tamayo P, Mootha VK, et al. Gene set enrichment analysis: a knowledge-based approach for interpreting genome-wide expression profiles. Proceedings of the National Academy of Sciences, USA 2005; 102: 15545–15550.

- Têtu B, Fortin JP, Gagnon MP, Louahlia S. The challenges of implementing a 'patient-oriented' telepathology network; the Eastern Québec telepathology project experience. Analytical Cellular Pathology (Amsterdam) 2012; 35: 11–18.

- Trapnell, C, Roberts, A, Goff, L, et al. Differential gene and transcript expression analysis of RNA-seq experiments with TopHat and Cufflinks. Nature: Protocols 2012; 7: 562–578.

- Wang F, Flanagan J, Su N, et al. RNAscope a novel *in situ* RNA analysis platform for formalin-fixed, paraffin-embedded tissues. Journal of Molecular Diagnosis 2012; 14: 22–29.

- Wang Q, Jia P, Li F, et al. Detecting somatic point mutations in cancer genome sequencing data: a comparison of mutation callers. Genome Medicine 2013; 5: 91.

- Warford A, Jasani B. Impact of analytical variables in breast cancer biomarker analysis. In: Badve S and Gökmen-Polar Y (Eds) *Molecular Pathology of Breast Cancer*, pp. 27–43. Basel, Springer International Publishing, 2016.

- Watkins AJ, Huang Y, Ye H, et al. Splenic marginal zone lymphoma: characterization of 7q deletion and its value in diagnosis. Journal of Pathology 2010; 220: 461–474.

- Wolff AC, Hammond ME, Hicks DG, et al. Recommendations for human epidermal growth factor receptor 2 testing in breast cancer: American Society of Clinical Oncology/College of American Pathologists clinical practice guideline update. Archives of Pathology & Laboratory Medicine 2014; 138: 241–256.

- Ye K, Schulz MH, Long Q, Apweiler R, Ning Z. Pindel: a pattern growth approach to detect break points of large deletions and medium sized insertions from paired-end short reads. Bioinformatics 2009; 25: 2865–2871.

Useful websites

- *GCRMA*: https://www.bioconductor.org/packages/3.7/bioc/vignettes/gcrma/inst/doc/gcrma2.0.pdf.

- *ChIp-seq*: www.illumina.com/Documents/products/datasheets/datasheet_chip_sequence.pdf.

- *Allen Brain Atlas*: www.brain-map.org.

- *The Human Protein Atlas*: www.proteinatlas.org.

Discussion questions

6.1 *Homogenate analysis*: A study for Braf V600E mutations from FFPE clinical biopsies of melanoma and colorectal cancer patients was carried out using semiconductor-based DNA sequencing.

(A)

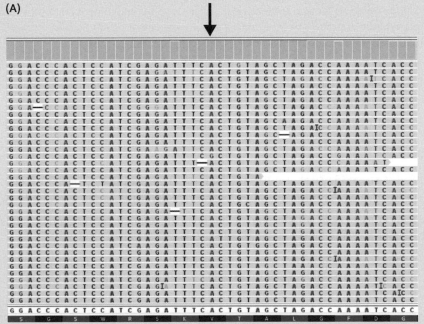

BRFA

(A) shows a mutation of CAC to CGC in Braf V600E in one of the tracks indicated by the arrow. On examining the quality of the G base it was found to be 13. Using the Phred Table (Table 5.1) decide whether you would call this a true Braf V600E mutation.

(B)

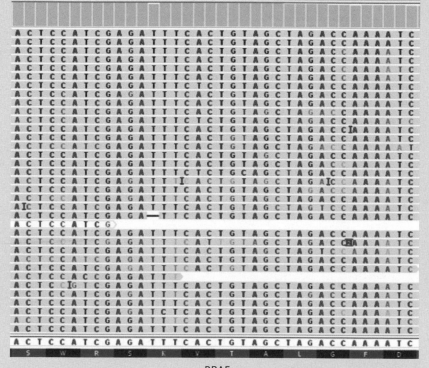

chr7:140,453,136

Total count:9177
A : 8870 (97%, 8870+, 0-)
C : 10 (0%, 10+, 0-)
G : 37 (0%, 37+, 0-)
T : 260 (3%, 260+, 0-)
N : 0

BRAF

(B) shows a CTC to CGC mutation in Braf V600E obtained from NGS of FFPE section from tissue biopsy. The pale yellow box is a summary of statistics of the base where the mutation occurs. The patient who had his biopsy sequenced for Braf V660E showed 3% of the bases were changed from A to T. The quality of all the T was examined and shown to be above 20. In (B) the 3 Ts had the following quality values; 29, 25, and 30.

Would you call this a true V600E Braf mutation? If not, what can you do to be absolutely certain whether or not it is a true mutation?

6.2 *Intact sample analysis*: The images shown here are from immunocytochemical stained preparations of cancer tissue.

(A)

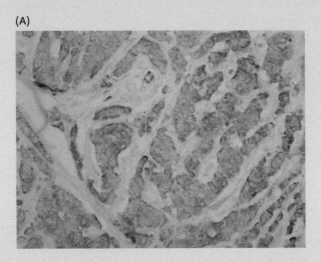

(A) Non-small cell carcinoma of lung ICC for cytokeratin 19. The grey to black staining is positive for the antibody.

(B)

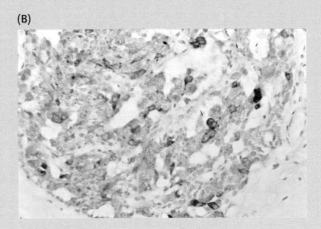

(B) Medullary carcinoma of thyroid ICC for calcitonin. The brown staining is positive for the antibody.

Showing your calculations, what scores do you get when you apply Allred and Quickscore semi-quantitative methods to the analysis of these images?

Use Table 6.2 to guide your calculations and ask colleagues to make their own calculations.

Answers to the self-check questions and tips for responding to the discussion questions are provided on the book's accompanying website:

 Visit: www.oup.com/uk/warford

7

Recent Technical Advances in Molecular Analysis

Nadège Presneau and Mary Alikian

Learning objectives

After studying this chapter you should be able to:

- Understand the recent technical advances in molecular analysis, especially liquid biopsy and digital PCR (dPCR).
- Define liquid biopsy and describe its clinical applications, particularly in cancer.
- Describe the technological advances in studying liquid biopsy.
- Define and understand the concepts of dPCR.
- Recognize different platforms and chemistries of dPCR.
- Understand the advantages and limitations of dPCR.
- Identify clinical applications for dPCR.

7.1 **Introduction**

This chapter focuses on two rapidly emerging technological advances in molecular analysis—the liquid biopsy and dPCR.

The **liquid biopsy** (LB) represents a minimally invasive procedure that can be used for screening, diagnosis, prognosis, and predictive testing of circulating biomarkers (proteins, DNA, RNA). Although LB can be applied to a wide range of conditions, the focus in this chapter is on its application in oncology where, in recent years, a tremendous effort has been made to develop new technologies and approaches to analyse circulating cancer biomarkers. LB may well represent the future of **precision medicine**, but it still has to circumvent some challenges in terms of its validation and standardization before it can be widely used in clinical practice.

Liquid biopsy
Collection, using minimally invasive methods, of samples for the molecular analysis of circulating tumour cells, cell-free nucleic acids, and proteins.

Precision medicine
The treatment of a patient with a targeted therapy based on the molecular profile of a disease.

Cross reference
See Chapter 4, Section 4.2.1.2 for further information on qPCR.

Copy number variation
The presence of variable numbers of copies of a particular sequence relative to a reference genome.

Since its introduction in the late 1990s, qPCR has become the gold standard for nucleic acid target quantification in genomics research and subsequently in clinical analysis. However, for applications such as low-level target monitoring, precise **copy number variation** (CNV), and rare mutation detection, qPCR has inherent limitations. dPCR is an adaptation of the qPCR technique, whose concept preceded that of qPCR. However, it was not adopted at the time due to technical and physical complications. With advances in microfluidics technologies, dPCR now promises to overcome the limitations of qPCR by allowing absolute quantification of target nucleic acids without the need for a standard curve, with higher precision and accuracy.

7.2 Liquid biopsy

LB is a new type of biopsy, as opposed to standard tissue biopsy (Table 7.1). Due to its minimally invasive property, it represents the ideal specimen to perform molecular analysis on a range of diseases (germline or somatic), including solid tumours. LB also represents a great alternative for organs that are difficult to reach or in situations where new invasive biopsy is not feasible.

The determination of protein-based cancer biomarkers in blood samples, such as carcinoembryonic antigen (CEA) and prostate-specific antigen (PSA), have been used for many years and should be regarded as a type of LB. These established applications are not considered further

TABLE 7.1 Standard tissue biopsy procedures used for cancer diagnosis

Type of biopsy	Brief description	Tissue sites
Bone marrow biopsy	Uses a long needle to draw a sample of bone marrow under local anaesthetic. Used to diagnose a variety of both non-cancerous and cancerous blood conditions like leukaemia, anaemia, infection, or lymphoma, but also to investigate metastasis from other primary organs	Bone marrow; often from hip bone
Endoscopic biopsy	Uses a thin, flexible tube (endoscope) with a light and a small camera on the end to see structures inside the body	Used for bladder, colon, or lung
Needle biopsies		
Fine needle aspiration (FNA)	Uses a very thin, hollow needle attached to a syringe to take out either a very small piece of tissue from a tumour or aspirate a small amount of fluid containing cells. It is relatively non-invasive, less painful and a quicker procedure than surgical biopsy	Breast lump or enlarged lymph node, cyst aspiration, and thyroid are common locations, but can also be performed on most parts of the body
Core needle biopsy	Involves making a small incision (cut) in the skin. A large needle is passed through the incision and several narrow samples from the suspected tumour are taken. This is a more invasive procedure than FNA. It is done under local anaesthetic	It can be performed on most parts of the body. Common locations include the lymph nodes, breast masses, bone lesions, and prostate
Vacuum-assisted biopsy	The biopsy procedure is performed under image guidance (mammogram, magnetic resonance imaging (MRI) or ultrasound). Through a small incision in the skin, a special biopsy needle is inserted into the lesion and, using a vacuum-powered instrument, several tissue samples are taken. The vacuum draws tissue into the centre of the needle and a rotating cutting device takes the samples. Local anaesthesia is used	This process can be used to remove completely benign breast lesions, for example, fibroadenomas
Image-guided biopsy	Image-guided biopsies taken using X-ray or CT scans	Used for breast, lung, and liver, but also other organs

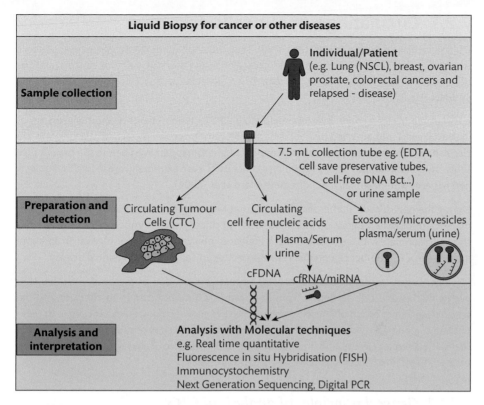

FIGURE 7.1

An overview of liquid biopsy. This figure depicts the general overview of the principle of liquid biopsy with its key steps of sample collection, preparation, and detection to analysis and interpretation of molecular results.

in this chapter. However, the application of LB, associated with recent developments to detect and enumerate circulating tumour cells (CTCs), to isolate circulating cell-free nucleic acids (cfDNA, cfRNA/miRNA), or to isolate exosomes/microvesicles containing fragments of DNA and/or RNA will be considered. Peripheral blood is the most common sample source for LB, but any human body fluids, such as urine, ascites, and pleural effusions can be used. The principle of the LB is summarized in Figure 7.1.

The LB has great potential for molecular and clinical diagnostic applications from detecting genetic abnormalities (gene mutations, or chromosomal rearrangements, or copy number), to establishing treatment options and monitoring treatment response, investigation of drug resistance, and disease progression, such as metastasis in cancer. Other applications include the monitoring of circulating endothelial cells in cardiovascular disease, foetal cells in the maternal circulation for prenatal diagnosis, cfDNA as a predictor of systemic lupus erythematosus severity and monitoring.

Key Point

LB is a rapid, minimally invasive procedure, and mostly painless for the patient to detect molecular biomarkers associated with diseases. In the context of cancer, it can be used for screening, prognosis, and predictive testing, patient stratification, and monitoring patient treatment.

7.2.1 Circulating tumour cells detection

CTCs were first described in 1869 (Ashworth, 1869) in a metastatic breast cancer autopsy and, since then, they have been isolated and characterized in a broad range of cancers, such as non-small cell lung cancer, breast, prostate, and colorectal cancers (Mavroudis, 2010; Pantel and Alix-Panabières, 2010).

CTCs are epithelial tumour cells (approximately 90% of human cancers are malignancies of epithelial cells) that are shed by the primary solid tumour or the metastatic tumour, and are circulating in the bloodstream. This process is thought to happen even before the patient presents any clinical symptoms of developing a cancer. CTCs are extremely rare in healthy subjects and patients with non-malignant diseases (Allard et al., 2004).

During the past decade, numerous clinical studies have demonstrated the use of CTCs as biomarkers for disease progression and metastasis. Specifically, high numbers of CTCs have been shown to correlate with aggressive disease, increased metastasis and poor survival. In addition, a change in CTC number can predict response to therapy and evaluate residual disease.

It is now widely accepted that the process of metastasis is due to a cascade of multi-step events that enable cells to detach from their primary tumour location to circulate into the vasculature (either blood or lymph vessels) and to eventually colonize a distant organ to recapitulate the primary tumour in the new niche. Based on this, it is reasonable to hypothesize that CTCs are the cells capable of metastasizing and, therefore, studying them will provide a greater insight into cancer aggressiveness and progression.

7.2.1.1 General principles of analysis of CTCs

Numerous academic and commercial technology platforms for isolation and analysis of CTCs have been reported in the last 10 years. More than 50 different approaches (technologies and platforms) have been published so far, in both research and clinical research setting (clinical trials) with highly variable detection rates, sensitivity, and specificity. Nonetheless, they all have in common that they need to discriminate, with high specificity and high sensitivity, a very rare cell population with its own phenotypic and genetic characteristics amongst the normal cells present in the sample. As an example, in metastatic cancer most of the patients have between 1 and 10 CTCs per mL of whole blood that contains 1 million white blood cells (WBC) and 1 billion red blood cells (RBC) per 1 mL. Hence, all of these technologies share the same objectives to:

- isolate ALL the CTCs in the sample (high capture efficiency);
- isolate only the CTCs and no other cell contaminants, such as WBCs (high isolation purity);
- do this in a relatively quick turnaround time (high throughput).

Although the threshold varies between studies, the accepted consensus defines the CTC count above a threshold of 5 CTCs per 7.5 mL blood for metastatic breast and prostate cancer, and more than three for metastatic colorectal cancer to be significant to correlate with disease progression (shortened overall survival and progression-free survival).

CTCs are mainly characterized and identified by their immunostaining pattern, cell morphology, size, electrical charge or shape (Figure 7.2).

Different approaches have been used to detect such rare cell populations (see Table 7.2). Of these CellSearch®, from Johnson & Johnson, is the sole platform cleared by the FDA for use for clinical diagnostics in breast cancer. This technology is based on isolating the CTCs utilizing the expression of a cell-surface marker specific to the tumour cells called epithelial cell adhesion

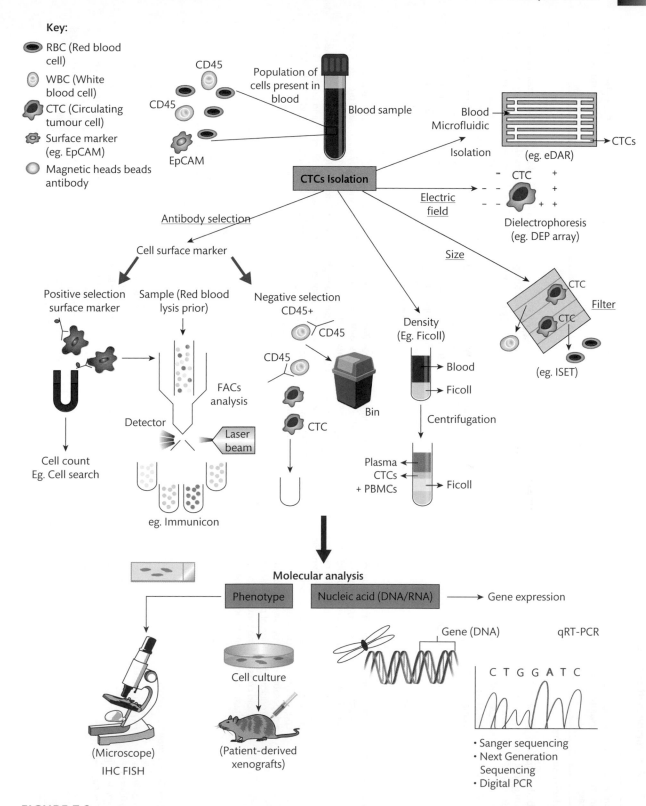

FIGURE 7.2

Illustration of the various methodologies that can be employed to detect, isolate, and analyse circulating tumour cells.

TABLE 7.2 Common methods of CTCs detection and analysis

Technology platforms or process	Assay-system examples	Principle of enrichment	Detection	Sensitivity	Comments	Examples of types of cancer studied
Benchtop instruments						
Antibody coupled with magnetic beads	CellSearch® system (formerly Veridex, Johnson & Johnson) AdnaTest (Adnagen AG)	Positive selection (EpCAM/CD326)-conjugated enrichment Positive selection using cocktails of antibodies specific for application, e.g. EpCAM and MUC-1 for breast cancer	Microscopic cell imaging using positive immunostaining of cytokeratins (CK), negative immunostaining of CD45, and DNA staining with DAPI and diameter larger than 5 µm. Captured CTCs are lysed and tested for expression patterns of various cancer-associated tumour markers, e.g., GA733-2, MUC-1, and HER2 for breast, using multiplex RT-PCR	Low sensitivity; blood sample size might be a limiting factor in sensitivity Sensitivity equal to CellSearch for metastatic breast cancer (Andreopoulou et al., 2012) and higher for metastatic colon cancer (Raimondi et al., 2014)	Use of FIXED cells Samples are defined as CTC-positive if the measured quantity of at least one of the tumour markers is above a defined threshold (Adnagen suggests >0.1 ng/mL)	(Metastatic) breast, prostate and colorectal cancers (Allard et al., 2004) Breast, prostate, ovarian, and colorectal cancers (Andreopoulou et al., 2012; Raimondi et al., 2014)
Flow cytometers	Immunicon Corporation coupled with FACS analysis ICeap- flow cytometer coupled with a disposable microfluidic chip to avoid cross-contamination	*Positive selection in two-steps*: immuno-magnetic ferrofluid enrichment with multiparameter flow cytometric and immunocytochemical analysis *Positive selection*: preparation of peripheral blood mononuclear cells, immunomagnetic CTC Separation (MACs column) and labelling	EpCAM+ followed by nucleic acid dye+, CD45–, and cytokeratin + (CAM5.2 mAb). Surface EpCAM double-positive selection/labelling method and anti-CKs + (8/7 &18) and anti-CD45–and nucleic acid +	Detection limit of rare cells using the cytometer is approximately 10^−5. Blood sample size might be a limiting factor in sensitivity	Use of LIVE cells. Further analysis by immunocytochemistry on Cytospins for morphology and other markers (cell surface; anti-mucin-1, anti-PSA or intracellular anti-CKs) to confirm their tumour-origin. Allows downstream experiments such as RT-PCR, chromosome aneuploidy and mutation analyses	(Metastatic) breast and prostate (Racila et al., 1998) (Metastatic) breast and prostate (Takao and Takeda, 2011)
LB Rare Cell isolation instrument						
High-definition (HD) fluorescence scanning microscopy	HD-CTC assay	No pre-enrichment step. Multiple-fluorescence staining followed by high quality images for morphology characterization	Immunofluorescence staining for anti-CKs, CD45 and nuclear staining (DAPI) on whole cell population to assess morphology of the CTCs	High sensitivity for detection and characterization of both early- and late-stage lung cancer CTCs (Wendel et al., 2012). Detects higher number of cells for same patients compared to CellSearch (Marrinucci et al., 2012)		Metastatic breast, prostate and pancreatic cancer (Marrinucci et al., 2012) Lung cancer (Wendel et al., 2012)

Technology platforms or process	Assay-system examples	Principle of enrichment	Detection	Sensitivity	Comments	Examples of types of cancer studied
Fibre-optic Array Scanning Technology (FAST™ PARC/ SRI Biosciences)— image cytometer	Laser-based techniques to scan broad fields of labelled blood cells immobilized on slides followed by Automated digital microscopy (ADM)	No pre-enrichment step (RBC lysis). Combination of FAST enrichment and ADM imaging	Immunofluorescence staining of up to 7 antigens including pan anti-CK, nuclear staining (DAPI) on whole cell population to assess morphology of the CTCs. Detection of fluorescent emissions using a fibre bundle with a large (50 mm) field of view	95% sensitivity and specificity of 3×10^{-6}	Subsequent tests using tissue-specific markers can be used for the characterization of the tissue of origin and for further molecular characterization. The use of a fixed substrate permits the re-identification and re-staining of cells allowing for additional morphologic and biologic information to be obtained from previously collected and identified cells	Breast and lung (Krivacic et al., 2004; Hsieh et al., 2006)
Laser scanning cytometers	MAINTRAC™	No pre-enrichment, only RBC lysis, no fixation or isolation	EpCAM magnetic beads with propidium red dye to determine if cells are dead or alive, detection made with use of fluorescence scanning microscope		Used in clinical trials to assess number of CTCs and if cells alive or dead. Used for guiding and monitoring chemotherapy treatments	Lung (Rolle et al., 2005)
Physical property-based methods	Screencell® Isolation by size of the epithelial tumour cells: ISET® (Vona et al., 2000) Flexible micro spring array: FSMA	No pre-enrichment.; single use (Desitter et al., 2011) RBC lysis followed by enrichment by blood filtration through filtering membranes with calibrated 8-µm pores No sample pre-processing	Use of a microporous membrane filter allowing size-selective isolation of CTCs Enriched cells are stained on the filter for cytomorphological examination or further characterized by immunocytochemistry Enrichment of viable CTCs and CTC clusters according to their sizes using etched parylene polymer layer (mechanical separation using micro spring gap structures). Peripheral blood is filtered through the FMSA device at pressures that allows isolation of viable cells		Further molecular characterization possible. Fixed cells for cytological characterization with biomarkers specific to the tumour type studied directly on the filter/ membrane, live cells for DNA/RNA extraction, tumour profiling (sequencing), FISH and cell culture to characterize the metastatic potential of the cells Isolation of viable CTCs that can be further characterized at cellular and molecular level	Lung cancer (Chudasama et al., 2017) Lung, breast, prostate, liver, kidney, cutaneous and uveal melanoma, pancreatic cancer and sarcoma, (Muller, 2005) Breast (Pinzani et al., 2006) Metastatic carcinomas of breast, prostate and lung. Comparison between cell search and ISET (Farace et al., 2011; Harouaka et al, 2014)

TABLE 7.2 *(continued)*

Technology platforms or process	Assay-system examples	Principle of enrichment	Detection	Sensitivity	Comments	Examples of types of cancer studied
CTC microdevices/microfluidics						
Microfluidics capture	CTC-chip EpCAM-coated microposts (Massachusetts General Hospital, USA) *Second generation CTC-chip:* Herringbone device HB-chip (Stott et al., 2010a)	No pre-enrichment or RBC lysis. Single step process	Surface area of 970 mm² containing an array of 78,000 microposts, chemically active by coating with antibodies to EpCAM. EpCAM-expressing CTCs bind to the microposts, whereas leucocytes and red blood cells are washed from the chip microfluidic channel with a herringbone pattern. Alternative strategy from CTC-chip that involves the use of surface ridges or herringbones in the wall of the device to disrupt streamlines, maximizing collisions between target cells and the antibody-coated walls themselves, thus not requiring the construction and functionalization of a complex micropost geometry.	Average capture efficiency 91.8% ± 5.2% (n = 6) for PC3 cells spiked into whole blood	*Imaging;* Lysis for DNA/RNA tumour profiling/genetics. The herringbone device allows *in-situ* immunostaining or cell lysis for further molecular characterization. In a subset of patient samples, the low shear design of the HB-Chip revealed microclusters of CTCs	Metastatic lung, prostate, pancreatic, breast and colon cancer in 115 of 116 (99%) samples, with a range of 5–1281 CTCs per mL and approximately 50% purity (Nagrath et al., 2007). Efficient cell capture was validated using defined numbers of cancer cells spiked into control blood, and clinical utility was demonstrated in specimens from patients with prostate cancer (Stott et al., 2010b)
Antibody-based Surface capture (affinity Chromatography)	*MGH device (Massachusetts General Hospital, USA) and second generation:* Cluster-Chip		Biomimetic Lipid coated microfluidics to Isolate viable CTCs and Microemboli (CTCs clusters)			Chen et al., 2016
Dielectric properties (polarizability) of cells	ApoStream® (licensed from MD Anderson Cancer Center, USA)	Peripheral blood mononuclear cell isolation containing CTCs	Microfluidic flow channel to isolate CTCs using a process called dielectrophoresis field flow assist. A combination of forces, dominated by the electrophoretic charge, attracts or repels cells to a charged electrode. Differential flow rates relative to distance from the electrode aid in fractionation of different cell types		Antibody-independent. Isolated cells are viable and therefore can be used for downstream analysis	Metastatic cancers including lung, breast and ovarian cancer (O'Shannessy et al., 2016)

Technology platforms or process	Assay-system examples	Principle of enrichment	Detection	Sensitivity	Comments	Examples of types of cancer studied
Combined approach	eDAR (Ensemble-Decision Aliquot Ranking)	No pre-enrichment	Positive selection with cell-surface markers, e.g. EpCAM or HER-2 that are labelled with fluorescent antibodies, ranked by aliquots (cells contained in droplets), and sorted. Aliquots are then purified and secondary labelling and microscopic imaging analysis undertaken		1 mL of whole blood analysed in less than 20 minutes. CTCs are enriched in a small field (<1 mm^2) for microscopic imaging. Allows isolation of cells for downstream analysis. Most of the steps, such as aliquot ranking, aliquot purification, and on-chip purification, are automated. Secondary labelling and imaging is semiautomated	Spiked breast cancer cell line SKBr3 in whole blood (Schiro et al., 2012; Zhao et al., 2013.)
Microelectronics and microfluidics technology-Dielectrophoresis	DEPArray	No pre-enrichment	Target cells are identified by combinations of intracellular and extracellular markers, as well as with the use of morphological features such as circularity or size (diameters). Image-based selection ensures recovery of intact cells of interest using the CellBrowser™ Software	1 tumour cell per 10^6 WBCs	Flexible sample type analysis. Can be used on whole blood, but also cell suspensions from frozen, fixed (FFPE) or fresh tissue from biopsies. Cell suspensions from fine needle aspirates, pleural fluid, and urine can also be used. Cells are gently manipulated and can be sorted in culture medium allowing them to be used for downstream cellular and molecular analysis	Colon (Fabbri et al, 2013) and breast (Peeters et al., 2013) cancer patients blood samples Spiked neuroblastoma samples (Carpenter et al., 2014)

Key: CD45, pan leucocyte antigen; CAM5.2, identifies principally CK8 and to some extent CK7; EpCAM, epithelial cell adhesion molecule; FFPE, formalin-fixed paraffin embedded molecule (EpCAM). EpCAM is a transmembrane glycoprotein that is found expressed at various levels in most of the normal epithelial cells, but is also found expressed in a variety of carcinomas, including colon carcinomas where it was first described, but is absent from haematological cells (Balzar et al., 1999).

The CellSearch® platform is based on *positive selection* with (EpCAM/CD326)-conjugated immunomagnetic enrichment (antibody-based capture) followed by microscopic cell imaging using positive immunostaining with additional epithelial markers, such as cytokeratins CK7/8 and 18; negative immunostaining of CD45 (a pan-leucocyte marker), DNA staining with DAPI (to assess cell viability), and inclusion of cells with a diameter larger than 5 μm. The method requires the cells to be fixed in order to be stained for the cytokeratins and, therefore, they cannot be used for other downstream functional analysis. Similarly, flow cytometry using pre-enrichment on positive selection of cell surface markers such as EpCAM have been developed. An example is the Immunicon Corporation assay, iCeap (CTC enumeration and analysis procedure). The advantage of the latter over CellSearch® is that the cells sorted are live, therefore allowing downstream molecular characterization and functional analysis.

Cross reference
See Chapter 3, Section 3.4 for more information on flow cytometry.

Other platforms use negative selection by depleting the blood sample lysing the red blood cells and retaining on the columns/device the cells expressing CD45 (CD45 depletion) to only keep a small fraction of cells in the eluent believed to be derived from the tumour being investigated. This approach has the advantage that not only does it select the EpCAM+ CTCs, but it also enriches for other circulating tumour cells with low or absent EpCAM that have possibly undergone an epithelial-to-mesenchymal transition (EMT) thought to be an important process during metastasis.

In addition to the EpCAM cell-surface markers, several platforms also utilize cytoplasmic markers such as the cytokeratins (18 and 19 for instance), or tumour type-specific cell-surface molecules, such as HER-2 in breast cancer.

Key Point

Epithelial cell surface markers such as EpCAM are used for selecting the CTCs, but other markers such as pan-cytokeratin and specific cytokeratin antibodies are also used for CTCs isolation.

There are several limitations to these approaches:

- Blood samples can be 'contaminated' with epithelial non-tumour cells expressing the molecule EpCAM, which will give a *false positive result*. Certain conditions, such as benign proliferative diseases, inflammation, tissue trauma, and surgical procedures, are also known to induce the circulation of epithelial cells in the bloodstream (Goeminne et al., 1999) and can also produce a false positive result.

- It has been suggested that during the process of metastasis, some of the tumour epithelial cells may change phenotype to a more mesenchymal-type (EMT) or even stem-like type, making expression of EpCAM protein low or absent. In these situations, a *false negative result* may occur. Mesenchymal markers such as vimentin could be used, but a certain number of leucocytes and macrophages have been shown to express mesenchymal markers, which would induce false positives.

To circumvent the issues of partial loss of expression of EpCAM by the CTCs, other methodologies have been developed based on the morphology, size, and/or deformability of the cells (Table 7.2). High-definition (HD)-CTC assay, fibre-optic array scanning technology (FAST), do not involve any pre-enrichment, and are based on immunofluorescence staining for anti-CKs (pan-CK staining, CK19), CD45, and DAPI (nucleus detection) on the whole cell population. The cells are imaged with high resolution technologies and, therefore, their presence is confirmed by visual analysis. As the cells do not require fixation, further molecular analysis with, for instance, PCR-based or NGS methods can be performed by extracting the DNA from the isolated cells.

Other platforms have based their CTC isolation size selection on using filter membranes such as Screencell® (microporous membrane filter) or ISET® (Isolation by Size of the Epithelial Tumour cells) that uses filtering membranes with calibrated pores of 8 μm in diameter. This is based on the fact that, usually, epithelial cancer cells are larger (median diameter 15 μm) than leucocytes. Both methods do not need a pre-enrichment step and, after RBC lysis, the blood is filtered to isolate cells that can be further characterized by cytological analysis with biomarkers specific to the tumour type, studied directly on the filter/membrane, or with DNA/RNA extraction for tumour profiling by sequencing, FISH, or cell culture to characterize the metastatic potential of the cells. Nonetheless, it should be noted that sorting CTCs by their size can present some limitations, as it will miss any CTCs that are smaller in diameter than the assay cut-off.

In the latest development of the technologies, microdevices (microfluidics platforms) have been developed to use minimal amount of peripheral whole blood sample and to avoid using any pre-enrichment step, cutting down on the preparation time. Such platforms include the 'CTC-chip', based on cell attachment to antibody (EpCAM)-coated microposts and antibody-based surface capture (affinity chromatography). These microdevices have the advantage of high sensitivity but tend to be of low specificity (see Box 2.2). They allow further characterization of the CTCs by either immunocytochemistry, cell culture, or cell lysis followed by DNA/RNA extraction.

One of the most promising developments is a combined approach of multiple CTC separation mechanism known as the eDAR (ensemble-decision aliquot ranking). Cells to be isolated are first labelled with a fluorescent tag, such as anti-EpCAM. The blood sample containing the fluorescent-tagged cells flows through a microfluidic channel. Non-fluorescent cells, which are smaller, will continue to flow towards a waste bin, whereas the fluorescent cells, which are bigger, will stumble on obstacles (collision) and be diverted through a control solenoid to an aliquot detection chamber; here, the CTCs will be captured and further characterized. This methodology has the advantage that multiple light sources can be used to allow multicolour sorting and, therefore, the combination of biomarkers. After the detection, the system allows for the cells to be analysed by further molecular analysis, such as PCR-based and NGS techniques.

Finally, the DEPArray™ system (Di-Electro-Phoretic Array system; Silicon Biosystems, Italy) is a semi-automated system that allows the isolation of rare cells, such as CTCs, from mixed-cell populations at the single-cell level (Fuchs et al., 2006). Based on the expression of specific cell-surface markers, cells are fluorescently labelled and introduced into a single-use microfluidic cartridge, which contains an array of individually controllable microelectrodes, each with embedded sensors. This circuitry creates dielectrophoretic (DEP) cages in which individual CTCs are trapped. An imaging system composed of six-channel fluorescent microscope and a CCD camera allows for the cells of interest to be detected and gently moved to specific locations on the cartridge to allow further analysis. The cells can then be recovered for cell culture to study cell-cell interactions or for DNA/RNA/protein profiling.

7.1 SELF-CHECK QUESTION

What are the main methods available to isolate CTCs?

7.2.1.2 Practical implementation in clinical diagnostic setting? Challenges and trends

As already mentioned above only the CellSearch® platform has received clearance from the FDA for diagnostic use. The other platforms or technologies have been used either in research settings or in clinical trials. They all have the objective to count CTCs for the prediction of disease progression and survival in metastatic and in certain cases in early-stage cancer (see Figure 7.3). Some are also looking at monitoring the real-time response from the patient to a given treatment.

The advantage of LB as a minimally invasive procedure is that serial samples can be obtained, which make monitoring treatment and recurrence possible, and can provide the clinician some insight into tumour resistance, as well as tumour heterogeneity. Where it is difficult to obtain a tissue biopsy, or biopsy is not feasible due to the invasiveness of the procedure, LB represents an ideal alternative.

Temporal changes in CTC numbers have been correlated reasonably well with the clinical course of disease in many cancers, such as lung, breast, prostate, and colon cancers (Cristofanilli et al., 2004; Cohen et al., 2006; Danila et al., 2007; Truini et al., 2014). Therefore, CTCs' changes in concentration and shape can be used as predictive and prognostic biomarkers. They may also provide information at the molecular level, such as genomic variation when the isolated cells are further analysed.

Biobanking

The regulated storage of samples with ethical consent under controlled conditions for analysis at a later date.

Currently, most CTC isolation platforms require that the whole blood is processed soon after collection, which makes **biobanking** samples for the long term difficult. In addition, CTCs are fragile and tend to degrade when collected in standard evacuated blood collection tubes. As an example, the CellSearch® platform requires the use of tubes called Cellsave that contain a preservative that can protect the CTCs for up to 96 hours at room temperature allowing for the tubes to be transported to the laboratory, where it will be analysed on the device.

Nonetheless, the measurement of CTCs is currently not yet recommended in cancer guidelines for diagnosis (UK National Institute for Health and Care Excellence; NICE) or to influence

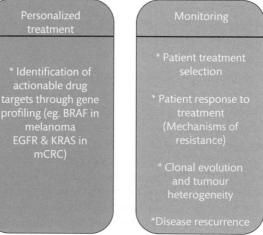

FIGURE 7.3
Potential clinical applications of liquid biopsy.
mCRC, metastatic colorectal cancer.

treatment decisions (Cree, 2015). There are few comparative studies and systematic reviews that show that most studies are too small to be considered to provide sufficient validation to permit clinical implementation (Gao et al., 2013). Accordingly, proof of clinical validity and utility holds this diagnostic tool back from routine use.

Due to the numerous challenges encountered by the identification and analysis of CTCs, circulating free nucleic acids and particularly circulating tumour DNA (ctDNA) have shown a lot of promise in the past few years and are developed in this chapter in the following section.

In conclusion, much remains to be learned about CTCs biology and their clinical potential as biomarkers for early screening, diagnosis, prognosis, and predictive testing.

7.2.2 Circulating cell-free nucleic acids detection

Circulating nucleic acids were first identified by French scientists Mandel and Métais in 1948 in human plasma (Mandel and Metais, 1948). Nonetheless, it was three decade later, in 1977, that the presence of cfDNA in cancer patient serum was first demonstrated by Leon and colleagues (Leon et al., 1977).

7.2.2.1 Circulating cell-free DNA (cfDNA)

It has been reported in the literature (Harber & Velculescu, 2014) that several thousand genome equivalents of DNA are typically present in 1 mL of circulating plasma, with more than 90% of healthy individuals having less than 25 ng cfDNA per mL (Stroun et al., 2001).

Certain conditions, such as inflammation, exercise, or tissue injury can substantially increase the amount of cfDNA levels found in blood (Lehmann-Werman et al., 2016).

Although this will not be discussed in detail in this chapter, it is of note that cfDNA has been also studied in other clinical settings, such as non-invasive prenatal testing to test for chromosome abnormalities, especially trisomy (an extra copy of a chromosome) or monosomy (a missing chromosome; see Norwitz and Levy, 2013).

In cancer, primary tumours shed tumour DNA via apoptosis and necrosis, and this DNA (ctDNA) circulates in the plasma of the patients among cfDNA from other sources. ctDNA represents between 0.1% and 10% of the total cfDNA molecules.

Two major levels are employed to study ctDNA in cancer patients. The first one is to estimate the amount of cfDNA in the circulation and the second one is to identify the genetic alterations, primarily found in the tumour tissue, in the cfDNA, such as points mutations, allele imbalances, microsatellite instability, loss of heterozygosity, and epigenetics events such as methylation (Diehl et al., 2005; Dawson et al., 2013).

Patients with similar cancer type may show different levels of ctDNA among themselves at the time of diagnosis, but the variations in concentration for the same patient have been shown to correlate with tumour progression/recurrences and response to treatment (Diehl et al., 2008).

Key Point

cfDNA is released into circulation by various pathologic and normal physiologic mechanisms, but in cancer a higher concentration has been shown to correlate with tumour progression and/or recurrences.

7.2.2.1.1 General principles of analysis of cfDNA

cfDNA can be detected in both plasma and serum, but it has been shown that there is a higher cfDNA concentration in serum that has been proposed to be due to 'contamination' from leucocyte lysis. Therefore, most studies use plasma for the analysis of cfDNA in cancer patients.

Blood collection, including transport and storage are critical steps in preparation for the analysis of cfDNA. Here, the major protocols established for blood collection, processing, and DNA extraction are presented.

Cross reference

See Chapter 1, Section 1.3.1 for more information on transport and storage, and Section 1.4.2.2 for more information on DNA extraction.

Issues encountered by researchers wanting to analyse cfDNA include the contamination of the sample with genomic DNA from WBCs, the low amount of ctDNA available for detection, and the fragmented nature of cfDNA. The size of cfDNA varies between 50 bp and 1000 bp, but usually within an average around 150–180 bp (Mouliere and Rosenfeld, 2015).

Standard evacuated blood collection tubes, such as EDTA blood draw tubes, are often used but an increasing number of studies are now using specifically designed draw tubes for cfDNA analysis such as the STREK tubes (cfDNA BCT®). The latter contain a preservative agent that stabilizes WBCs to avoid WBC lysis and release of gDNA and inhibitors to prevent nuclease-mediated degradation of cfDNA. These 'specialized' types of tubes permit the storage of the blood specimen for up to 14 days as opposed to a traditional EDTA blood draw tube, which would require the immediate processing of the sample, usually within an hour of collection. This allows the batching of samples and transport from remote locations to the analysis laboratory.

For EDTA tube collection, as soon as possible after blood taking and within an hour of the blood draw, the sample is processed to obtain the plasma that will contain the cfDNA. Most of the protocols are based on a series of two consecutive centrifugations. First, the whole blood is centrifuged at 1500 g for 10 minutes to remove RBCs, the supernatant containing the plasma is then removed taking care not to disturb the buffy coat containing WBCs. The supernatant is then centrifuged at a much higher force >10,000 g for 10 minutes to remove any remaining cells and stored at $\geq$−20°C for further DNA extraction.

There are a number of kits on the market that can be used for cfDNA extraction from plasma. Publications have discussed the main issues in analysing cfDNA, including DNA extraction. The main message here is that the use of a SOP is recommended that is easy to implement for either manual or automated methods. For routine laboratory use, automated methods are recommended in order to achieve reproducibility and cost effectiveness.

As an example, a kit that is often cited in publications and has been shown to perform well is the Qiagen kit: QIAamp circulating nucleic acid. However, many other companies have developed their own kits for cfDNA extraction, for example, NucleoSpin™ plasma XS, EpiGenTek's FitAmp™ kit (EpiGentek, Farmingdale, NY), and Norgen's Plasma/Serum Cell-Free Circulating DNA Kit.

Cross reference

See Chapter 7, Section 7.3 for coverage of digital PCR.

After extraction, cfDNA can be analysed with a range of molecular techniques, such as PCR-based methods and NGS methods for methylation analysis.

Cross reference

See Chapter 6, Section 6.2.3 for more information on NGS methods for methylation analysis.

7.2.2.1.2 Practical implementation in clinical diagnostic setting? Challenges and trends

Several studies have demonstrated that cfDNA contains the same somatic mutations detected in the primary tumour tissue and that are absent in normal cells. This is important as it can help to identify actionable drug targets, such as EGFR and KRAS mutations, in metastatic colorectal cancers, or BRAF V600E mutation in melanoma as examples using LB. In addition, the somatic mutation(s), once identified in the ctDNA, can be monitored to evaluate response to treatment for the patient, residual disease, and tumour progression.

The major challenge is, therefore, to detect genetic alterations present in a small amount of cfDNA, but this is being circumvented by the development of highly sensitive technologies such as PCR-based methods and dPCR.

Cross reference
See Chapter 4 for coverage of PCR-based methods and Chapter 5 for coverage of NGS.

Like CTCs, analysis of ctDNA is attractive as it is a minimally invasive procedure and allows serial sampling for monitoring a patient's disease. This may represent the future of precision medicine, whereby existing, but also *de novo* mutations will be able to be identified not only for genotyping the tumours, but also for monitoring response to treatment, detecting residual disease, and eventually screening for early detection (Figure 7.3). As with CTCs, use of cfDNA for clinical diagnostic is currently not yet recommended in cancer guidelines for diagnosis (NICE) or to influence treatment decisions (Cree, 2015). Lack of standardization remains one of the greatest hurdles, and sensitivity and specificity still need to be validated in a bigger cohort.

7.2 SELF-CHECK QUESTION

How can ctDNA be used?

7.2.2.2 Circulating cell-free RNAs and non-coding RNAs (microRNAs (short) and long non-coding RNAs)

Identical to cfDNA, circulating gene transcripts are also detectable in plasma and serum. The first study identifying circulating RNA was in melanoma, which showed expression of the transcripts for the melanocyte-specific tyrosinase gene by RT-PCR (Smith et al., 1991).

As for circulating gene transcripts, circulating microRNAs (miRNAs), as well as non-coding RNAs are also detectable in plasma/serum. MiRNAs, in general, are non-coding, short RNA molecules that are tissue-specific and play an important role in the regulation of gene expression and subsequent protein synthesis (Esteller, 2011). The cellular processes they are involved in can be deregulated in diseases and, in cancer, this may influence cell proliferation, differentiation, and apoptosis. Circulating miRNAs were identified in 2008 as potential biomarkers in patients with solid tumours (Mitchell et al., 2008). Since then, numerous studies have identified and studied circulating miRNAs in diverse types of cancers, as diagnostic tools, as well as prognostic and predictive biomarkers (Schwarzenbach et al., 2014). Equally, long non-coding RNAs (lncRNAs) have also been examined in cancer as new diagnostic and prognostic biomarkers, and are considered potential therapeutic targets (Qi and Du, 2013).

PCR-based methods are used to look for miRNA in plasma and several reports suggest that this is a robust method for the detection of relapse, even early cancer detection (Zandberga et al., 2013; Ulivi et al., 2013; Sanfiorenzo et al., 2013).

Detection of circulating RNA/miRNA/lncRNA molecules is still in its infancy compared with cfDNA. For instance, there are some questions regarding the origin of the transcripts circulating in the blood of cancer patients. Some have suggested that they could be coming from the haematopoietic cells in response to the disease condition.

Key Point

In addition to cfDNA, other nucleic acids such as RNA, miRNA, and long non-coding RNA can be isolated from the plasma of cancer patients.

7.2.3 Exosomes and microvesicles

Exosomes and microvesicles can be found circulating in the bloodstream and biological fluids, and they can contain diverse macromolecules from the originating cells, such as proteins, DNA, mRNAs, miRNAs, and lncRNAs. They are collectively called extracellular vesicles (EV). Exosomes are membrane-bound phospholipid vesicles ranging in size between 30–150 nm in diameter that are actively secreted by a variety of mammalian cells. Microvesicles are bigger entities ranging in size from 100 to 1000 nm in diameter. Apoptotic bodies are also part of the EV, are shed by dying cells, and are the biggest vesicles ranging in size between 50 and 5000 nm.

Despite the fact that they have been known for more than 20 years, little is known about their role in normal physiology and, therefore, why they are present in cancer and involved in many different disease states is still largely unknown. Nonetheless, it is thought that tumour-derived exosomes are involved in cell-to-cell communication, and could contribute to the progression of the disease.

Exosome levels are found to be elevated in the plasma of some cancer patients compared with healthy controls. A recent pilot study conducted on pancreatic cancers has shown that a protein called glypican-1 (GPC1) was present in exosomes secreted by pancreatic cancer cells, but not on other exosomes secreted by other cell types (Melo et al., 2015). This could point to a new source of diagnostic biomarkers for cancer.

Several protocols have been published for exosomes isolation, which are based on differential centrifugation at increasing g-force and duration, to pellet sequentially the smallest particles. Indeed, the bigger vesicles, the apoptotic bodies, are collected at ~2000 g, whereas the exosomes are collected by ultra-centrifugation above 100,000 g.

Presently, there is no technical standardization on how to collect samples, and how to isolate and analyse EV (Witwer et al., 2013). Therefore, the study of EV, and particularly exosomes in cancer, is still confined to research settings. However, with technological advances, it is possible that EV will become another source of diagnostic, prognostic, and predictive markers in cancer, and be used to deliver drug targets.

7.2.4 Conclusion

LB is a rapidly advancing field and the subject of intense research. As opposed to standard biopsy, LB as a minimally invasive procedure can provide a comprehensive real-time picture of the tumour burden, response to treatment (emergence of mechanism of resistance), and progression (recurrences and metastasis). It is tantalizing to think that LB could represent the future for early detection of biomarkers associated with cancer in asymptomatic individuals. Nevertheless, there are still several limitations, such as lack of standardization in protocols for collecting, processing, and analysing samples that are holding back the introduction of LB in the clinical setting.

7.3 Digital PCR (dPCR)

Since its introduction in the late 1990s, real-time qPCR has become the gold standard for nucleic acid target quantification in genomics research and, subsequently, in clinical analysis. However, for applications such as low-level target monitoring, determination of precise CNV and rare mutation detection, qPCR has inherent limitations. dPCR is an adaptation of the qPCR technique whose concept preceded that of qPCR. However, it was not adopted due to technical and physical complications. With the advances in microfluidics technologies, dPCR promises to overcome the limitations of qPCR by allowing absolute quantification of target nucleic acids without the need for a standard curve, with higher precision and accuracy (Figure 7.4).

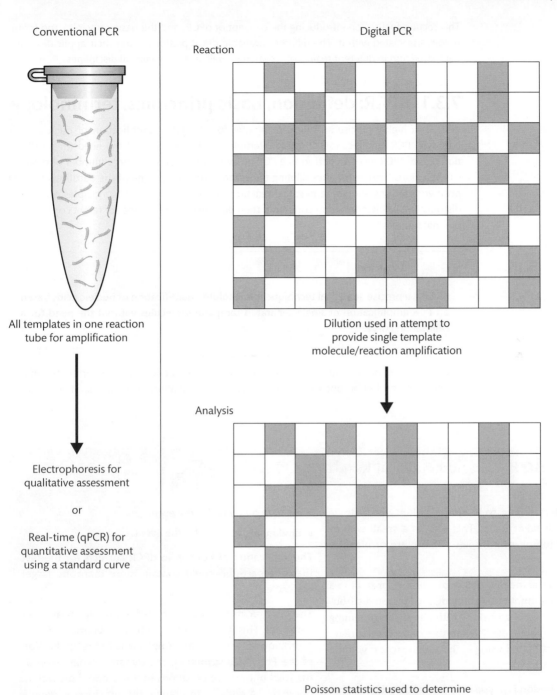

Conventional PCR

All templates in one reaction
tube for amplification

Electrophoresis for
qualitative assessment

or

Real-time (qPCR) for
quantitative assessment
using a standard curve

Digital PCR

Reaction

Dilution used in attempt to
provide single template
molecule/reaction amplification

Analysis

Poisson statistics used to determine
probability of single template
partition (green boxes) versus more
than one template (red boxes) to
give copies of template/μL

FIGURE 7.4
Comparison between conventional and digital PCR.

This section focuses on introducing the concept of dPCR, and the basic principles and terminology associated with it. The different available dPCR platforms, and their applications are compared and in different realms of research as well as diverse clinical disciplines.

7.3.1 dPCR: definition, basic principles, terminology

dPCR is a highly precise analytical technique for absolute quantification of nucleic acids based on PCR amplification of a single template molecule without the need for a calibration curve. dPCR involves performing PCR on single molecules in a large number of separate sub-reactions (partitions) after diluting the sample, such that some of the partitions contain positive reactions and some negative reactions, purely following the principles of Poisson distribution. Applying Poisson correction then determines the number of target copies in the original sample.

Key Point

dPCR is a precise analytical technique for absolute quantification of nucleic acids based on PCR amplification of single separated template molecules without the need for a calibration curve.

The most common terminologies that accompany dPCR applications are partitions, lambda (λ), Poisson distribution, and the dynamic range or the quantification 'sweet spot' (Box 7.1).

BOX 7.1 Summary of key terms used in dPCR

Partition refers to the fixed space within which single molecule PCR takes place. This can be a small well or water-in-oil (emulsion) droplet of nanolitre or picolitre volume.

Lambda (λ) is the mean target copy number present in a partition. It is estimated by applying the Poisson distribution to account for a positive partition initially containing more than one molecule. The number of copies per reaction can be estimated using λ, the total reaction volume and total partition number.

Poisson distribution is a type of binomial distribution that describes the probability of a rare event (target molecule) in a fixed partition size. Assumptions are:

• large population (partitions) of fixed size;

• a rare event;

• a binary outcome for the event;

• random distribution for the event.

The application of Poisson in dPCR corrects for the fact that one partition could contain more than one target molecule.

The *dynamic range* of dPCR is defined by the number of partitions. This is also influenced by the volume and concentration of target in the sample. Due to the application of the Poisson distribution, the dynamic range exceeds the total number of partitions in a reaction; however, at the extreme ends of the range, the precision is greatly reduced. The most accurate quantification is reached when $\lambda = 0.6$–1.6. Hence, the 'sweet spot' of a platform is defined by the range of λ values that can be accurately quantified with acceptable precision.

A partition is the fixed space within which the single molecule PCR occurs. This can be a small well or water-in-oil emulsion droplet of nanolitre or picolitre volumes. Lambda (λ) represents the mean target copy number present per partition. It is estimated applying Poisson distribution to the number of positive partitions (k) per reaction or to the number of negative partitions (r) per reaction. The number of copies per reaction can be estimated using λ, the reaction volume, and the total number of partitions (n).

$$\lambda = -\ln[1-(k/n)] \text{ or } \lambda = -\ln(r/n)$$

$$\text{Copies per reaction} = \lambda * n$$

where copies in the sample = copies per reaction * the dilution factor.

The application of the Poisson statistics in dPCR corrects for the fact that a positive partition can contain more than one target molecule (Figure 7.4). Therefore, the dynamic range of dPCR exceeds beyond the number of partitions per reaction with the most precise quantification achieved with λ= 0.6–1.6. This concept is considered when deciding the optimal sample dilution to be included in the reaction. For example in a dPCR reaction of 20,000 partitions, the ideal number of cells to be loaded for precise quantification is 32,000 cells (20,000 × 1.6 = 32,000), which is equivalent to 107 ng of DNA. This way, each partition in the reaction would receive 1.6 templates.

Several interrelated factors affect the dynamic range of dPCR, including the number of partitions, the volume of sample interrogated, and the original concentration of the target in the sample. When the sample volume is not limiting, increasing the number of partitions increases the sensitivity. Conversely, sample availability dictates the theoretical sensitivity that could be achieved. On the other hand, the quantification of a rare target dictates the need for greater partition numbers, whereas samples with highly abundant targets could be quantified with a smaller number of partitions provided the sample is adequately diluted.

7.3.2 dPCR attributes

- *Absolute quantification*: dPCR technology provides an absolute count of target nucleic acid copies in samples without the need for running standard curves. This makes dPCR ideal for applications that require absolute quantities of target copy numbers, such as minimal residual disease (MRD) monitoring, viral load analysis, and genetically modified organism (GMO) detection. Neither calibration standards nor reference genes are required for absolute quantification. Poisson statistics provides a simple estimation for target copy numbers in a sample by taking the ratio between positive and negative partitions.

- *Unparalleled precision*: The massive sample partitioning afforded by dPCR enables the reliable measurement of small fold differences (<2-fold) in nucleic acid sequence copy numbers among samples in addition to the precise detection of low-abundance target copy numbers (<35 ct).

- *Increased signal-to-noise ratio*: Partitioning allows high-background copy numbers to be diluted, increasing the odds for low-level templates to stand out, thus permitting for the sensitive detection of rare targets and enabling a ±10% precision in quantification.

- *Removal of PCR bias*: Error rates are reduced by removing the amplification efficiency reliance of qPCR, enabling the detection of small (1–2-fold) differences.
- *Higher tolerance to PCR inhibition*: Since dPCR data collection occurs at endpoint (after 40 cycles), the reaction is less susceptible to PCR inhibition. Matrices that would, in other systems, have led to lower target quantification can be tested by dPCR with reduced effect on the final result.

7.3.3 dPCR platforms

Today, there is a plethora of dPCR platforms that differ in a range of characteristics including partition type, number of partitions (influencing dynamic range), reaction volume and loss in reaction volume, chemistries they support, throughput, different 'hands-on time', and costs of consumables. Therefore, the choice of instrument is application dependent taking into account several factors, including desired sensitivity, precision, throughput, and budget. A comparison between different dPCR platforms and between dPCR versus qPCR is provided in Table 7.3 and Table 7.4, respectively, and discussed later in more detail.

There are, effectively, two main methods for generating the partitions for a dPCR: prefabricated reaction wells (in a chip or plate) or droplets (water-in-oil emulsions). Prefabricated platforms include the BioMark® HD (Fluidigm, San Francisco, CA), QS®3D (ThermoFisher Scientific, Santa Clara, CA) and Constellation (Formluatrix, Bedford, MA) with the Clarity® (JN Medsys, Cambridge) and Naica Crystal dPCR (Stilla Technologies, Paris). The emulsion-based technologies include the QX200® droplet dPCR system (BioRad Laboratories, Hercules, CA) and the RainDrop® (RainDance Technologies, Billerica, MA).

TABLE 7.3 Comparison of commercial digital PCR platforms

Platform specifications	Fluidigm	QS3D	BioRad	RainDance
	IFC & BioMark HD	QS3D	BioRad	RainDrop
dPCR	Chip	Chip	Droplet	Droplet
PCR	Real-Time	End Point	End Point	End Point
# of partitions	48 × 770 Array 12 × 765 Array	20,000	20,000	10 million
Platform cost	£200,000	£40,000	£80,000	£95,000
Cost per reaction	£30–£20	£10	£8	£20
Partition vol (nL)	0.48 nL–6 nL	0.865 nL	0.85 nL	0.005 nL
Reaction vol (µL)	0.65 µL–4.6 µL	15 µL	20 µL	50 µL
Number of multiplex reactions	2	2	2–5	2–5
Automation	Not available	Not available	Available	Not available

TABLE 7.4 Quantitative PCR (qPCR) versus digital PCR (dPCR)

qPCR	dPCR
Relative quantification to a reference gene or a standard curve	Absolute quantification. No need for standard curve and reference genes. However, reference gene quantification is still required to evaluate the quality of the samples and the efficiency of the pre-dPCR steps, particularly when quantifying RNA
Compromised sensitivity and precision at the lower end of the dynamic range	The sensitivity increases with the increasing number of partitions and the volume of sample interrogated. The precision of measurement is predictable due to the application of Poisson binary distribution. However, it reduces outside the 'sweet spot' with the most precise quantification reached when $\lambda = 0.6–1.6$
Competitive amplification, which masks low abundance target quantification	Single molecule amplification and increased signal-to-noise ratio

7.3 SELF-CHECK QUESTION

How many types of partitions are available for dPCR and how do they differ from each other?

In terms of number of partitions, several platforms, particularly the ones that were introduced earlier to the market, allow 20,000 partitions of wells or droplets (QS®3D and BioRad, respectively). In theory, droplet partitions are more heterogeneous in size compared with solid reaction wells due to inherent technicalities of droplet generation. Variable partition size has implications for the accuracy of measurements as Poisson statistics mandates a fixed partition size for optimal estimation of target copy numbers; otherwise, underestimation occurs due to multiple occupancy of partitions. This becomes particularly apparent when quantifying high abundance target molecules. Therefore, it is important that this information is provided by the manufacturer, so that, if necessary, the variability can be factored into the Poisson calculations. However, experimental evidence has shown that the variability in the number of partitions also affects the final results as the number of partitions directly influences the dynamic range. The RainDrop® platform allows interrogation of a larger volume of sample per individual reaction than the BioMark® HD, QS®3D and BioRad® (50 μL versus 5, 15 and 20 μL, respectively). The larger capacity and greater number of partitions per reaction makes this platform theoretically the most sensitive platform at the present.

The BioMark® HD is the only platform that supports a real-time PCR amplification of single molecules. All other platforms support the endpoint system where positive and negative populations are separated into two clusters at the end of the reaction resembling cell sorting on a FACS machine. Although real-time curves look familiar as they resemble the curves produced on a qPCR instrument, they are not necessary for obtaining accurate quantification. Most dPCR platforms support the same chemistries applied on qPCR, i.e. both the TaqMan hydrolysis probe and the EvaGreen chemistries. Most platforms allow the use of different master-mixes and reagents used on regular qPCR instruments despite the provision of their own proprietary solutions. Multiplexing is easier on the droplet-based platforms compared with the chip-based platforms. Chip-based platforms are limited by the number of fluorescent detection channels suited within the system, which usually does not exceed three channels, one of which is used for detecting the passive reference dye leaving only two for duplex detections. However, the droplet-based platforms can multiplex 5–8 targets, utilizing not only the number of channels, but also the intensity of the fluorescent dyes included in the reaction. In the latter scenario, the same dye can detect three different targets simply by manipulating the concentration of the dye per target.

Up-front machine cost, cost per experiment, hands-on time, experimental set-up, and throughput are also important factors when considering dPCR instruments. Some platforms are more affordable than others. The QS®3D has the lowest upfront machine cost; however, it has a lower throughput and is more labour intensive compared with BioRad®. QS®3D allows 24 reactions per run, with a maximum of three runs per day, whereas BioRad® can perform 96 reactions per run, and three runs a day (288 reactions in total, daily). RainDrop® allows eight reactions per run, and 24 reactions per day making this platform the least high-throughput. However, since QS®3D and BioRad® platforms require multiple reactions per sample to reach optimal sensitivity, throughput is counter-balanced by sensitivity and cost of consumables. In summary, therefore, the choice of instrument is application-dependent and should take into account several factors, including desired sensitivity, precision, throughput, and budget.

In their current format, dPCR platforms are limited by the amount of sample that can be analysed as they cannot compete with the absolute sensitivity of qPCR. Furthermore, a large reaction volume is required to reach absolute sensitivity using limiting dilutions. Hence, if simplified quantification by dPCR is to have an impact on patient care, then versatile higher throughput instruments that facilitate large reaction volumes (>50 µL) and larger partition numbers will be required. The industry is aware of this limitation and is already moving rapidly toward providing innovative solutions.

7.4 SELF-CHECK QUESTION

What are the main differences between the commercially available dPCR platforms?

7.3.4 dPCR applications

7.3.4.1 Gene expression

The increased precision of dPCR can provide higher resolution in many aspects of gene expression measurement. It enables to precisely quantify finer changes in expression levels (<2-fold). dPCR also provides a greater sensitivity when quantifying rare targets or RNA from very limited material. It is used for the analysis of gene expression at both the DNA and RNA levels in the following areas:

- *Detection and measurement of genomic DNA methylation*: Due to its great capacity to discriminate alleles, digital PCR can provide more sensitive detection and accurate measurement of methylation events. Its extreme precision also makes it a method of choice when operating with low amounts of starting material (such as chromatin immunoprecipitation experiments).

- *Increased sensitivity in transcriptional analyses with absolute quantification*: dPCR can increase the detection of rare transcripts, and provide easier and more accurate quantitation. Either one-step or two-step RT-PCR can be used in dPCR.

- *Detection of rare mRNAs and miRNAs with rapid turnover*, including those present in complex matrices such as blood.

- *Detection of targets using EvaGreen*, which allows users to perform gene expression analysis without TaqMan probes.

7.3.4.2 Rare allele detection

dPCR allows for increased performance in the detection and quantitation of rare sequences as target quantification is independent of the number of amplification cycles. Moreover, when detecting related sequences (SNPs, allelic variants, edited RNA), partitioning reduces competition with the more abundant background (wild-type) species. This technique is particularly useful for low abundance targets, targets in complex backgrounds, allelic variants (SNPs) and for monitoring of subtle changes in target levels. Expanding uses of dPCR for rare sequence detection include:

- Detection of cancer below the level detectable by current tests.
- Monitoring for new mutations and duplications in cancer as they arise.
- Detection of viral loads, such as HIV, below those detectable by current testing.
- Non-invasive testing in bodily fluids for infectious diseases and cancer, including cfDNA.
- Non-invasive prenatal testing using cell-free foetal DNA (cffDNA).
- Detection of transplant rejection in circulating DNA.

7.3.4.3 Copy number variation (CNV)

CNVs are found throughout the genome, with a relatively higher abundance in non-coding regions that may or may not be associated with a detectable phenotype change. Disease-associated CNVs can be inherited or generated *de novo*. Until recently, qPCR assays and microarray hybridization have been the main methods used to determine CNV in the genome. dPCR technology overcomes a number of inherent limitations of qPCR and microarray techniques for CNV analysis.

Performing comparative genomic hybridization (CGH), the test and reference are differentially labelled and then hybridized to the clones on the array. The sensitivity of CGH in detecting fine changes in CNV depends on the representation of the genomic sequence of the clones, probe characteristics, and signal-to-noise ratios. Multiple replicates or several probes can be needed to increase sensitivity, requiring the averaging of data obtained with multiple probes, which reduces the resolution of an individual array. In qPCR assays, many replicates are needed to accurately discriminate CNVs. The precision and sensitivity of dPCR technology allows the system to distinguish small changes much more readily. Additionally, the accuracy of dPCR is less sensitive to changes in amplification efficiency, a major cause of inaccuracy in qPCR measurements. The advent of dPCR now permits very high-resolution determination of CNV, often using smaller sample and reagent volumes.

7.3.4.4 NGS library quality control

Digital PCR can increase the efficiency and accuracy of NGS, saving both money and time. Integration of digital PCR into an NGS workflow can occur at several steps:

- Amplification of target libraries, ensuring better representation of low abundance species.
- Accurate quantification of NGS libraries.
- Validation of sequencing results.

7.3.4.5 Single-cell analysis

Single-cell PCR is challenging. Not only can the isolation of an intact single cell be difficult, but getting accurate results from the low levels of starting material has been technically challenging. The following characteristics of dPCR make it a sensitive and robust tool for single-cell analysis:

- Precise detection and amplification of targets at low template levels.
- No requirement for a standard curve or housekeeping genes.
- Reduced sensitivity to PCR-inhibiting components in crude cell lysates.
- Simultaneous detection of four targets through multiplexing.
- Can be combined with other techniques such as NGS.

7.5 SELF-CHECK QUESTION

What are the main applications for dPCR?

7.3.5 Considerations when planning for dPCR-based experiments

Several factors need to be considered in the application of dPCR in both research, as well as in clinical settings as a diagnostic testing tool.

The successful translation of the new methodology into clinical practice requires that pre-clinical research agrees on using specific terminology to reduce confusion, follows good experimental design and adequately reports experimental details. The digital Minimum Information for Publication of Quantitative Digital PCR Experiments (dMIQE) guideline has been developed to facilitate uniform terminology for dPCR and identify the parameters needed to assist the independent assessment of experimental data. RT-dPCR is used when the quantified target is reverse-transcribed transcript molecule (gene expression). dPCR is used to refer to the technology in general and when used without the preceding RT, it inherently implies the quantification of genomic DNA molecules.

Another concept to consider is the clinical definition of sensitivity as opposed to the analytical sensitivity. Analytical sensitivity is expressed as the LoD of an analyte denoting the lowest concentration that can be accurately detected with 95% certainty. In the clinic, however, the clinical sensitivity of an assay is often defined by the ability of the test to detect the log reduction of the ratio between the target and the reference gene, compared with the baseline at diagnosis. The definition of clinical sensitivity dictates that both dPCR and RT-dPCR remain susceptible to errors associated to upstream processing factors, such as sampling, RNA or DNA extraction, and the efficiency of the RT and cDNA synthesis steps. Reference gene quantification, therefore, is required to assess sample and pre-PCR processing quality.

The other consideration is the importance of assay standardization. The fact that no calibration curve is required to quantify target molecules does not imply that assay standardization and platform performance evaluation are discounted. The detection of any rare target by any PCR method mandates the accurate description of the assay's LoD that defines its sensitivity and specificity. dPCR is no different. Therefore, both assay design and standardization are paramount. dPCR can be more sensitive, but remains susceptible to poor assay design in addition to pre-PCR processes and molecular dropout. Appropriate positive and negative controls are paramount to assess false positivity rates and aid the establishment of accurate quantification amplitude thresholds.

Key Point

The fact that no calibration curve is required to quantify target molecules using dPCR does not mean that assay standardization and platform performance evaluation are not required.

The purpose for the dPCR application dictates the number of partitions required. For example, when performing a genotyping assay, a dPCR reaction with 20,000 partitions is more than enough to provide an accurate estimation. However, when the aim is to detect a rare allele or a low-level mutation, then the template type investigated dictates the type of calculation. When DNA is the template used, estimating the amount of DNA included in the dPCR reaction as nanograms provides an estimation of the number of cells being investigated. For example, 33 ng of DNA equates to about 10,000 cells, when using DNA from a diploid human genome. This means that the theoretical sensitivity reached is 0.01%, and a 20,000 partition size would be adequate. However, when the expression of a rare allele is investigated, then the sensitivity is defined by the number of the allele detected from the reference gene as a quality control measurement for the entire quantification process starting from RNA extraction down to the dPCR quantification. For example, if the required expression level for the reference gene is 100,000 copies per sample, then a reaction with at least 60,000 partitions is required to ensure that the platform is capable of quantifying the required amount of reference gene reaching the desired sensitivity, which is 0.001% in this example.

Key Point

When quantifying a highly abundant target, the sample needs to be diluted adequately so that a precise quantification is possible. However, factors that affect the sensitivity, while quantifying a rare target include increased number of partitions, larger volume of investigated sample, and the presence of the target in the sample in the first place.

Another factor to consider is the multiplexing dilemma for rare-allele detection, posed by the high reference background and low target copy numbers in one reaction. This is particularly an issue when quantifying RNA molecules. To express the disease levels as a percentage ratio between the transcript levels of the target and reference genes, the copy numbers of both genes are measured in a sample containing an unknown copy number of the target gene (0–10,000 copies), but an almost fixed number of the reference gene (10^4–10^5) indicating a good quality sample. When duplexing, a balance must be achieved between the accurate quantification of both the highly abundant reference gene and the low abundance target gene, without unduly compromising the sensitivity of the assay. To meet this requirement a platform with a larger number of partitions would be of particular advantage. For example, when aiming to quantify 1 copy in 100,000 molecules on an RT-dPCR platform with a 20,000 partition, at least 3 or 9 reactions per sample would be required to reach a comparable sensitivity per reaction or triplicate reactions on RT-qPCR, respectively [100,000/1.6 = 62,500 partitions (~3 reactions)]. On the other hand, a partition size of 10 million allows one reaction per sample to reach the sensitivity of the triplicate RT-qPCR reactions combined without risking reaction saturation by the reference gene.

Additional aspects to understand are potential sources of error. Accuracy of quantification is influenced by bias and variance. Systematic bias leading to underestimation can be associated with dPCR particularly when quantifying highly abundant target quantities. Underestimation could be due to poor assay design, inhibitors, non-random distribution of the sample due to linkage or sample lack of inhomogeneity, molecular drop-out or non-uniform partition size. Therefore, the inclusion of internal positive controls is important for qPCR, especially when reporting negative case results. Regarding precision, the random nature of nucleic acid molecules distribution across the partitions makes the precision of measurement both predictable and precise compared with qPCR. However, the accuracy is difficult to assess due to the lack of methods that are capable of verifying dPCR results. This highlights the importance of the use of certified reference materials to assess the performance of dPCR platforms.

Key Point

Considerations when planning for dPCR-based experiments or clinical tests include:

- **The clinical definition of sensitivity as opposed to the analytical sensitivity.**
- **The importance of assay standardization.**
- **The number of partitions required.**
- **Multiplexing when quantifying low target copy numbers in a background of a highly abundant reference gene in one reaction.**
- **Identification of sources of quantification error.**

7.3.6 dPCR applications including clinical applications

7.3.6.1 Genetically modified organisms

A GMO is an organism whose genetic material has been altered using genetic engineering techniques, i.e. organisms that contain a novel combination of genetic material not present naturally. 'Transgenic organisms' are a subtype of GMOs whose genetic make-up has been altered by the addition of genetic material from an unrelated organism. GMOs are used to produce many medications and foods, and are widely used in scientific research.

A popular example of GMOs is genetically modified crops (GMCs) or biotech crops. Since their introduction to agriculture in 1996, GMCs are becoming increasingly popular with new crops being approved for commercialization every year. Hence, the regulation of the development (production), cultivation, and import of these new crops is a key concept accompanying the advances in the technology. Therefore, it is increasingly becoming important to be able to detect and identify GMO-derived product and distinguish between authorized and unauthorized use of modified crops. Additionally, there is a pressing requirement for monitoring and validation of product labelling. The current regulation in the USA and Canada does not require compulsory labelling of GMO products, whereas the EU regulation is more stringent and requires mandatory labelling up to a threshold GM content level of 0.9% (expressed as the percentage of event-specific DNA copy numbers in relation to the **taxon**-specific DNA copy numbers, calculated in terms of haploid genomes; EU recommendation 2004/787/EC).

At present, qPCR is the most commonly accepted for detection, identification, and quantification of GMOs. However, challenges in GMO detection using PCR methods are numerous:

- Although it is true that the methodology has exquisite sensitivity, reproducibility, and accuracy, the divergence in protocols, instruments, enzymes, buffers, and taxons mandates extensive harmonization amongst laboratories to ensure reliable and comparable quantification.

- The appropriate estimation of the measurement uncertainty associated with an analytical result is crucial for decisions on the compliance of the sample tested. This is a particular challenge with the qPCR method when the target is present at very low concentrations, particularly at borderline to the recommended 0.9%.

- The insert within the GMO needs to be of a known sequence. Added to this is the increasing number of GMOs on the market worldwide, which leads to time-consuming and more costly GMO quantification in cases where there are multiple GMOs in a sample.

- The detection of one GMO at a time.

- The inability to distinguish transgenic crops with stacked traits.

This refers to transgenic cultivars derived from crosses between transgenic parent lines, combining the transgenic traits of both parents, i.e. GMCs with more than one genetic insert.

In comparison with qPCR, dPCR has several advantages when it comes to the quantification of GMOs:

- dPCR enables the determination of absolute target copy numbers present in a reaction. This eliminates multiple issues including the bias of amplification efficiency between the taxon and target when quantifying using qPCR.

- dPCR is more precise in quantifying low target quantities which increases the reliability of low-level GMC quantification.

- Droplet dPCR allows multiplexing of 2–8 targets in one reaction allowing multiple GMO detection in one reaction.

- Assays that work on qPCR platforms are easily transferred to dPCR platforms.

- The only drawback that persists with dPCR is the inability to detect unknown GMO sequences. However, combining the quantitative capabilities of dPCR with NGS circumvents this limitation.

Taxon

A group of one or more populations of an organism or organisms seen by taxonomists to form a unit.

7.6 SELF-CHECK QUESTION

What are the main advantages of dPCR in GMO testing?

7.3.6.2 Virology

The development of quantitative molecular methods has characterized the evolution of clinical virology in the last two decades. Using these methods, the *in vivo* role of viral load, viral replication activity, and viral transcriptional profiles has been investigated and correlated with disease outcome and progression. In addition, the pathophysiology of viruses involved in human disease has been highlighted. From a medical point of view, quantitative methods have provided the rationale for therapeutic intervention and monitoring in medically challenging viral diseases.

Viruses are capable of persisting indefinitely after infection as the viral genome integrates itself to the host's genome and uses the host's replication machinery to replicate itself. Therefore, sequences specific to the viral genome are ideal markers to use for viral disease diagnosis, monitoring response to antiviral therapy, and the detection of viral reservoirs. Quantitation of virus-specific nucleic acid in a sample using qPCR has become an essential methodology in clinical and molecular microbiology laboratories. Disease progression, prognosis, selection of antivirals, and response to therapy have been linked to the initial viral load or changes in load observed during continuous monitoring for HIV, cytomegalovirus (CMV), Epstein–Barr virus (EBV), human herpesvirus 6 (HHV-6), and BK virus in addition to disease diagnosis.

At present, qPCR is the most commonly used quantitative method in virology. However, several in-house PCR methods have been used with a high degree of result variability when compared with each other. Although the WHO has recently made available some international convention calibrators for EBV, CMV, HIV-1, HBV, and HCV, their values are arbitrarily defined, and not based on a calibration hierarchy corresponding to an independent and stable reference system such as the International System of Units (SI) as described in ISO 17511. The subsequent generation of secondary commercial standards has added another layer to the source of variability. Concordance in quantitative viral DNA results between laboratories and between serial samples from a patient within the same laboratory is a prerequisite for definition of generally accepted clinical thresholds for viral infection, and for monitoring disease initiation and progression. Therefore, standardization, accuracy, and precision remain challenging in clinical virology testing with qPCR.

dPCR in this context is a valuable tool for the development of certified reference materials, whose values are traceable to higher-order standards and reference measurement procedures, which will further contribute to the understanding of analytical performance characteristics and promote clinical data comparability. dPCR has been shown to be a highly precise measurement technique, especially for low concentrated viral materials, and has demonstrated good accuracy in discriminating different genotypes of the same virus. Furthermore, it has been successfully applied in a one-step format for the quantification of RNA viruses showing increased tolerance to inhibitors and higher accuracy at lower concentrations than qPCR. Some limitations in the clinical sensitivity of dPCR have been observed that may be linked to restricted sample volume inputs for the currently available dPCR platforms, but the fast-advancing technology is already evolving towards addressing this issue. Similar to GMO detection, the inability to identify new variants within a fast evolving viral genome is a main limitation for dPCR. However, coupled with NGS the two technologies complement each other in this application.

7.7 SELF-CHECK QUESTION

What are the main advantages of dPCR in viral load testing?

7.3.6.3 *Pharmacogenomics*

Pharmacogenetics is the study of the role of genetics in drug response. It deals with the influence of acquired and inherited genetic variation on drug response in patients by correlating gene expression or SNPs with drug pharmacokinetics and patient response to the drug. In this sense, pharmacogenomics feeds into the precision medicine in which drugs and drug combinations are optimized for each individual's unique phenotype. Pharmacogenomics is similar in principle to pharmacogenetics except that it takes the genomic and epigenetic profile, and the effect of multiple genes or genetic networks on drug response. For providing recommendations that inform the choice of treatment for a given disease, two possible approaches are used—genotyping of single genes or exome/whole genome sequencing. Examples are the polymorphism in the *G6PD* gene, which causes favism, mutations in enzymes that are responsible for variances in drug metabolism, and responses such as polymorphisms in the cytochrome P450 enzymes, which account for the metabolism of approximately 80–90% of currently available prescription drugs.

The main advantage for dPCR in pharmacogenetics is that it is capable of providing a single or multiple gene genotyping in a timely and cost-effective fashion. In addition, its ability to perform absolute quantification simplifies the process of validating new markers identified by the ongoing pharmacogenomics research. The precision of digital PCR is also useful as more data emerges on the role of germline CNVs in pharmacogenetics.

7.3.6.4 *Haematological malignancies*

In most haematological-derived malignancies, patients are monitored frequently to observe responses to treatment, to detect minimal residual disease, and to give the 'all clear', leading to discontinuation of therapy. Indeed, for patients with chronic myeloid leukaemia (CML) patients who have reached sustainable undetectable levels of MRD can consider coming off therapy, and continuing their life medication-free. In acute myeloid leukaemia (AML), MRD monitoring is particularly relevant in three scenarios:

- to assess the reduction of the leukaemic burden while on therapy;
- to make sure that leukaemic clones are undetectable before allogenic transplantation;
- whether the leukaemia is reappearing after either chemotherapy or transplantation consolidation.

In scenarios, the absence of MRD or its decline has been linearly correlated with good prognosis and low risk of relapse.

Research and clinical trials have identified different types of biomarkers in different types of leukaemia suitable for MRD monitoring. These fall into three main categories:

1. Fusion gene transcripts such as *PML-RARA* in AML, *MLL* gene rearrangements in acute lymphoblastic lymphoma (ALL), and *BCR-ABL1* in CML.

2. Single mutations in certain genes such as mutations in the *FLT3, RUNX1, CEBPA, TET2, NPM1* and *DNMT3A* genes in AML, and the *JAK2, MPL* and *CALR* genes in myeloproliferative neoplasms (MPN).

3. Overexpressed genes, such as *WT1* in AML.

RT-qPCR is the gold standard for monitoring MRD and is routinely applied for this purpose. However, it has inherent limitations related to reduced precision at the lower end of the calibration curve, in addition to significant variation in assay performance between different laboratories, emphasizing the need for standardization. The use of conversion factors and the introduction of international reference materials to act as calibration standards have helped mitigate this problem, but the former requires a burdensome system of sample exchange, while the latter are difficult to produce, thus limiting their availability. RT-dPCR has unique

Cross reference

See Chapters 8 and 9 for further details on haemopoeitic diseases.

Cross reference

See Chapter 8, Section 8.3 for more details about chronic myeloid leukaemia.

Cross reference

See Chapter 8, Section 8.2 for more details about acute myeloid leukaemia.

Cross reference

See Chapter 8, Section 8.4 for more details about acute lymphoblastic leukaemia.

advantages for MRD monitoring in haematological malignancies as it offers to simplify the standardization procedure and improves on both the sensitivity and precision of measurement. Another application for RT-dPCR is to value assign reference materials, which can be used either for the calibration of secondary 'in-house' control materials or for directly quantifying transcript copy numbers.

In the field of transplant medicine, dPCR has been used to quantify cell-free donor-specific DNA molecules in the peripheral blood of heart transplant recipients on the premise that the quantity of donor DNA in the circulation would reflect cellular rejection of the graft. dPCR can therefore be used for quantitation of specific loci in consecutive clinical samples.

7.8 SELF-CHECK QUESTION

What is the current method used for molecular monitoring of MRD in haematological malignancies? What are the drawbacks of this? List the main advantages of dPCR in the investigation of these malignancies.

7.3.6.5 Precision medicine applied to solid tumours

Precision medicine is a key healthcare enterprise that is gathering momentum. PCR strategies, particularly dPCR, have a unique role in the successful and efficient delivery of this initiative. Precision medicine requires the identification of meaningful molecular biomarkers and/or profiles to assist in the appropriate clinical management of patients and is core to its success. These biomarkers will assist the rational matching of patients to effective therapies, and will facilitate the use of molecular stratification to inform prognosis and clinical decision-making.

Current technological advances, particularly NGS, have played a pivotal role in identifying potentially novel molecular biomarkers. Solid organ tumours are now routinely screened

CASE STUDY 7.1 The use of dPCR to produce reference material used to assign copy numbers for targets within clinical samples

CML is one of very few cancers where the disease is kept under control indefinitely, so long as the therapy is continuously administered. Therefore, frequent monitoring of patients' response to therapy is an integral part of patient management. The level of the disease causing fusion transcript, the major *BCR-ABL1* transcript, is the biomarker monitored using RT-qPCR application and requires the acquisition of a standard curve with known serially diluted concentrations. Historically, the CML community had relied on laboratory-specific standard curves and normalized their values by exchanging a range of patient samples at different disease levels with a reference laboratory. However, dPCR is now used to generate a reference material with dilution points that have been precisely quantified using

the technology. The introduction of this reference material has simplified the process of normalizing results from different laboratories and is widely adopted through the CML community.

In addition to the major *BCR-ABL1* transcript, which is present in 90% of CML patients, 10% carry six rarer transcript types. In these patients, it is very difficult to monitor their responses to therapy as the generation and standardization of standard curves is a laborious and non-cost effective procedure. dPCR, however, has circumvented this, where reference standard curves, specific to each rare transcript type, are generated and value assigned with high precision using dPCR quantification.

for mutations in oncogenes such as *EGFR* in lung cancer, *KRAS* in colorectal cancer, *BRAF* and *NRAS* in melanoma, and *BRCA1* and *BRCA2* in breast cancer that are known to predict response to specific types of therapies. CNVs also have unique clinical significance as they have been associated with different cancers, such as the amplification of the *HER-2* gene in breast cancer. As research progresses more genes are being identified as causative or prognostic in different cancer types. However, defining the clinical relevance and the validation of these biomarkers are the main challenges. Other considerations include significant biological, clinical, logistical, and economic challenges. Absolute quantification, rare-variant detection, and CNV estimation in this context can be uniquely addressed using dPCR. In addition to validating NGS findings, dPCR itself can also be used as a diagnostic, prognostic, and monitoring test.

> **Cross reference**
> More information about breast cancer can be found in Chapter 10 and more information about epithelial tumours and melanoma can be found in Chapter 11.

Particular challenges that exist while testing biomarkers in solid tumours are related to the sample type, sample amount, tumour heterogeneity, and the sensitivity of the methods used for testing the biomarkers. In solid organ tumours, the nucleic acid is often extracted from either biopsies or fine needle aspirates embedded in FFPE blocks leading to its degradation/fragmentation as a result of fixation. On the other hand, they could be investigated in blood samples where a large amount of WBC background is present. A further challenge is that cancers often consist of multiple subclones. It is, therefore, clinically relevant to identify the presence of low-level drug-resistant subclones at the start of treatment, as this would stratify patients into those who are likely to relapse early after therapy, or those who may benefit from combination targeted therapy. This proven tumour heterogeneity presents a direct challenge to variant allele frequency reporting as the frequency of the mutated allele is reliant on the percentage of cancer cells that carry a specific druggable mutation. dPCR, coupled with semi-automated laser capture microdissection technology, allows the analysis of as pure a population of cancer cells as is feasible, reducing the contamination from surrounding stroma cells for accurate variant allele frequency reporting.

> **Cross reference**
> See Chapter 1 Box 1.3 for more details on fixation.

In patients treated for different tumour types, particularly in lung cancer, monitoring of biomarkers as surrogates for response to therapy using dPCR and/or NGS has recently become a hot topic in monitoring the response of these cancers to different therapies. The debate is as to which of the two methodologies is more suitable for this purpose. Although the former is quicker, cheaper and potentially more sensitive than NGS, it's is only able to quantify, at best, a few biomarkers at a time, whereas NGS can provide a complete profile for as many biomarkers as required. The cost/benefit of the two is, therefore, not yet completely established and for the time being the use of both technologies is appropriate.

7.9 SELF-CHECK QUESTION

What are the main advantages of dPCR in solid tumours?

7.3.6.6 Non-invasive prenatal testing

Since the discovery of foetal DNA in maternal plasma and the developments in precise measurement technologies non-invasive prenatal diagnosis has witnessed remarkable developments. Testing for aneuploidy of chromosomes 13, 18, and 21 is currently approved for clinical testing using qPCR for prenatal risk assessment. qPCR is also used for the risk assessment of other common diseases, such as sickle cell anaemia, haemophilia, and inherited metabolic disorders. dPCR coupled with circulating free-foetal DNA testing adds unique advantage for prenatal testing as it reduces the risks an invasive procedure poses on the foetus. Furthermore, it allows the non-invasive detection of foetal aneuploidies by the analysis of cffDNA and RNA in maternal plasma or serum. This topic is further discussed in Chapter 13, Section 13.3.3'.

Key Point

Current clinical applications for dPCR include virology, pharmacogenomics, investigation of haematological malignancies, precision medicine of solid tumours, and non-invasive prenatal testing.

7.3.6.7 Automated near-patient testing

The use of walk-away fully automated closed systems for qPCR, such as the Cepheid cartridges by GeneXpert for *BCR-ABL1* transcript level quantification in CML and testing for tuberculosis, has provided a practical alternative for low-throughput laboratories and for use in countries where assay standardization and the monitoring of performance is particularly challenging. The input is a peripheral blood sample where RNA or DNA extraction, RT and qPCR steps are automated inside the cartridge with the transcript levels reported, based on the internal algorithm, without the need for a standard curve. The system is reproducible and has a quick turn-around time of less than 2 hours. dPCR could be a potential future development that may improve on the cartridge sensitivity. The cost per cartridge, however, remains a considerable factor that might influence wider uptake of such systems.

 # Chapter summary

This chapter has:

- Defined the liquid biopsy (LB) and contrasted it with the standard tissue biopsy.

- Introduced the main LB technologies available to study circulating tumour cells and cell-free nucleic acids (DNA, RNA, microRNAs, non-coding RNAs).

- Described the main molecular characteristics that are used to define what the phenotype of circulating tumour cells is likely to be.

- Presented the main applications of studying LB and potential benefit for molecular diagnostics.

- Defined and outlined the concepts of digital PCR (dPCR).

- Introduced different platforms and chemistries of dPCR.

- Discussed the advantages and limitations of dPCR.

- Explored various clinical applications for dPCR.

 # Further reading

Liquid biopsy

- Hong B, Zu Y. Detecting circulating tumor cells: current challenges and new trends. Theranostics 2013; 3: 377–394.

- Krishnamurthy N, Spencer E, Torkamani A, Nicholson L. Liquid biopsies for cancer: coming to a patient near you. Journal of Clinical Medicine 2017; 6: 3–13.

- Perakis S, Speicher MR. Emerging concepts in liquid biopsies. BMC Medicine 2017; 15: 75.

dPCR

- Devonshire AS, Honeyborne I, Gutteridge A, et al. Highly reproducible absolute quantification of Mycobacterium tuberculosis complex by digital PCR. Analytical Chemistry 2015; 87: 3706–3713.

- Huggett JF, Cowen S, Foy CA. Considerations for digital PCR as an accurate molecular diagnostic tool. Clinical Chemistry 2015; 61: 79–88.

- Huggett JF, Foy CA, Benes V, et al. The digital MIQE guidelines: minimum information for publication of quantitative digital PCR experiments. Clinical Chemistry 2013; 59: 892–902.

- Milavec M, Dobnik D, Yang L, et al. GMO quantification: valuable experience and insights for the future. Analytical and Bioanalytical Chemistry 2014; 406: 6485–6497.

- Sanders R, Mason DJ, Foy CA, Huggett JF. Evaluation of digital PCR for absolute RNA quantification. PLoS One 2013; 8: e75296.

- Whale AS, Cowen S, Foy CA, Huggett JF. Methods for applying accurate digital PCR analysis on low copy DNA samples. PLoS One 2013; 8: e58177.

- Nancy BY, Tsui Rezan A, Kadir KC, et al. Noninvasive prenatal diagnosis of hemophilia by microfluidics digital PCR analysis of maternal plasma DNA. Blood 2011; 117(13): 3684–3691.

References

- Allard WJ, Matera J, Miller MC, et al. Tumor cells circulate in the peripheral blood of all major carcinomas but not in healthy subjects or patients with nonmalignant diseases. Clinical Cancer Research 2004; 10: 6897–6904.

- Andreopoulou E, Yang LY, Rangel KM, et al. Comparison of assay methods for detection of circulating tumor cells in metastatic breast cancer: AdnaGen AdnaTest BreastCancer Select/Detect™ versus Veridex CellSearch™ system. International Journal of Cancer 2012; 130: 1590–1597.

- Ashworth TR. A case of cancer in which cells similar to those in the tumors were seen in the blood after death. Australian Medical Journal 1869; 14: 146–149.

- Balzar M, Winter MJ, de Boer CJ, Litvinov SV. The biology of the 17-1A antigen (Ep-CAM). Journal of Molecular Medicine (Berlin) 1999; 77: 699–712.

- Carpenter EL, Rader J, Ruden J, et al. Dielectrophoretic capture and genetic analysis of single neuroblastoma tumor cells. Frontiers in Oncology 2014; 31: 201.

- Chen W, Allen SG, Reka AK, et al. Nanoroughened adhesion-based capture of circulating tumor cells with heterogeneous expression and metastatic characteristics. BMC Cancer 2016; 16: 614.

- Chudasama D, Barr J, Beeson J, et al. Detection of circulating tumour cells and survival of patients with non-small cell lung cancer. Anticancer Research 2017; 37:169–173.

- Cohen SJ, Alpaugh RK, Gross S, et al. Isolation and characterization of circulating tumor cells in patients with metastatic colorectal cancer. Clinical Colorectal Cancer 2006; 6: 125–132.

● Cree I. Liquid biopsy for cancer patients: principles and practice. Pathogenesis 2015; 2 (1–2): 1–4.

● Cristofanilli M, Budd GT, Ellis MJ, et al. (2004) Circulating tumor cells, disease progression, and survival in metastatic breast cancer. New England Journal of Medicine 2004; 351: 781–791.

● Danila DC, Heller G, Gignac GA, et al. Circulating tumor cell number and prognosis in progressive castration-resistant prostate cancer. Clinical Cancer Research 2007; 13: 7053–7058.

● Dawson SJ, Tsui DW, Murtaza M, et al. Analysis of circulating tumor DNA to monitor metastatic breast cancer. New England Journal of Medicine 2013; 368: 1199–1209.

● Desitter I, Guerrouahen BS, Benali-Furet N, et al. A new device for rapid isolation by size and characterization of rare circulating tumor cells. Anticancer Research 2011; 31: 427–441.

● Diehl F, Li M, Dressman D, et al. Detection and quantification of mutations in the plasma of patients with colorectal tumors. Proceedings of the National Academy of Sciences, USA 2005; 102: 16368–16373.

● Diehl F, Schmidt K, Durkee KH, et al. Analysis of mutations in DNA isolated from plasma and stool of colorectal cancer patients. Gastroenterology 2008; 135: 489–498.

● Esteller M. Non-coding RNAs in human disease. Nature Reviews: Genetics 2011; 12: 861–874.

● Fabbri F, Carloni S, Zoli W, et al. Detection and recovery of circulating colon cancer cells using a dielectrophoresis-based device: KRAS mutation status in pure CTCs. Cancer Letters 2013; 335: 225–231.

● Farace F, Massard C, Vimond N, et al. A direct comparison of cell search and ISET for circulating tumour-cell detection in patients with metastatic carcinomas. British Journal of Cancer 2011; 105: 847–853.

● Fuchs AB, Romani A, Freida D, et al. Electronic sorting and recovery of single live cells from microlitre sized samples. Lab on a Chip 2006; 6: 121–126.

● Gao P, Jiao SC, Bai L, et al. Detection of circulating tumour cells in gastric and hepatocellular carcinoma: a systematic review. Journal of International Medical Research 2013; 41: 923–933.

● Goeminne JC, Guillaume T, Salmon M, et al. Unreliability of carcinoembryonic antigen (CEA) reverse transcriptase-polymerase chain reaction (RT-PCR) in detecting contaminating breast cancer cells in peripheral blood stem cells due to induction of CEA by growth factors. Bone Marrow Transplantation 1999; 24: 769–775.

● Harber DA, Velculescu V E. Blood-based analyses of cancer: circulating tumor cells and circulating tumor DNA. Cancer Discoveries 2014; 4: 650–661.

● Harouaka RA, Zhou MD, Yeh YT, et al. Flexible micro spring array device for high-throughput enrichment of viable circulating tumor cells. Clinical Chemistry 2014; 60: 323–333.

● Hsieh HB, Marrinucci D, Bethel K, et al. High speed detection of circulating tumor cells. Biosensitivity and Bioelectronics 2006; 21:1893–1899.

- Krivacic RT, Ladanyi A, Curry DN, et al. A rare-cell detector for cancer. Proceedings of the National Academy of Sciences, USA 2004; 101:10501–10504.

- Lehmann-Werman R, Neiman D, Zemmour H, et al. Identification of tissue-specific cell death using methylation patterns of circulating DNA. Proceedings of the National Academy of Sciences, USA 2016; 113: 1826–1834.

- Leon SA, Shapiro B, Sklaroff DM, Yaros MJ. Free DNA in the serum of cancer patients and the effect of therapy. Cancer Research 1977; 37: 646–650.

- Mandel P, Métais P. Les acides nucléiques du plasma sanguin chez l'homme. Comptes Rendus des Seances de la Societe de Biologie et des ses Filiales 1948; 142: 241–243.

- Marrinucci D, Bethel K, Kolatkar A, et al. Fluid biopsy in patients with metastatic prostate, pancreatic and breast cancers. Physics & Biology 2012; 9: 016003.

- Mavroudis D. Circulating cancer cells. Annals of Oncology 2010; 21: 95–100.

- Melo SA, Luecke LB, Kahlert C, et al. Glypican-1 identifies cancer exosomes and detects early pancreatic cancer. Nature 2015; 523:177–182.

- Mitchell PS, Parkin RK, Kroh EM, et al. Circulating microRNAs as stable blood-based markers for cancer detection. Proceedings of the National Academy of Sciences, USA 2008; 105: 10513–10518.

- Mouliere F, Rosenfeld N. Circulating tumor-derived DNA is shorter than somatic DNA in plasma. Proceedings of the National Academy of Sciences, USA 2015; 112: 3178–3179.

- Nagrath S, Sequist LV, Maheswaran S, et al. Isolation of rare circulating tumour cells in cancer patients by microchip technology. Nature 2007; 450: 1235–1239.

- Norwitz ER, Levy B. Noninvasive prenatal testing: the future is now. Reviews in Obstetrics and Gynecology 2013; 6: 48–62.

- O'Shannessy DJ, Davis D W, Anderes K, Somers EB. Isolation of circulating tumor cells from multiple epithelial cancers with ApoStream® for detecting (or monitoring) the expression of folate receptor alpha. Biomark Insights 2016; 11: 7–18.

- Pantel K, Alix-Panabières C. Circulating tumour cells in cancer patients: challenges and perspectives. Trends in Molecular Medicine 2010; 16: 398–406.

- Peeters DJ, De Laere B, Van den Eynden GG, et al. Semiautomated isolation and molecular characterisation of single or highly purified tumour cells from CellSearch enriched blood samples using dielectrophoretic cell sorting. British Journal of Cancer 2013; 108: 1358–1367.

- Pinzani P, Salvadori B, Simi L, et al. Isolation by size of epithelial tumor cells in peripheral blood of patients with breast cancer: correlation with real-time reverse transcriptase-polymerase chain reaction results and feasibility of molecular analysis by laser microdissection. Human Pathology 2006; 237: 711–718.

- Qi P, Du X. The long non-coding RNAs, a new cancer diagnostic and therapeutic gold mine. Modern Pathology 2013; 26: 155–165.

- Racila E, Euhus D, Weiss AJ, et al. Detection and characterization of carcinoma cells in the blood. Proceedings of the National Academy of Sciences, USA 1998; 95: 4589–4594.

- Raimondi C, Gradilone A, Naso G, Cortesi E, Gazzaniga P. Clinical utility of circulating tumor cell counting through CellSearch(®): the dilemma of a concept suspended in Limbo. Oncology Targets & Therapy 2014; 7: 619–625.

- Rolle A, Günzel R, Pachmann U, et al. Increase in number of circulating disseminated epithelial cells after surgery for non-small cell lung cancer monitored by MAINTRAC(R) is a predictor for relapse: A preliminary report. World Journal of Surgical Oncology 2005; 3: 18.

- Sanfiorenzo C, Ilie MI, Belaid A, et al. Two panels of plasma microRNAs as non-invasive biomarkers for prediction of recurrence in resectable NSCLC. PLoS One 2013; 8: e54596.

- Schiro PG. Zhao M, Kuo JS, et al. Sensitive and high-throughput isolation of rare cells from peripheral blood with ensemble-decision aliquot ranking. Angewandte Chemie International Edition in English 2012; 51: 4618–4622.

- Schwarzenbach H, Nishida N, Calin GA, Pantel K. Clinical relevance of circulating cell-free microRNAs in cancer. Nature Reviews: Clinical Oncology 2014; 11: 145–156.

- Smith B, Selby P, Southgate J, et al. Detection of melanoma cells in peripheral blood by means of reverse transcriptase and polymerase chain reaction. Lancet 1991; 338: 1227–1229.

- Stott SL, Hsu CH, Tsukrov DI, et al. Isolation of circulating tumor cells using a microvortex-generating herringbone-chip. Proceedings of the National Academy of Sciences, USA 2010a; 107: 18392–18397.

- Stott SL, Lee RJ, Nagrath S, et al. Isolation and characterization of circulating tumor cells from patients with localized and metastatic prostate cancer. Science of Translational Medicine 2010b; 2: 25ra23.

- Stroun M, Lyautey J, Lederrey C, Olson-Sand A, Anker P. About the possible origin and mechanism of circulating DNA: Apoptosis and active DNA release. Clinica Chimica Acta 2001; 313: 139–142.

- Takao M, Takeda K. Enumeration, characterization, and collection of intact circulating tumor cells by cross contamination-free flow cytometry. Cytometry A 2011; 79: 107–117.

- Truini A, Alama A, Dal Bello MG, et al. Clinical applications of circulating tumor cells in lung cancer patients by CellSearch system. Frontiers in Oncology 2014; 4: 242.

- Ulivi P, Foschi G, Mengozzi M, et al. Peripheral blood miR-328 expression as a potential biomarker for the early diagnosis of NSCLC. International Journal of Molecular Sciences 2013; 14: 10332–10342.

- Vona G, Sabile A, Louha M, et al. Isolation by size of epithelial tumor cells: a new method for the immunomorphological and molecular characterization of circulating tumor cells. American Journal of Pathology 2000; 156: 57–63.

- Wendel M, Bazhenova L, Boshuisen R, et al. Fluid biopsy for circulating tumour cell identification in patients with early- and late-stage non-small cell lung cancer: a glimpse into lung cancer biology. Physical Biology 2012; 9: 016005.

- Witwer KW, Buzás EI, Bemis LT, et al. Standardization of sample collection, isolation and analysis methods in extracellular vesicle research. Journal of Extracellular Vesicles 2013; 2.

- Zandberga E, Kozirovskis V, Abols A, et al. Cell-free microRNAs as diagnostic, prognostic, and predictive biomarkers for lung cancer. Genes Chromosomes Cancer 2013; 52: 356–369.

- Zhao M, Nelson WC, Wei B, et al. New generation of ensemble-decision aliquot ranking based on simplified microfluidic components for large-capacity trapping of circulating tumor cells. Analytical Chemistry 2013; 85: 9671–9677.

 Discussion question

1.1 What are the challenges that liquid biopsy is facing to its implementation in clinical routine?

Answers to the self-check questions and tips for responding to the discussion question are provided on the book's accompanying website:

Visit: www.oup.com/uk/warford

8

Haemopoietic Diseases 1: Leukaemias

Georgina Ryland and Piers Blombery

Learning objectives

After studying this chapter, you should confidently be able to:

- Describe the differences between acute and chronic leukaemias.
- Discuss the role of genomics in leukaemia risk stratification and the implications on prognosis.
- Understand the principles of B-cell and T-cell cell receptor clonality analysis and how this is used to make a diagnosis of malignancy.
- Describe different molecular methods for monitoring minimal residual disease in haematological malignancies.

8.1 An introduction to leukaemia classification

Cross reference
See Chapter 9 for more details about lymphoproliferative disorders.

The term 'leukaemia' is used to describe the group of cancers characterized by an elevated number of malignant haematopoietic cells in the peripheral blood and/or bone marrow. A distinction between leukaemias and lymphomas is made, but sometimes precise categorization is difficult (see Box 8.1).

BOX 8.1 Leukaemia versus lymphoma

There are no precise definitions for the terms leukaemia and lymphoma and in certain contexts they can be interchanged, which can lead to confusion. Leukaemia generally describes malignancies derived from haematopoietic cells (either myeloid or lymphoid), which have sufficient involvement of the blood or bone marrow. The term lymphoma refers to lymphoid malignancies that present in secondary lymphoid tissues, such as the lymph nodes; however, most lymphomas can also involve the blood or bone marrow. Likewise, leukaemia (particularly the lymphoid leukaemias) can involve the lymph nodes. For example, diffuse large B-cell lymphoma usually presents in lymph nodes, but can involve the blood/bone marrow, whereas T-prolymphocytic leukaemia typically presents in the blood and bone marrow, but can involve lymph nodes. Chapter 9 examines the molecular aspect of lymphomas and how this is used in diagnostic pathology in more detail.

Historically, the first complete description of leukaemia is credited to the German pathologist Rudolf Virchow in 1845, who observed an abnormally large number of white blood cells (WBCs) in a patient and termed this *weisses blut*—white blood—a literal description of the vast excess of WBCs he observed down the microscope. Later he termed the disease 'leukämie' from the Greek words *leukos* meaning 'white' and *haima* meaning 'blood'. This seminal observation by Virchow provided a foundation for understanding leukaemia as a malignant proliferation (or neoplasia) of white cells in the blood, rather than a proliferation of blood cells secondary to infection as had been postulated by numerous physicians before him. Today, leukaemia is recognized as a complex disease with many different types that can be distinguished based upon morphological differences in maturation stages, lineage commitment, and the genetic aberrations contributing to leukaemogenesis, the combination of which predicts clinical behaviour.

As can be seen in Figure 8.1 the common morphological feature of all leukaemias is an elevated number of white blood cells in the peripheral blood and/or bone marrow. Clinically and pathologically, leukaemia can be subdivided into a number of broad groups. The first of these divisions is based on the lineage from which the malignant blood cell is derived, either myeloid or lymphoid, as illustrated in Figure 8.2.

(A)   (B)

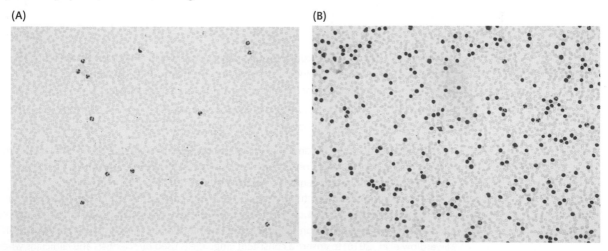

FIGURE 8.1

Normal and leukaemic blood films. The first indication that a patient has leukaemia is often made from the morphological assessment of a blood or bone marrow specimen. These peripheral blood smears show the number of haematopoietic cells seen in a healthy person (A) compared with a patient with chronic lymphocytic leukaemia (B). An increased number of lymphocytes in the blood is known as a lymphocytosis.

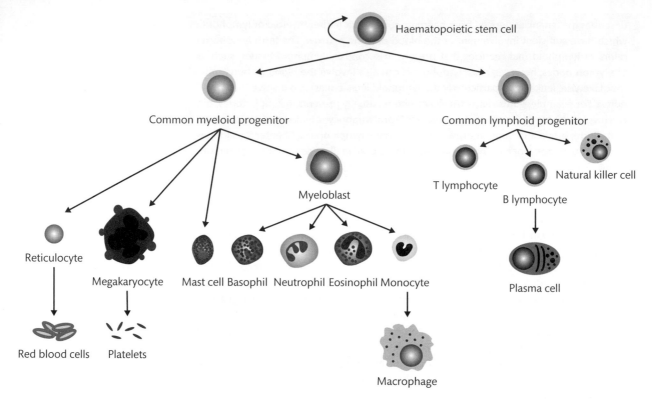

FIGURE 8.2

Haematopoietic cell development. Haematopoiesis begins with multipotent haematopoietic stem cells (HSCs), which reside in the bone marrow. HSCs can differentiate into lineage-specific stem cells (either myeloid or lymphoid), which will ultimately give rise to each of the different types of mature blood cells. The leukaemias can be broadly classified into those derived from cells of the myeloid or lymphoid lineage.

- *Myeloid leukaemia*: arises from myeloid progenitor cells, which under normal circumstances give rise to erythroid (red blood cells), granulocytic (neutrophils, eosinophils, and basophils), monocytic (monocytes, macrophages), mast cells, or megakaryocytic (platelets) lineages.

- *Lymphoid leukaemia*: arises from lymphoid progenitor cells, which under normal circumstances give rise to B lymphocytes, T lymphocytes, plasma cells, and natural killer cells.

While the categorization of some cases of leukaemia into myeloid or lymphoid lineages is not always clear cut, this distinction, nevertheless, serves as a useful framework for understanding the clinical behaviour and genomic changes in these diseases. The importance of the cell of origin as a starting point for tumour diagnosis is reflected in the *World Health Organization (WHO) Classification of Tumours of Haematopoietic and Lymphoid Tissues*. This is a unified and international classification scheme for all haematopoietic neoplasms (Arber et al., 2016; Swerdlow et al., 2016) to ensure a consistent language is used by scientists and clinicians when describing these malignancies. A number of WHO entities are defined, in part, by specific genetic abnormalities, including gene rearrangements and gene mutations, and so molecular haematology and cytogenetic laboratories play an important role in assigning the correct diagnosis to a patient's disease. Other features that are also important in the WHO classification scheme are morphology, immunophenotype, and clinical behaviour.

In addition to being distinguished by their lineage, the leukaemias can also be categorized based on their clinical aggressiveness into either:

- *Acute leukaemia*: these disease entities tend to be derived from more primitive haematopoietic precursors (e.g. lymphoblasts). Acute leukaemias are clinically aggressive and generally rapidly fatal without urgent treatment.
- *Chronic leukaemia*: these entities tend to be derived from the more mature counterparts (e.g. mature B lymphocytes). Chronic leukaemias develop more slowly and may not require treatment unless causing symptoms.

The clinical presentations that distinguish between acute and chronic leukaemia is provided in Box 8.2.

BOX 8.2 Clinical presentation and routine investigation of acute and chronic leukaemia

The clinical signs and symptoms of acute or chronic leukaemia are typically related to the underproduction of normal blood cells that is associated with bone marrow their replacement by the abnormal leukaemic cells. Therefore, patients may present with signs and symptoms related to:

- *Anaemia*: shortness of breath, fatigue, pallor, syncope (loss of consciousness caused by a fall in blood pressure).
- *Neutropenia*: life-threatening sepsis, atypical infections, recurrent infections.
- *Thrombocytopenia*: spontaneous bruising, bleeding (e.g. epistaxis, gastrointestinal bleeding).

In addition, patients may also have signs and symptoms related to leukaemic infiltration of various organs. Common symptomatic sites of infiltration include spleen (splenomegaly), skin (leukaemia cutis), and gum.

The typical diagnostic work-up for a patient with suspected acute or chronic leukaemia includes a bone marrow aspirate and trephine. The aspirate is reviewed for the morphological evidence of excessive primitive appearing haemopoietic cells (blasts) in the case of acute leukaemia or an excess of mature lymphoid or myeloid cells in the case of the chronic leukaemias. Immunophenotyping by flow cytometry is then performed (guided by the cytological features) in order to define the lineage, maturation, and immunophenotypic aberrancy of the abnormal cells further. Typically, morphological and immunophenotypic data is used to establish the initial diagnosis. Cytogenetic and molecular data provides further subclassification of disease type and prognostic information to guide treatment strategies.

Combining these two classifications provides a total of four main categories:

1. Acute myeloid leukaemia.
2. Chronic myeloid leukaemia.
3. Acute lymphoblastic leukaemia.
4. Chronic lymphoproliferative disorders.

Cancer Research UK statistics show that there were around 9500 new cases of leukaemia in the UK in 2014, with more than half of these cases diagnosed in people aged 70 and over. Chronic lymphocytic leukaemia and acute myeloid leukaemia accounted for the majority of diagnoses (around 3500 and 3000 cases, respectively). While this disease terminology appears to denote single disease entities (e.g. 'acute myeloid leukaemia'), there is marked clinical and genomic heterogeneity within each disease category. Characterization of this molecular heterogeneity in routine molecular diagnostic practice can have dramatic implications for diagnosis, treatment, and prognosis.

This chapter will discuss the genomic changes of acute and chronic leukaemias that are most relevant both to understanding the underlying pathobiology of these diseases and to impacting patient outcomes. The methods commonly used in the molecular diagnostic laboratory to identify these genomic lesions will also be introduced.

8.1 SELF-CHECK QUESTION

Outline the main subgroups of leukaemia with regard to cell of origin and clinical behaviour.

Key Point

The leukaemias are a group of blood cancers characterized by an elevated number of white blood cells in the blood and/or bone marrow. They are further subdivided, based on their cell of origin (myeloid or lymphoid) and clinical aggressiveness (acute or chronic). The WHO classification is an international classification scheme used by haematologists and scientists to describe these malignancies.

8.2 Acute myeloid leukaemia

Acute myeloid leukaemia (AML) is an aggressive haematological malignancy resulting from the clonal expansion of immature myeloid lineage cells. In most cases, a diagnosis of AML is made when myeloid lineage blasts comprise at least 20% of the nucleated cells in the peripheral blood or bone marrow. The notable exception to this rule is that when particular recurrent cytogenetic abnormalities are detected [e.g. t(8;21), inv(16), or t(15;17)], the diagnosis of AML can be made regardless of blast percentage. AML is a clinically, morphologically, and biologically heterogeneous disease. However, there is a well-established role for using genetic information to simplify some of this disease complexity and identify subgroups of patients with genomic changes relevant to clinical practice.

The diagnostic karyotype is the cornerstone of the diagnostic work-up and initial subcategorization of patients with AML. A commonly used hierarchical cytogenetic classification system was originally developed by analysing close to 6000 patients and was recently updated by the European LeukaemiaNet (http://www.leukemia-net.org/; Grimwade et al., 2010; Döhner et al., 2017). This classification system separates patients into three prognostic groups—favourable, intermediate, and adverse. Cytogenetic abnormalities characterizing these groups are listed in Table 8.1. The leukaemogenic mechanisms resulting from these translocations, copy number changes, or inversion events are considered the critical founding abnormality in the pathogenesis of a significant proportion of AML.

Beyond cytogenetic risk classification the presence of mutations in numerous genes, including *FLT3*, *NPM1*, *CEBPA*, *DNMT3A*, and *KIT*, have clinically significant effects on response to

TABLE 8.1 Cytogenetic and molecular abnormalities used in the risk-stratification of AML

Risk group	Cytogenetic/molecular abnormality
Favourable	t(15;17)(q24.1;q21.2) resulting in *PML-RARA* fusion
	t(8;21)(q22;q22.1) resulting in *RUNX1-RUNX1T1* fusion
	inv(16)(p13.1q22) and t(16;16)(p13.1;q22) resulting in *CBFB-MYH11* fusion
	Mutations in *CEBPA* (biallelic), *GATA2* or *NPM1*
Intermediate	Cytogenetic and molecular genetic abnormalities not classified as favourable or adverse including a normal karyotype
Adverse	inv(3)(q21.3q26.2) and t(3;3)(q21.3;q26.2) resulting in overexpression of *MECOM* and haploinsufficiency of *GATA2*
	add(5q), del(5q) or -5
	t(6;9)(p23;q34.1) resulting in *DEK-NUP214* fusion
	add(7q), del(7q) or −7
	t(v;11q23.3) *KMT2A* rearranged
	t(9;22)(q34.1;q11.2) resulting in *BCR-ABL1* fusion
	−17 or abn(17p)
	Complex or monosomal karyotype
	FLT3-ITD
	Mutations in *DNMT3A, TP53, RUNX1* or *ASXL1*

Key: abn, abnormal; add, addition; del, deletion; inv, inversion; t, translocation.

therapy and survival within specific cytogenetic subgroups. Detection of mutations in these genes does not supersede cytogenetic testing, but rather helps to clarify the heterogeneity existing between patients with the same primary chromosomal abnormality that are unexplained by the pattern of additional cytogenetic changes.

A key decision in the treatment pathway for patients with AML is whether to perform an allogeneic stem cell transplant (alloSCT) in first remission (i.e. after induction/consolidation therapy). Due to the significant risk of morbidity/mortality from the transplant itself only patients with a high risk of relapse are considered for this treatment. Therefore, the most important clinical implication of cytogenetic/molecular risk stratification is to predict patients who will benefit from transplant (i.e. those at high risk) from those who will not (i.e. those with favourable risk).

The next section outlines the main subtypes of AML, and describes the cytogenetic and molecular basis of these entities.

8.2 SELF-CHECK QUESTION

Describe the role of karyotype in the prognostication of AML.

8.2.1 Favourable-risk AML

8.2.1.1 *Acute promyelocytic leukaemia*

Acute promyelocytic leukaemia (APML) is a rare subtype of AML (comprising <10% of diagnoses), which characteristically presents with severe and potentially life-threatening coagulopathy (impairment of blood clotting). Morphologically, the disease is characterized by the proliferation of abnormal promyelocytes in the bone marrow and peripheral blood. Left untreated, patients with APML have a median survival of less than 1 month.

More than 95% of APMLs are characterized by a reciprocal translocation involving the downstream sequences of the *retinoic acid receptor alpha* (*RARA*) gene on chromosome 17q21.2 and the promoter and upstream sequences of the *promyelocytic leukaemia* (*PML*) gene on chromosome 15q24.1. Both genes play an important role in normal haematopoiesis, with *PML* involved in multiple tumour suppressive processes and *RARA* functioning as a transcription factor that functions as a regulator of granulocytic maturation beyond the promyelocyte stage. The t(15;17) PML–RARA fusion protein observed in APML impairs both the pro-apoptotic and growth suppressive activity of PML, while also gaining an enhanced ability to inhibit retinoic acid responsive genes, blocking terminal differentiation of promyelocytes in the bone marrow.

Conventional metaphase cytogenetics/FISH can detect the t(15;17) *PML–RARA* translocation and is the most important initial test to definitively establish the diagnosis. Molecular testing, such as quantitative real time (RT)-PCR, can be used to confirm and quantify mRNA produced from the fusion gene and is also a more sensitive method for the identification of rare complex or cryptic *PML–RARA* rearrangements (accounting for approximately 5% of *PML–RARA* positive cases). As illustrated in Figure 8.3, the breakpoint is located in intron 2 of *RARA*, while there are three breakpoint cluster regions localized to intron 6 (bcr1), exon 6 (bcr2), and intron 3 (bcr3) of *PML* producing long, variant, and short isoforms of the resulting fusion.

t(15;17) positive APML (classic, complex, and cryptic rearrangements) predicts a favourable response to the retinoid-differentiating agent ATRA (all-trans retinoic acid). Treatment with ATRA overcomes the differentiation arrest in leukaemic promyelocytes and results in rapid promyelocyte maturation into granulocytes. The combination of ATRA with arsenic trioxide (ATO), which accelerates the degradation of *PML–RARA*, is associated with disease cure in over 90% of patients. The prognosis for treated patients is better than that for any other subtype of AML.

A small number of patients with the clinical and morphological presentation of APML do not have an identifiable t(15;17). In these cases, rare variant translocations involving *RARA* (but not *PML*) have been identified. The most common of these include:

- t(11;17) resulting in a *ZBTB16-RARA* fusion.
- t(5;17) resulting in a *NPM1-RARA* fusion.
- t(11;17) resulting in a *NUMA1-RARA* fusion.
- t(17;17) resulting in a *STAT5B-RARA* fusion.
- t(17;17) resulting in a *PRKAR1A-RARA* fusion.
- t(4;17) resulting in a *FIP1L1-RARA* fusion.
- t(X;17) resulting in a *BCOR-RARA* fusion.

Identifying these alternative translocations has important therapeutic implications as different *RARA* fusion partners alter the sensitivity of leukaemic promyelocytes to ATRA therapy. For example, fusions involving *ZBTB16* and *STAT5B* are insensitive to ATRA-ATO therapy, whereas cells containing *NPM1–RARA* and *NUMA1–RARA* fusions are sensitive.

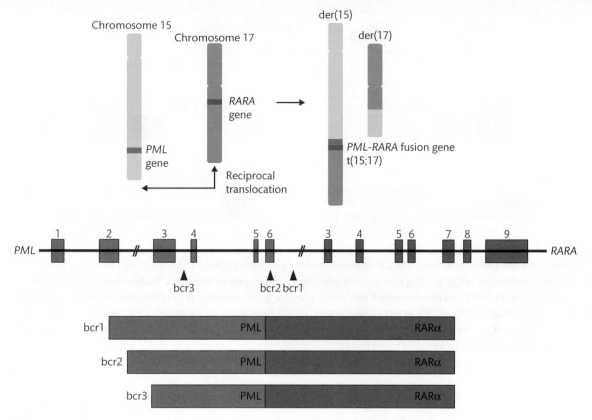

FIGURE 8.3

Generation of the PML-RARA fusion protein by t(15;17) in APML. A balanced translocation between the long arms of chromosomes 15 and 17 generates a fusion between *PML* and *RARA* retaining functional domains of both genes. Depending on the location of the breakpoint within *PML* three main isoforms are generated including long (bcr1), variant (bcr2), and short (bcr3). Various methods are used to detect the fusion protein, including cytogenetics, FISH, and RT-PCR. Key: der, derivative chromosome.

8.2.1.2 Core-binding factor acute myeloid leukaemia

There are two subtypes of core-binding factor leukaemia (CBF-AML), so named because the cytogenetic abnormalities detected in these diseases involve the core-binding factor complex genes *RUNX1* and *CBFB*. Working through Figure 8.4 it can be seen that the CBF complex is a heterodimeric transcription factor comprised of RUNX1 (CBFα) and CBFB (CBFβ), which allows for active transcription of genes necessary for haematopoiesis upon binding to CBF enhancer sites upstream of target genes. The *RUNX1–RUNX1T1* fusion results from a translocation involving chromosomes 8 and 21 (t(8;21)), while pericentric inversion inv(16) or, less frequently, translocation of two chromosome 16 homologues (t(16;16)), give rise to the *CBFB–MYH11* fusion gene. Figure 8.4 shows that the gene fusion inhibits the formation of a stable DNA binding complex, preventing the expression of CBF target genes and a block in myeloid differentiation.

The t(8;21) is typically detected by G-banded karyotyping and FISH at diagnosis using locus-specific probes for the *RUNX1–RUNX1T1* fusion when the leukaemic burden in the specimen is high. After treatment RT-PCR can be used as a more sensitive method to detect and confirm the presence of the fusion gene. While G-banded karyotyping is also used to detect the *CBFB–MYH11* fusion, the inv(16) and t(16;16) can be difficult to identify when chromosome

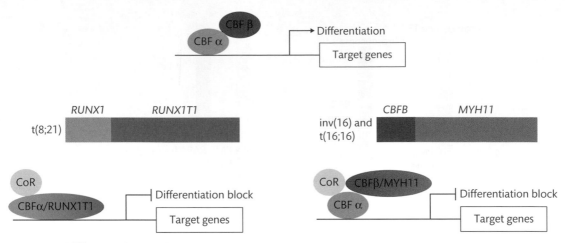

FIGURE 8.4

Disruption of the CBF pathway in AML. The CBF complex consisting of *RUNX1* (CBFα) and *CBFB* (CBFβ) facilitates gene transcription upon binding to CBF binding sites and recruiting other transcriptional activators. Translocation of *RUNX1* on chromosome 21 and *RUNX1T1* on chromosome 8 produces a chimeric gene consisting of the DNA binding domain of *RUNX1* fused to the full length of *RUNX1T1*. The *CBFB-MYH11* fusion protein results from either pericentric inversion or translocation. Disruption of *RUNX1* by t(8;21) or *CBFB* by inv(16) or t(16;16) recruits transcriptional repressors (CoR) to CBF binding sites, converting the CBF complex from an activator to a repressor of transcription.

morphology is suboptimal and either FISH or RT-PCR may be required to detect the fusion transcript.

In CBF-AML concomitantly mutated *KIT* (a member of the type III receptor tyrosine kinase family encoding for the proto-oncogene c-KIT) has been associated with a higher incidence of relapse and worse outcome. In normal cells c-KIT sends proliferation cues to intracellular messengers following stimulation by its ligand stem cell factor. Gain-of-function mutations in *KIT* result in proliferation signals without the need for ligand binding and an overproduction of white blood cells. As seen in Figure 8.5, *KIT* mutations cluster most frequently within exon 17, which encodes the activation loop of the second kinase domain, or in exon 8, which encodes the extracellular domain involved in receptor dimerization. These mutations occur either by point mutation at Asp816 or Asn822 (exon 17), or by small deletions (often with subsequent insertion) involving codon Asp419 causing its loss or replacement (exon 8). *KIT* mutations are found in 20-45% of CBF-AML.

In both APML and CBF-AML the presence of a fusion transcript provides a molecular target for monitoring patients during treatment. Quantification of the relative number of *PML–RARA*, *CBFB–MYH11* or *RUNX1–RUNX1T1* transcripts can be used to pre-empt disease relapse in patients that are not yet symptomatic. The methodology used is discussed later in the section on monitoring *BCR-ABL1* in chronic myeloid leukaemia (Section 8.3.2').

8.3 SELF-CHECK QUESTION

What are the types of AML associated with a favourable risk, and what cytogenetic and molecular abnormalities characterize these entities?

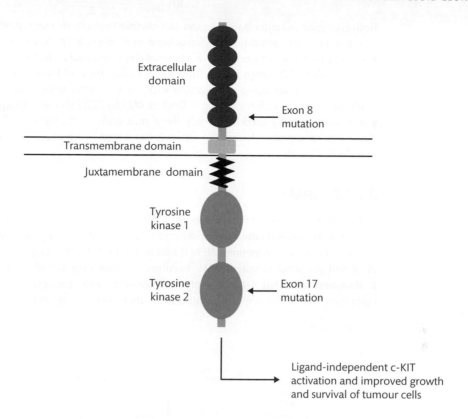

FIGURE 8.5

KIT **mutations in CBF-AML.** Activating *KIT* mutations cluster within exon 8 encoding for the extracellular domain and exon 17 located within the second tyrosine kinase domain. These mutations result in uncontrolled cell division and are associated with a poor prognosis in CBF-AML.

8.2.2 Intermediate-risk acute myeloid leukaemia

Approximately half of patients presenting with AML have no clonal chromosomal aberrations at diagnosis detectable by conventional cytogenetic analysis. These cases, termed cytogenetically normal AML or normal karyotype AML, comprise a heterogeneous group with an intermediate risk of progression and relapse. Acquired gene mutations can be used to further stratify these patients. These genes comprise either those that activate signal transduction pathways and increase proliferation and/or survival of haematopoietic progenitor cells (such as *FLT3*), or genes involved in gene transcription (transcription factors or components of the transcriptional co-activation complex) that cause impaired differentiation (such as *CEBPA* and *NPM1*).

8.2.2.1 FLT3

Mutation of the receptor tyrosine kinase *fms-like tyrosine kinase 3* (*FLT3*) are either in-frame duplications (commonly referred to as internal tandem duplications, *FLT3*-ITD) involving the juxtamembrane and first tyrosine kinase domains encoded by exons 14 and 15, or, less frequently, point mutations within the second tyrosine kinase domain (*FLT3*-TKD). While both classes of mutation lead to ligand-independent constitutive activation of *FLT3* signalling, only *FLT3*-ITDs are definitively associated with poorer prognosis, particularly when the allelic ratio (i.e. the number of ITD-mutated alleles compared with the number of wildtype alleles) is high. ITDs range in size from a few to hundreds of base pairs. Moreover, only about two-thirds are actual duplications with the remainder being insertions or complex duplications and insertions. Testing of DNA or RNA by PCR followed by fragment size analysis is most typically used to identify these mutations as illustrated in Figure 8.6, while direct sequencing or allele-specific methods can detect TKD mutations at hotspot residues Asp835 and Ile836.

8.2.2.2 NPM1

In the absence of a *FLT3*-ITD, mutations in *NPM1* or *CEBPA* are associated with complete remission and survival rates not dissimilar to that of patients with cytogenetically favourable risk leukaemia. *Nucleophosmin* (*NPM1*) mutations are the most frequent molecular lesion identified in normal karyotype AML, occurring in more than half of cases. *NPM1* encodes a phosphoprotein that helps to shuttle ribosomal proteins between the nucleus and the cytoplasm, and also has roles in centrosome duplication during mitosis, cell proliferation,

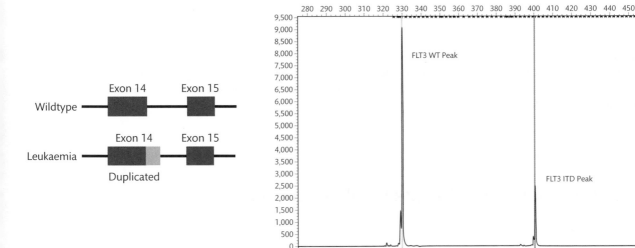

FIGURE 8.6

***FLT3* cytokine receptor mutations.** Internal tandem duplications (ITDs) are the most common type of activating mutations in the *FLT3* cytokine receptor. These in-frame insertions occur in exons 14 and 15 encoding for the juxtamembrane domain and tyrosine kinase domain 1. ITDs in the juxtamembrane domain destabilize its negative regulatory function, thus promoting constitutive signalling to downstream effectors including the MAPK, STAT5 and PI3K pathways critical for cell survival, proliferation, and differentiation. The oncogenic mechanism of kinase domain ITDs is less well understood, but may result in a different conformational change. Because *FLT3*-ITDs are heterogeneous in their size, location, and allelic ratio, fragment size analysis remains the most reliable method for their detection. In this assay genomic DNA is amplified using fluorescently labelled primers flanking exons 14 and 15, and the fragments are separated by capillary gel electrophoresis and analysed using fragment analysis software. In the example shown, a distinct secondary peak is detected in addition to the wildtype peak indicating the presence of an in-frame insertion.

and apoptosis. The most commonly observed *NPM1* mutation is a duplication of four base pairs (c.860_863dupTCTG), however more than 50 different frameshift mutations have been described and cluster within the C-terminal portion of the protein. Figure 8.7 shows that these mutations lead to loss of important tryptophan residues (Trp288 and Trp290 in particular) that remove a nucleolar localization signal and generate an additional nuclear export signal motif which results in NPM1 sequestration in the cytoplasm and prevents its function as a transport protein.

NPM1 mutations are useful to follow as a **minimal residual disease marker** in AML as this mutation is acquired relatively early in AML pathogenesis and, therefore, tends to be present in the entire leukaemic compartment. Moreover, *NPM1* mutations tend to be stable at relapse.

8.2.2.3 *CEBPA* and *GATA2*

Mutations in the myeloid transcription factor *CEBPA* (*CCAAT/enhancer binding protein* α) mainly occur in normal karyotype AML with an incidence of approximately 10%, but have also been associated with other specific chromosomal lesions such as del9q. The typical pattern observed is cooperating *CEBPA* mutations located in the N- and C-terminal regions and this is illustrated in Figure 8.8. N-terminal mutations, which are typically frameshift insertions or deletions, lead to loss of the full-length p42 isoform, while maintaining translation of the short p30 isoform that acts as a dominant negative protein. In-frame C-terminal mutations occur within the basic leucine zipper (bZIP) domain and disrupt dimerization and DNA-binding activities

> **Minimal residual disease**
> Term used to describe the situation when small numbers of malignant cells remain in the patient during treatment or after treatment when the patient is in remission. MRD is the major cause of relapse in cancers. The demonstration of MRD can distinguish between patients who need intensive and potentially more toxic therapy from those who do not.

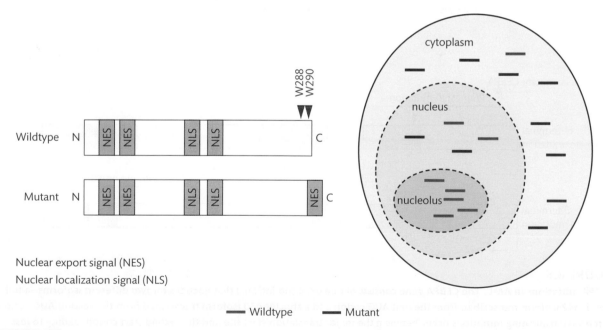

Nuclear export signal (NES)
Nuclear localization signal (NLS)

── Wildtype   ── Mutant

FIGURE 8.7

Nucleophosmin mutations in AML. *Nucleophosmin* mutations in AML are typically frameshift occurring almost exclusively in the C-terminal region. The aberrant cytoplasmic accumulation of mutant NPM1 results from (i) loss of tryptophan residues normally involved in the nucleolar localization of the protein, and (ii) the generation of a new nuclear export signal motif.

of *CEBPA*. Frequently, biallelic *CEBPA* mutations are detected (double mutant, *CEBPA*^{DM}) and, in most cases, consist of a combined N-terminal frameshift of one allele and a C-terminal in-frame mutation of the other. Alternative combinations of *CEBPA*^{DM} are noted including non-sense and missense mutations (atypical *CEBPA*^{DM}), as well as mutations occurring between the N-terminus and bZIP domain. Wildtype CEBPα protein expression is absent in the context of *CEBPA*^{DM}, in contrast to tumours with a single mutated *CEBPA* copy (*CEBPA*^{SM}), which retain expression of the wildtype allele. Importantly, favourable outcome is restricted to AML with *CEBPA*^{DM}. *CEBPA*^{DM} are a transcriptionally homogeneous group with a gene expression profile that is distinct from that observed in AML with *CEBPA*^{SM} and correlates with sensitivity to JAK-STAT pathway inhibition *in vitro*.

Reinforcing their distinction as separate entities, concurrent mutation of commonly mutated AML genes is frequently observed in *CEBPA*^{SM} (notably mutant *NPM1* and *FLT3*-ITD), whereas the mutational spectrum seen in CEBPA^{DM} is relatively unique and involves mutations of *GATA2* and *CSF3R* in 20–40% of cases. *GATA2* is a zinc finger transcription factor, important in haematopoietic stem cell proliferation and normal megakaryocyte development. Heterozygous missense mutations cluster in the zinc finger domains of *GATA2*, which is the protein region that contributes to the stability and specificity of DNA binding, as well as interactions with transcription cofactors. The relationship between co-occurring

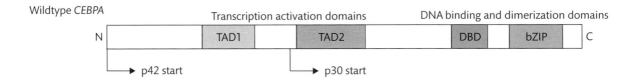

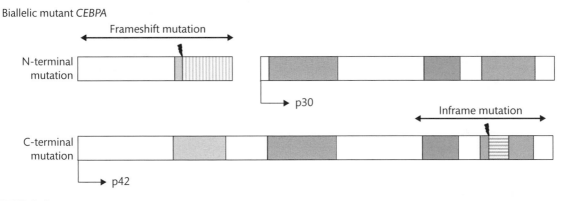

FIGURE 8.8

CEBPA mutations in AML. The *CEBPA* gene consists of one exon (no introns) that encodes for two different isoforms—a full-length p42 isoform transcribed from the first AUG codon and a shorter p30 isoform transcribed from the second AUG codon. N-terminal truncating mutations occur between the major translation start site and the second start codon leading to loss of the full-length isoform, but expression of a dominant negative p30 isoform. C-terminal in-frame mutations disrupt the dimerization activity of *CEBPA* and, consequently, DNA binding. Most *CEBPA*-mutated AML carry two mutations on separate alleles with a specific combination of an N-terminal frameshift mutation on one allele and a C-terminal in-frame mutation on the other.

$CEBPA^{DM}$ and *GATA2* is unclear. Mutations in *CSF3R* (encoding the granulocyte colony-stimulating factor (G-CSF) receptor) cluster at threonine 618, which occurs immediately proximal to the transmembrane domain. The recurrent activating mutation Thr618Ile renders the receptor ligand-independent resulting in constitutive JAK-STAT pathway signalling. Importantly, the Thr618Ile mutation predicts sensitivity *in vitro* to inhibition of the JAK-STAT pathway downstream of the receptor providing the rationale for exploring tyrosine kinase inhibitors that target JAK proteins as a potential therapeutic avenue in these patients.

Germline *CEBPA* mutations are associated with predisposition to AML, with a second mutation (often in the C-terminal) acquired at presentation. Likewise, germline *GATA2* mutations also increase susceptibility to AML. It is, therefore, important to determine the somatic or germline nature of *CEBPA* and *GATA2* mutations in AML, as this information has important implications for family members.

8.2.2.4 DNMT3A

The tumour suppressor gene *DNMT3A*, a DNA methyltransferase, is mutated in ~30% of normal karyotype AML. These mutations are typically heterozygous and fall within the catalytic domain at the arginine amino acid at position 882 (Arg882). Arg882 mutations are thought to act in a dominant negative manner; the mutant protein restricts the formation of active tetramers that are usually comprised of wildtype protein, which leads to reduced enzyme activity and focal hypomethylation. In contrast, other loss-of-function mutations may not fully inactivate the protein in isolation and are, therefore, often biallelic. However, these are rare and how they contribute to loss of tumour suppressive activity is less well understood.

In AML mutations in *DNMT3A* frequently co-segregate with mutations in *NPM1* and *FLT3*-ITDs, and are associated with a poor prognosis. *DNMT3A* mutations occur early in leukaemogenesis and contribute to a pre-leukaemic state that predisposes to (but is insufficient for) the development of overt haematological malignancy. The poor prognosis associated with *DNMT3A* may be attributed to a high relapse rate as non-leukaemic *DNMT3A* mutant progenitors are difficult to eradicate with chemotherapy and thus persist to reinitiate disease if a new oncogenic mutation is acquired.

8.2.2.5 RUNX1

RUNX1 is a transcription factor essential for normal haematopoiesis. In addition to its involvement in recurrent balanced rearrangements such as t(8;21) *RUNX1-RUNX1T1* and t(3;21) *RUNX1-MECOM*, somatic mutations in *RUNX1* are observed in approximately 10% of patients with AML where they are associated with a particularly poor prognosis. *RUNX1* missense mutations typically occur in the Runt domain and tend to be monoallelic, whereas nonsense or frameshift mutations often result in loss of all or part of the transactivation domain. Functionally, both types of mutations alter amino acids essential for DNA binding and heterodimerization. Mutations in *RUNX1* infrequently co-occur with *NPM1* mutations, but are associated with *MLL*-PTD (partial tandem duplication). They also display a unique pattern of gene expression that is shared by other adverse-risk AML, such as those with del(7q)/monosomy 7, inv(3)/t(3;3), and a complex karyotype.

It is important to confirm the somatic nature of *RUNX1* mutations in AML as germline mutations in this gene predispose to familial platelet disorder, a rare autosomal dominant disease characterized clinically by thrombocytopenia, and functional platelet defects causing easy bleeding and bruising. These patients also have a propensity to develop myeloid malignancies. Identifying AML patients with inherited *RUNX1* mutations is important to avoid alloSCT from related donors who also carry the familial mutation. Active monitoring of at-risk family members is also recommended.

8.2.2.6 IDH1 and IDH2

Isocitrate dehydrogenase 1 and 2, encoded for by the *IDH1* and *IDH2* genes, are enzymes involved in glucose metabolism. Working through Figure 8.9 it can be seen that wildtype *IDH1* and *IDH2* catalyse the production of α-ketoglutarate (α-KG) from isocitrate, whereas the mutant forms gain neomorphic activity and produce the oncometabolite 2-hydroxyglutarate (2-HG). 2-HG inhibits numerous α-KG-dependent histone demethylases and DNA hydroxylases resulting in widespread epigenetic changes and a block in cellular differentiation. In AML heterozygous substitution mutations in the *IDH* genes occur at critical arginine residues including codons 132 (*IDH1*) and 140 and 172 (*IDH2*), and occur in up to 20% of patients. Although these mutations are not themselves prognostic in AML, selective small molecule inhibitors of mutant IDH induce cell differentiation in *in vitro* and *in vivo* models, and are currently being evaluated in clinical trials of advanced haematological malignancies including AML. Identifying these mutations, therefore, has significant therapeutic relevance.

8.4 SELF-CHECK QUESTION

Imagine you are a haematologist and receive molecular testing results for a patient with AML that demonstrate a *FLT3*-ITD, and mutations in *NPM1* and *DNMT3A*. What is the impact of these results on prognosis and treatment?

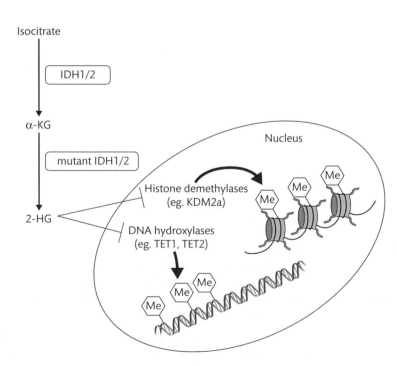

FIGURE 8.9

Effects of mutations in *IDH1* and *IDH2*.
Mutations in *IDH1* and *IDH2* result in simultaneous loss of their normal catalytic activity (the production of α-ketoglutarate (α-KG) from isocitrate) and gain of a new function (the production of 2-hydroxyglutarate, 2-HG). 2-HG is an α-KG antagonist and competitively inhibits multiple α-KG dependent histone demethylases and DNA hydroxylases. This alters epigenetic control of stem and progenitor cell differentiation, contributing to leukaemogenesis in AML.

8.2.3 Adverse-risk acute myeloid leukaemia

Approximately one-third of AML patients present with adverse-risk cytogenetics at diagnosis. AML in this group likely represents many different rare biological syndromes, with a number of specific cytogenetic abnormalities comprising this subgroup. For example, t(6;9) that results in the chimeric fusion gene between *DEK* (6p23) and *NUP214* (9q34.1), and inv(3) and t(3;3) involving *MECOM* (3q26.2) and *GATA2* (3q21.3) are each seen in about 1% of AML. In this setting, mutation profiling does not offer additional diagnostic or prognostic information for patients with adverse-risk cytogenetics. Rather, mutational profiling may identify specific genetic lesions associated with targeted therapies that may serve as an alternative treatment approach in these patients. A good example of this would be *IDH1* and *IDH2* mutations, which may be targeted by small molecule inhibitors as noted previously.

One-third to one-half of all AML classified as adverse-risk disease have multiple cytogenetic abnormalities seen in a single karyotype. If the number of cytogenetic abnormalities is three or more, such cases are defined as having a complex karyotype. Instead of balanced translocations, complex karyotypes are characterized by loss or gain of chromosomal material, predominantly loss of chromosomes 5, 7, and 17, and gain of chromosomes 1, 8, 11, and 21. The term monosomal karyotype is used to define the 60% of complex karyotypes that have two or more autosomal monosomies or one autosomal monosomy plus at least one additional structural abnormality. Monosomal karyotype, which commonly involves chromosomes 5 and 7, predicts for a particularly poor prognosis with the lowest proportion of AML patients attaining a complete remission and the highest rate of relapse. Deletions or mutations at 17p, involving the well-known tumour suppressor gene *TP53*, which regulates cell cycle in response to cellular stress, are closely related with complex and monosomal karyotypes. *TP53* abnormalities also commonly co-occur with loss of part or all of chromosomes 5 or 7.

Myeloid lineage malignancies involving aberration of 3q appear to define a specific molecular entity of adverse-risk disease, which is among the most refractory to therapy. This group includes karyotype abnormalities affecting the *MECOM* locus on chromosome 3q, such as inv(3)(q21.3q26.2) and t(3;3)(q21.3;q26.2), seen in AML, as well as myelodysplastic syndrome and chronic myeloid leukaemia. Consequent to chromosome 3 rearrangements a distal *GATA2* enhancer is repositioned to the *MECOM* locus with dual pathogenic consequences:

- oncogenic activation of the transcription factor *MECOM*;
- monoallelic expression of *GATA2*.

Recent high-throughput sequencing has also identified secondary genetic events that contribute to leukaemic transformation (Groschel et al., 2015). Mutations within the zinc finger domains of the non-rearranged *GATA2* allele are common and are often present concurrently with a mutation in the splicing factor *SF3B1*. Also, the majority of AML with aberrant *MECOM* contain changes predicted to activate receptor-tyrosine kinase activity, through *NRAS* and *KRAS* hotspot mutation or loss of *NF1*. In addition to their common genetic background, all 3q aberrant myeloid malignancies display a similar gene expression profile suggestive of shared biology and classification as a single disease entity.

8.5 SELF-CHECK QUESTION

Define a monosomal karyotype. How does this affect prognosis in AML?

> ## Key Point
>
> The most important initial distinction in acute myeloid leukaemia is the identification/exclusion of acute promyelocytic leukaemia. After this, in the upfront setting, molecular and cytogenetic testing are used to determine which patients require an allogeneic transplant in first remission.

8.3 Chronic myeloid leukaemia

Chronic myeloid leukaemia (CML) is a stem cell derived myeloproliferative neoplasm characterized by the expansion of myeloid progenitors and accumulation of mature effector cells in the blood and bone marrow. Unlike AML, CML is not associated with maturation arrest, but rather a failure of apoptosis, which contributes to the increased white cell count observed in these patients.

From a historical perspective CML is an important entity as it was the first cancer in which a hallmark chromosome abnormality, the Philadelphia chromosome (Ph), was identified and a pathological correlation suggested. This discovery by Peter Nowell and David Hungerford in 1960 (Nowell and Hungerford, 1960) was the foundation for further seminal work culminating in the identification of the reciprocal t(9;22) translocation by Janet Rowley in 1973 (Rowley, 1973). It is now known that the t(9;22)(q34.1;q11.2) juxtaposes the *ABL1* proto-oncogene (a non-receptor tyrosine kinase) on chromosome 9 and the *breakpoint cluster region* (*BCR*) gene on chromosome 22, giving rise to the pathogenic *BCR-ABL1* fusion transcript on chromosome 22 along with a non-functional *ABL1-BCR* on chromosome 9 as illustrated in Figure 8.10.

The translated BCR-ABL1 protein contains many important functional domains of its parent proteins, the most important functional consequence of which results from the collocation of the BCR dimerization domain and the ABL1 tyrosine kinase domain. The constitutive tyrosine kinase activity of the BCR-ABL1 oncoprotein provides a positive signal to the RAS/MAPK, PI3K, and JAK-STAT pathways leading to increased proliferation and reduced response to apoptotic stimuli.

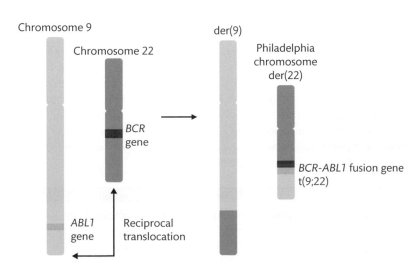

FIGURE 8.10

Generation of the Ph chromosome.
Translocation of *BCR* on chromosome 22 and *ABL1* on chromosome 9 results in a fusion gene, which is transcribed and translated into the BCR-ABL1 chimeric protein. Key: der, derivative chromosome.

The natural history of untreated CML is characterized by a tri-phasic clinical course—chronic phase, accelerated phase, and blast crisis. Most patients present in chronic phase after experiencing mild non-specific symptoms or being diagnosed incidentally after a blood test performed for other reasons. Chronic phase is associated with a hypercellular bone marrow and peripheral leukocytosis, and the patient will often have an enlarged spleen.

In the absence of treatment CML progresses through an accelerated phase characterized by increasing blast count to acute leukaemia or 'blast crisis'. While the majority of blast crisis manifest as acute myeloid leukaemia, in approximately 20–30% of cases the immunophenotype of the blasts are lymphoid. This is termed a lymphoid blast crisis and is a manifestation of the BCR-ABL1 fusion being present in a stem cell which precedes lymphoid/myeloid differentiation (see Figure 8.2). Blast crisis CML is very difficult to treat being highly refractory to chemotherapy with a poor overall survival. The accelerated phase may bridge chronic phase and blast crisis, however not all patients follow this linear progression and blast crisis can develop abruptly from chronic phase or be present at diagnosis.

In addition to being the first malignancy to be associated with a particular chromosomal abnormality, CML is also notable as the first malignancy to show remarkable success with a molecular targeted therapy—imatinib. This first-generation tyrosine kinase inhibitor (TKI) has an inhibitory effect by binding to the ATP-binding site of the hyperactive chimeric protein and preventing downstream signalling that gives rise to the leukaemic transformation in CML. Second-generation tyrosine kinase inhibitors may be useful in the upfront setting and in the context of acquired resistance as discussed in Section 8.3.3.

8.6 SELF-CHECK QUESTION

What is the fundamental molecular abnormality in CML?

8.3.1 Methods for detecting the t(9;22) at diagnosis

Diagnostic assays for suspected cases of CML are based on the standard tests of G-banded karyotyping, FISH, and RT-PCR, all of which aim to specifically detect the Ph chromosome or BCR-ABL1 fusion abnormality. Karyotyping will detect 95% of patients with the Ph chromosome. Other structural abnormalities are seen in around 10% of patients, frequently Ph chromosome duplication, isochromosome 17q and trisomy 8 and 19, and may be associated with significantly inferior response to treatment or disease evolution to acute leukaemia.

The point at which the BCR and ABL1 genes fuse is heterogeneous and cannot be distinguished by cytogenetic analysis, but rather is best detected with molecular methods. In the majority of hybrid transcripts, the breakpoint in ABL1 arises in the intron between exon 1b and exon 2 (a2) resulting in the body of ABL1 being juxtaposed to variable portions of BCR. The most common breakpoints in BCR result in either BCR exon 13 (e13, also called b2) or BCR exon 14 (e14, also called b3) fused to ABL1 exon 2. These fusions result in a 210-kDa protein (p210) that is detected in the vast majority (99%) of patients with CML. Alternative rarer BCR breakpoints involve BCR exon 1 (e1) resulting in a smaller 190-kDa (p190) protein that has greater transforming potential or BCR exon 19 (e19, previously called c3) resulting in a larger 230-kDa protein (p230). Figure 8.11 illustrates the differences between the various BCR-ABL1 fusion transcripts.

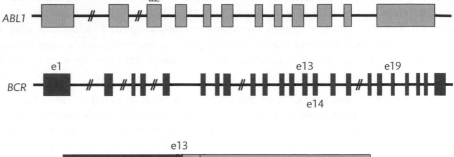

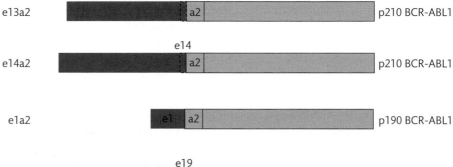

FIGURE 8.11

Generation of BCR-ABL1 fusion proteins. Different breakpoints in the *BCR* gene generate different sized BCR-ABL1 fusion proteins. Breakpoints within intron 2 of *ABL1* and introns 13 and 14 of *BCR* generate the e13a2 and e14a2 variants, which are detected in the majority of patients. Less common breakpoints can generate e1a2 and e19a2 variants, which produce fusion proteins of a different size.

In a small proportion of cases (5%), the Ph chromosome is cytogenetically cryptic. In these cases, a diagnosis can be made by FISH or PCR to identify the *BCR-ABL1* fusion gene or mRNA transcript, respectively. These variant Ph rearrangements appear to respond equally well to imatinib.

8.7 SELF-CHECK QUESTION

List the common breakpoints that generate a *BCR-ABL1* fusion. What are the different methods used to detect the Ph chromosome, and what are the relative advantages and disadvantages of each?

8.3.2 Monitoring response to treatment

A reduction in the number of leukaemic cells in CML is associated with improved survival. Therefore, monitoring treatment response is essential to identify patients at high risk of treatment failure and disease progression that may benefit from a change in therapy. Moreover, patients that have undetectable disease by sensitive molecular methods for sufficient duration may be able to stop tyrosine kinase inhibitor (TKI) therapy without relapsing. Monitoring in CML takes three forms—haematological, cytogenetic, and molecular, which are outlined in Table 8.2. Given that conventional chromosome analysis is time and labour intensive, and up to 5% of specimens are unsuitable, combined with the superior detection limit of molecular methods, RT-PCR is now preferentially used to monitor for the emergence of residual leukaemic cells. However, cytogenetic/FISH analysis is still used in the early stages of treatment with multiple well-established treatment milestones based on these analyses. The European LeukaemiaNet (http://www.leukemia-net.org/) has developed definitions of treatment failure, optimal response, and warning signs following the use of a TKI as first-line therapy (outlined in Table 8.3). Patients who have

TABLE 8.2 Different classifications for monitoring response in CML

Response by type	Definition
Haematologic response	
Complete (CHR)	Leukocyte count $<10 \times 10^9$/L, platelet count $<450 \times 10^9$/L, normal differential, no splenomegaly
Cytogenetic response	
No response	>95% Ph+ metaphases in bone marrow
Minor (Minor CyR)	36–95% Ph+ metaphases in bone marrow
Major (Major CyR)	1–35% Ph+ metaphases in bone marrow
Complete (CCyR)	0% Ph+ metaphases in bone marrow
Molecular response	
Major (MMR)	3-log reduction of *BCR-ABL1* on the IS ($\leq$0.1%)
Deep (MR^4, MR$^{4.5}$, MR^5)	Undetectable *BCR-ABL1* with an assay sensitivity $\geq$4, 4.5, or 5.0 logs

Key: IS, International Scale.

large reductions in leukaemic cell burden within the first year of treatment have excellent long-term outcome with few relapses.

When monitoring CML with molecular methods, an RNA template extracted from the peripheral blood or bone marrow is amplified using reverse transcriptase enzyme and random oligonucleotide primers to form cDNA. This cDNA is then used in RT-PCR to quantitate both the *BCR-ABL1* transcript and a control gene (typically either *ABL1* or *BCR*), with the result expressed as a ratio of *BCR-ABL1* to control. Molecular monitoring is generally performed every 3 months.

TABLE 8.3 Definitions of response to TKI as first-line treatment in CML

	Optimal	Warning	Failure
3 months	*BCR-ABL1* IS $\leq$10% and/or Ph+ $\leq$35%	*BCR-ABL1* IS >10% and/or Ph+ 36–95%	Non-CHR and/or Ph+ >95%
6 months	*BCR-ABL1* IS <1% and/or Ph+ 0%	*BCR-ABL1* IS 1–10% and/or Ph+ 1–35%	*BCR-ABL1* IS >10% and/or Ph+ >35%
12 months	*BCR-ABL1* IS $\leq$0.1%	*BCR-ABL1* IS >0.1–1%	*BCR-ABL1* IS >1% and/or Ph+ >0%
Then, at any time	*BCR-ABL1* IS $\leq$0.1%		Loss of CHR Loss of CCyR Confirmed loss of MMR New Ph+ clonal cytogenetic abnormalities *ABL1* TKD mutation

Key: IS, International Scale.

As quantitative assays can vary in their sensitivity when performed by different laboratories, attempts have been made to standardize *BCR-ABL1* RT-PCR, including recommendations for specimen handling, RNA preparation and storage, cDNA synthesis, and PCR protocol. The establishment of an International Scale (IS), whereby parallel testing of samples with a reference laboratory is used to produce a laboratory-specific conversion factor that corrects for differing sensitivities, has further improved the reproducibility and comparability of quantitative results (Hughes et al., 2006). A more recent approach for standardization is the development of reference calibration reagents, which allow each laboratory to determine their calibration factor internally (White et al., 2010; Cross et al., 2016). Given the technical variability of this assay, changes in a single molecular finding alone are not used to define treatment failure, but rather increasing PCR signal over time.

8.8 SELF-CHECK QUESTION

Why is it important for patients with CML to achieve a deep molecular response?

8.3.3 Mechanisms of resistance

A subset of patients with CML exhibit either primary (failure to achieve a significant therapeutic response) or secondary (disease reappearance after an initial response) resistance to initial TKI therapy. The most commonly identified mechanism of secondary resistance is the development of point mutations in the *ABL1* tyrosine kinase domain, whereas *ABL1* mutations are rare in primary resistance suggesting the involvement of other mechanisms (such as Ph chromosome duplication). Like other resistance mutations, pre-existing mutant clones are selected for based on their ability to survive and expand in the presence of drug, rather than being caused by drug treatment. Once primary resistance is established there is a higher risk of developing kinase domain mutations that further impair response to TKI therapy.

The kinase domain of *BCR-ABL1* consists of four major components—the phosphate binding-loop, ATP-binding site, catalytic domain, and an activation-loop. Two major types of resistance mutations are recognized:

- Those at positions directly involved in imatinib binding. For example, Thr315Ile occurs at the ATP-binding site and prevents the formation of hydrogen bonds with imatinib.
- Those that change the conformation of *BCR-ABL1* and prevent the kinase from adopting a conformation compatible with imatinib binding. Mutations in the phosphate binding-loop (such as Met244Val, Gly250Glu, Gln252His, Tyr253Phe/His, Glu255Lys/Val), and activation-loop (His396Arg/Pro) are examples of this category of resistance mutation.

To date, close to 100 different resistance mutations have been detected, the most common of which are illustrated in Figure 8.12 (Mughal et al., 2016). Changing between available TKIs can be used in cases with resistance-associated mutations. For example, ponatinib is useful in the setting of a Thr315Ile resistance mutation, while Tyr253His, Glu255Lys, and Phe359Val mutations are resistant to nilotinib, but sensitive to dasatinib. Therefore, mutation analysis via direct sequencing is recommended in the case of failure to respond and suboptimal response as it will strongly influence the choice of an alternative TKI.

8.9 SELF-CHECK QUESTION

What is the difference between primary and secondary resistance to TKIs? How can these develop?

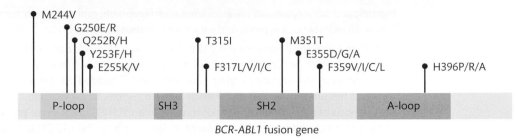

FIGURE 8.12

BCR-ABL1 resistance mutations in CML. These mutations occur within the tyrosine kinase domain of *ABL1* with over 100 different mutations described to date. Those accounting for resistance in more than two-thirds of CML patients are illustrated. Mutations either occur at positions which directly contact imatinib or occur at positions involved in the structural conformation of the protein.

Key Point

All aspects of the treatment of CML are centred around the quantitative testing of *BCR-ABL1* transcripts detectable in the peripheral blood and bone marrow aspirate. These assays are amongst the most sensitive of all assays in molecular diagnostics. A rise in *BCR-ABL1* transcripts while a patient is on TKI treatment can be due to failure to take the drug, but can also be a sign of the development of resistance mutations leading to changes in the BCR-ABL1 fusion protein.

8.4 Acute lymphoblastic leukaemia

ALL is an aggressive, primitive haematological malignancy derived from lymphoblast progenitor cells, which in their normal state ultimately give rise to B and T lymphocytes. ALL is the most common paediatric malignancy and with current treatment strategies is generally associated with very favourable outcomes. In contrast, ALL is uncommon in adults and is associated with much poorer outcomes, in part due to treatment factors (i.e. the ability for older patients to tolerate intense treatment), but also due to the enrichment of poorer risk genetic abnormalities in these patients.

8.4.1 B-acute lymphoblastic leukaemia

The most common types of genetic lesions seen in B-ALL are the loss/gain of whole chromosomes (aneuploidies) and recurrent chromosomal translocations. Treatment choices in B-ALL are heavily influenced by patient risk stratification based on the detection of numerous recurrent genetic lesions, particularly in the paediatric setting. The common genetic lesions observed can be divided into those which impart a favourable outcome and those that are associated with unfavourable outcomes.

Examples of favourable genetic lesions include:

- *Hyperdiploid karyotype*: the gain of chromosomes in a non-random fashion is frequently observed in childhood ALL. These gains tend to involve chromosomes X, 4, 6, 10, 14, 17, 18 and 21 in particular. Cases that have over 50 chromosomes in total are termed 'high hyperdiploid' karyotypes and are associated with favourable outcomes with current treatments. An example of a hyperdiploid karyotype is shown in Figure 8.13.

- *t(12;21)*: the reciprocal translocation t(12;21) results in a chimeric transcription factor that consists of *RUNX1* (also mutated and dysregulated in some cases of AML) and *ETV6*. This translocation is cytogenetically cryptic as conventional G-banded cytogenetics cannot easily detect the juxtaposition of the similarly banded regions of *RUNX1* and *ETV6*. Instead, the diagnosis of a *ETV6-RUNX1* translocation relies on FISH. t(12;21) occur predominantly in paediatric ALL and become less frequent as age increases. The t(12;21) is an early event in the development of B-ALL and may occur *in utero*, and requires the development of other genetic lesions (mutations/copy number changes) for the full malignant phenotype to manifest.

Examples of unfavourable risk lesions include:

- *Ph+ ALL*: in addition to CML, the Philadelphia chromosome can also occur in ALL. By cytogenetic/FISH analysis the Ph chromosomes in these two entities are indistinguishable; however, in Ph+ ALL the p190 isoform (as a result of the e1a2 fusion transcript) is most frequently observed. The proportion of ALL that is Ph+ increases with age, being infrequent in children but observed in approximately a quarter of cases of adult ALL. While Ph+ ALL has been consistently associated with inferior outcomes, use of TKIs (e.g. imatinib) has led to a recent improvement in outcomes.

- *'Ph-like ALL'*: when gene expression profiling is performed in B-ALL a group of patients that do not have the Philadelphia chromosome detectable cluster with patients that have Ph+ ALL. These cases have been termed 'Ph-like' and, similar to their Ph+ counterparts, are associated with a poor prognosis. These cases have been shown to harbour translocations and mutations in cytokine receptors and tyrosine kinases (and related genes). The most common lesions seen in this group involve *CRLF2*, and result in fusion to either the immunoglobulin heavy chain (IGH) locus by translocation or *P2RY8* by copy number change of *PAR1*. Other commonly involved fusions involve *CSF1R*, *EPOR*, and *JAK2*. Identification of this group of patients is important due to their inferior outcomes, but also because of the ability to target these kinase fusions with currently available TKIs (Roberts et al., 2014). A list of fusions found in Ph-like ALL is provided in Table 8.4.

TABLE 8.4 Kinase activating fusions in B-ALL with a Ph-like gene expression profile

Kinase	Partner genes identified to date
CRLF2	IGH, *P2RY8*
JAK2	*ATF7IP, BCR, EBF1, ETV6, PAX5, PPFIBP1, SSBP2, STRN3, TERF2, TPR*
ABL1	*ETV6, NUP214, RCSD1, RANBP2, SNX2, ZMIZ1*
ABL2	*PAG1, RCSD1, ZC3HAV1*
EPOR	IGH, IGK
CSF1R	*SSBP2*
PDGFRB	*EBF1, SSBP2, TNIP1, ZEB2*

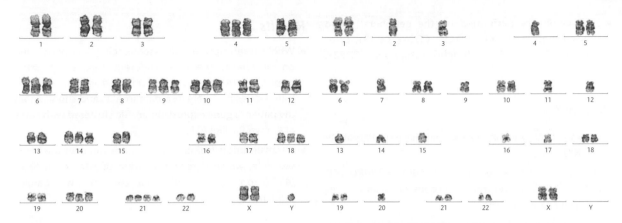

FIGURE 8.13

Aneuploid karyotypes in B-ALL. The hyperdiploid karyotype involves gains of chromosomes X, 4, 6, 10, 14, 17, 18, and 21. In this example of B-ALL in a male (shown on the left) extra copies of chromosomes X, 4, 6, 9, 10, 14, 18, 20, and 21 were detected by G-banded karyotyping. A hypodiploid karyotype involves loss of chromosomes, as demonstrated in the karyotype on the right with loss of 2, 3, 4, 7, 9, 11, 12, 13, 14, 15, 16, 17, and 20 in a female patient. This is an example of low hypodiploid ALL (32–39 chromosomes), which are often associated with germline *TP53* mutations.

Image courtesy of Dr Meaghan Wall, Victorian Cancer Cytogenetic Service, St Vincent's Hospital, Melbourne, Australia.

- *iAMP21*: intrachromosomal amplification of chromosome 21 is a rare but high-risk subtype of B-ALL, which also tend to have concomitant *RAS* pathway mutations.

- *Hypodiploid karyotype*: hypodiploid ALL (<44 chromosomes) can be further separated into either near haploid cases (approximately 24–31 chromosomes) or low hypodiploid (32–39 chromosomes). Low hypodiploid cases are highly associated with *TP53* mutations which have been noted to be germline in almost 50% of cases. An example of a hypodiploid karyotype is shown in Figure 8.13.

- *Abnormalities of* IKZF1 *(Ikaros)*: approximately 15% of B-ALL have either deletions or deleterious mutations in Ikaros, a transcription factor essential for normal lymphoid development. Ikaros abnormalities are highly enriched in Ph+ ALL (and to a lesser degree Ph-like ALL) being present in approximately 70–80% of cases.

8.10 SELF-CHECK QUESTION

Outline the genetic lesions in B-ALL that determine prognosis.

CASE STUDY 8.1 Ph-like acute lymphoblastic leukaemia

Patient history

- An 8-year-old boy was taken to the GP with increasing shortness of breath and lethargy. Physical examination revealed only pallor.

- FBC and blood film showed a haemoglobin of 65 g/L, WBC 32.3 × 10⁹/L consisting predominantly of small- to medium-sized primitive appearing cells with high nuclear to cytoplasmic ratio and basophilic agranular cytoplasm.

- Flow cytometry performed on the peripheral blood showed these cells to be CD45+, CD34+, CD19+, cytoplasmic CD79a+, CD10+ (bright), cytoplasmic CD22+, and TdT+.

Results 1

- G-banded karyotyping showed a normal male karyotype (46, XY).
- FISH on interphase cells for BCR/ABL1, ETV6/RUNX1, KMT2A (MLL), and centromere probes for chromosomes 4, 10, and 17 was normal.
- A microarray analysis was performed, which showed a homozygous deletion on chromosome 9p, which included the *CDKN2A* locus.

Significance of results 1

The immunophenotype is consistent with B-acute lymphoblastic leukaemia; however, none of the common recurrent abnormalities were detected by karyotype and FISH analysis. G-banded karyotyping can miss abnormalities in ALL due to poor chromosomal morphology and, therefore, techniques such as microarray are useful to pick up genetic lesions at a greater sensitivity.

The patient responded to chemotherapy initially, but relapsed soon after completion.

Results 2

- Whole transcriptome RNA-sequencing was performed on the relapse specimen. Differential gene expression from the RNA-seq data was compared against an established dataset of RNA-seq data in ALL, and showed that the patient's gene expression profile clustered with cases of known Ph-like ALL.
- A translocation/fusion algorithm was run over the RNA-seq data, which detected a fusion of *PAX* (exon 5) to *JAK2* (exon 19). This finding was confirmed with Sanger sequencing.

Significance of results 2

Ph-like ALL is difficult to diagnose using 'conventional' methodologies due to the wide variety of genetic lesions that can be seen. Whole transcriptome RNA-seq in this disorder has the benefit of both being able to detect the characteristic gene expression profile of this subgroup, as well as being able to detect any fusions that may be present.

The patient received chemotherapy with the addition of ruxolitinib and he achieved a complete remission, after which he received an allogeneic stem cell transplantation.

8.4.2 T-acute lymphoblastic leukaemia

T-ALL can be divided into different subgroups based on the hypothesized stage of T-cell development from which the malignant T lymphoblasts have arisen, with each stage having relatively unique immunophenotypic, clinical, and genomic characteristics. From a molecular point of view, T-ALL can be divided into early T-precursor and cortical-type T-ALL.

Early T-precursor ALL (ETP-ALL) is an aggressive and poor prognosis malignancy derived from one of the earliest T lymphoblast precursors. Genomically, ETP-ALL is characterized by numerous mutations that are more typically associated with myeloid malignancies such as *FLT3*-ITDs and *DNMT3A* mutations. Dysregulation of the *RAS* pathway (by activating mutations in *NRAS*, *KRAS*, etc.) is also observed more frequently in ETP-ALL than in other T-ALL subtypes.

Cortical-type T-ALL can be further divided into an 'early' group (corresponding to normal T-cell precursors undergoing the early cortical stage of thymocyte differentiation) characterized by aberrant *TLX1* (HOX11) expression and a 'late' group characterized by aberrant

TAL1 expression. One mechanism leading to aberrant expression of *TLX1* or *TAL1* in these groups is from either translocation or copy number changes resulting in the juxtaposition of the enhancers of the T-cell receptor alpha/gamma loci with *TLX1/TAL1*. Both early and late cortical groups have a high incidence of aberrant activation of the NOTCH1 signalling pathway, which is central to normal T-cell fate decision. Enhanced NOTCH1 signalling in T-ALL is typically the result of either activating mutations in *NOTCH1* or deleterious mutations of the negative NOTCH regulator *FBXW7*. Another genomic lesion common to both early and late groups is copy number loss of *cyclin-dependent kinase inhibitor 2A (CDKN2A)* resulting in dysregulated cell cycling.

Other molecular lesions seen across the subtypes include mutations in tumour suppressor genes (including *ETV6*, *WT1*, and *GATA3* in ETP-ALL), copy number loss of *PTEN* resulting in aberrant PI3K-AKT activation and mutations in epigenetic regulators (e.g. *EZH2*, *SUZ12*).

8.11 SELF-CHECK QUESTION

What are the associations between T progenitor maturation stage and genetic lesion?

Key Point

ALL therapy is highly dependent on the baseline genetic lesions present (particularly karyotype), but also the response to therapy as measured by minimal residual disease testing (typically of either the T-cell receptor (TCR) or IGH sequence). The individual leukaemias TCR/IGH rearranged sequence can be used as a unique 'barcode' for the patient's disease, as will be discussed further in Section 8.5.1.

8.5 Chronic lymphocytic leukaemia (and other chronic lymphoid leukaemias)

Chronic lymphocytic leukaemia (CLL) is the most common form of adult leukaemia, representing around one-third of all leukaemia cases. This low-grade B-cell malignancy results from the clonal proliferation of differentiated mature B lymphocytes that accumulate in the blood, bone marrow, and other lymphoid tissues. The clinical course of CLL is highly variable, ranging from an indolent disease never requiring therapy, to a rapidly progressing disorder leading to early death. The transformation of CLL into an aggressive lymphoma, most commonly diffuse large B-cell lymphoma, is termed Richter's syndrome or Richter's transformation (see Box 8.3). The disparity in disease course in CLL is partly reflected by the existence of two main disease subtypes defined by the mutational status of the variable region of the immunoglobulin gene.

BOX 8.3 Richter's transformation

Richter's transformation is defined as the transformation of CLL into an aggressive lymphoma, most commonly DLBCL. Approximately 2–10% of patients with CLL will progress to Richter's transformation. The molecular pathogenesis of Richter's transformation is not well understood, but various genetic characteristics, including cell cycle de-regulation via inactivation of *TP53* and *CDKN2A/CDKN2B*, and activation of *NOTCH1* and *MYC*, together with an unmutated immunoglobulin heavy variable gene (*IGHV*) are associated with an increased risk of transformation. The most important factor determining outcome in patients with Richter's transformation is the clonal relationship between the DLBCL and underlying CLL, with clonal transformation (observed in approximately 80% of patients) associated with poorer outcome and a median survival of 1 year. Sequencing of the *IGHV* gene in both the CLL and lymphoma tissue is a simple assay that can make this distinction.

8.5.1 Immunoglobulin heavy chain gene mutational status

B-cells are lymphocytes that can differentiate and secrete antibodies (immunoglobulins) upon antigen recognition. A unique B-cell receptor (BCR) is found on the surface of each B lymphocyte, and consists of two identical heavy chains (IGH) and two identical light chains (IGL), with each chain divided into constant and variable regions. The variable region of IGH is encoded by the rearrangement of three genes denoted *IGHV* (variable, V), *IGHD* (diversity, D), and *IGHJ* (joining, J), while the variable region of IGL comprises a V and J region. Development and diversity in antigen recognition arises from rearrangements of the BCR via three genetic mechanisms—V(D)J recombination, somatic hypermutation, and immunoglobulin class switch recombination.

The IGH locus (located at 14q32.3) contains approximately 123–129 V regions, 27 D regions and 9 J regions, which are separated in the germline sequence. During B-cell development in the bone marrow, one of each of these regions is juxtaposed to create a contiguous VDJ sequence in the DNA. Figure 8.14 demonstrates this process. Firstly, a J and D region are joined, followed by the joining of a V region to the D-J complex. In addition, random insertion and deletion of nucleotides can occur at the junction sites during rearrangement, further increasing the receptor repertoire. Once in the germinal centre of a lymph node mature B-cells undergo clonal expansion in response to T-cell activation and enhance their antigen affinity via somatic hypermutation, a process which introduces random mutations into the rearranged *IGHV* gene.

Malignant lymphocytes in CLL are clonally derived from a single malignant B-cell and will thus all contain identical (monoclonal) immunoglobulin rearrangements. These can evolve either from pre-germinal or post-germinal centre B-cells, dividing CLL into two genetically distinct groups:

- IGHV-*mutated*: where B-cells have passed through the germinal centre and have thus undergone somatic hypermutation of the *IGHV* locus.

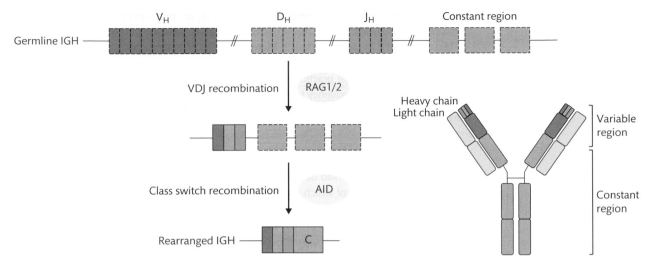

FIGURE 8.14

Generation of antigen receptor diversity in B-cells. VDJ recombination results in the generation of antigen receptor diversity in B-cells (in the form of immunoglobulins) and in T-cells (in the form of T-cell receptors). As a representative example, the immunoglobulin heavy chain (IGH) in its germline configuration is illustrated. In lymphocytes the RAG complex (comprised of RAG1 and RAG2) mediates recombination among *IGHV*, *IGHD*, and *IGHJ* segments, while the AID enzyme facilitates recombination of the constant region to give the immunoglobulin isotype (for example, IgM or IgG). The near random fashion of this process generates an enormous antibody repertoire, which is an important component of adaptive immunity. In contrast, clonally derived malignant lymphocytes will all share the same B- or T-cell receptor. Receptor clonality assessment is an important molecular test in suspected cases of lymphoid leukaemia and lymphoma.

- IGHV-*unmutated*: originates from B-cells that have not yet passed through the germinal centre and thus have an unmutated *IGHV* locus.

This biological difference translates into differences in clinical outcome, with *IGHV*-unmutated CLL representing a more aggressive form of the disease with a higher tumour burden that is often refractory to treatment. *IGHV* mutation status is, therefore, important for clinical management of CLL.

IGHV mutation testing involves PCR to amplify the clonal B-cell population in a blood or bone marrow specimen, followed by direct sequencing. Nucleotide sequences are then compared with publicly available reference databases of germline *IGHV* sequences, such as the International Immunogenetics Information System (IMGT, www.imgt.org), to determine the specific *IGHV* family and mutation frequency. Somatic hypermutation is defined as a 2% or more difference from the germline variable gene sequence, whereas less than 2% is defined as unmutated.

8.12 SELF-CHECK QUESTION

What is the difference between a mutated and an unmutated *IGHV*? What are the clinical and biological implications of this distinction?

8.5.2 Cytogenetic abnormalities in CLL

There are no cytogenetic abnormalities that are exclusive to CLL and so cytogenetics cannot be used to positively identify CLL cases. However, once a diagnosis of CLL has been established by

8.5.3 Gene mutations in CLL

Recent large-scale sequencing studies of CLL have identified other acquired mutations which are also important prognosticators (Baliakas et al., 2015; Landau et al., 2015). Mutations in these genes, including *NOTCH1*, *MYD88*, and *SF3B1* (summarized in Table 8.5), often occur in parallel to the aforementioned cytogenetic aberrations. As asymptomatic patients are managed with watch and wait until the development of symptoms, mutation testing of these genes is not recommended at diagnosis but rather at the point of requiring treatment.

The *NOTCH1* transmembrane receptor is mutated in 10–17% of newly diagnosed CLL and a third of relapsed or refractory CLL. Binding of the *NOTCH1* ligand to the receptor results in a series of events leading to proteolytic cleavage of its intracellular domain (termed the intracellular domain of Notch, ICN), which can then migrate to the nucleus and regulate expression of genes involved in pro-survival and anti-apoptotic signals. A recurrent 2 bp deletion in exon 34 (c.7541_7542del) accounts for 80% of all observed mutations and generates a premature stop codon (Pro2514Argfs*4) in the PEST domain of NOTCH1. PEST domain mutations prolong the half-life of ICN through interference with FBXW7-SCF ubiquitin ligase complex mediated ubiquitination and degradation. Other frameshift and nonsense mutations disrupting the PEST domain have also been observed, which, like the recurrent Pro2514Argfs*4, increase the stability of an active NOTCH1 isoform. Mutant *NOTCH1* is associated with the *IGHV*-unmutated subgroup that often exhibit early progression, chemotherapy refractoriness and Richter's transformation. Moreover, approximately 40% of *NOTCH1*-mutated CLL have a trisomy 12 suggesting functional synergy between these two genetic events.

SF3B1 mutations are associated with adverse prognostic features, including Richter's transformation and chemorefractory disease, and an unmutated *IGHV* locus. They also often co-occur with *ATM* deletion or mutation. This gene encodes for a central component of the U2 spliceosome that orchestrates the excision of introns from pre-mRNA to form a mature mRNA transcript. 10–15% of CLL have heterozygous missense mutations at evolutionary conserved hotspots within the C-terminal repeat HEAT domains of *SF3B1*, the most common of which results in a Lys to Glu at codon 700 (Lys700Glu) and accounts for approximately 50% of all events. Other recurrently targeted residues include Lys666 and Gly742. Mutations in this gene lead to defective RNA processing of multiple transcripts.

A hotspot mutation in *myeloid differentiation primary response 88* (*MYD88*) is mutated in ~5% in *IGHV*-mutated CLL. This missense mutation, resulting in a Leu to Pro at amino acid 265 enforces ligand-independent binding of MYD88 to IL-1 receptor-associated kinase 1 (IRAK1) and constitutive signalling to STAT3 and the NF-κB pathway. *MYD88* mutations are enriched among cases with isolated deletion of chromosome 13q.

TABLE 8.5 Frequently mutated genes in CLL

Gene	*IGHV* association	Prognostic significance	Richter's transformation or chemorefractory disease
ATM	*IGHV*-unmutated	Poor	Yes
TP53	*IGHV*-unmutated	Poor	Yes
NOTCH1	*IGHV*-unmutated	Poor	Yes
MYD88	*IGHV*-mutated	Unknown	No
SF3B1	*IGHV*-unmutated	Poor	Yes

High-throughput sequencing has also identified a long list of biologically and clinically uncharacterized genes mutated at lower frequencies in pathways known to be important in disease pathogenesis (for example, B-cell signalling genes *TRAF2*, *TRAF3* and *CARD11*, and DNA damage response genes *CHEK2* and *BRCC3*), as well as less well-characterized putative drivers (for example, *RPS15* and *IKZF3*). In addition, the use of drugs targeted to the inhibition of BCR signalling has led to the uncovering of previously unrecognized resistance mutations in BCR pathway genes. For example, mutations at the ibrutinib binding site of *BTK* or gain-of-function mutations in *PLCG2*. Although these mutations have not yet gained the qualification of predictive factors it is possible that in the future some of these molecular markers will be used to predict the highly variable clinical course of CLL beyond that of established factors such as *IGHV* mutation status and cytogenetic abnormalities.

8.14 SELF-CHECK QUESTION

Describe the function of NOTCH1. How does mutation alter this function?

8.5.4 Hairy cell leukaemia

Hairy cell leukaemia (HCL) is an indolent B-cell malignancy comprising 2% of lymphoid leukaemias. Morphologically, the leukaemic cells are recognizable for their abundant cytoplasm, reniform (kidney shaped) nucleus, and 'hairy' cytoplasmic projections. The disease-defining genetic event in essentially all HCL is a somatic amino acid substitution of glutamic acid for valine at position 600 of the RAF serine/threonine protein kinase *BRAF* (Val600Glu; Tiacci et al., 2011). The Val600Glu substitution occurs within the activation loop of *BRAF*, thereby leading to oncogenic signalling through the MAPK pathway and ultimately increased cell proliferation, survival, and neoplastic transformation.

Up to 80% of patients with HCL achieve a complete remission following treatment with purine analogue chemotherapeutics. In the subgroup of HCL that are refractory or relapsed the finding of *BRAF* mutations is particularly important—targeted inhibitors of the Val600Glu mutation have been developed for use in other *BRAF* mutant cancers (notably melanoma), but probably represent a novel therapeutic option in HCL patients as well.

8.15 SELF-CHECK QUESTION

What are the cellular effects of a *BRAF* Val600Glu mutation?

8.5.5 T-cell large granular lymphocytic leukaemia and T-cell prolymphocytic leukaemia

Like the BCR in B-cells, the T-cell receptor (TCR) gene loci contain many different *V*, *D*, and *J* gene segments that undergo rearrangement during normal T lymphocyte differentiation. Each TCR consists of two different chains coupled together, with the majority of T-cells expressing an alpha (TRA) and beta (TRB) chain, and a minority expressing gamma (TRG) and delta (TRD) chains. Each chain consists of one variable and one constant region. The variable coding regions of TRA and TRG are generated by the recombination of VJ regions, whereas those that form TRB and TRD are generated by VDJ recombination.

In the case of T-cell malignancy, where all malignant cells originate from a founding neoplastic clone, the TCR population will be homogeneous and thus analysis of the TCR repertoire is used to facilitate the diagnosis of a neoplastic T-cell process. As the TRG locus is rearranged in the vast majority of T-cell malignancies, this is the first choice of locus for PCR analysis. Amplification of the J and V segment families is performed using fluorescently labelled consensus primers which allows fragments to be visualized by fragment size analysis. A clonal population is represented by a single prominent peak like that shown in Figure 8.15, whereas polyclonal populations are represented as a Gaussian distribution of

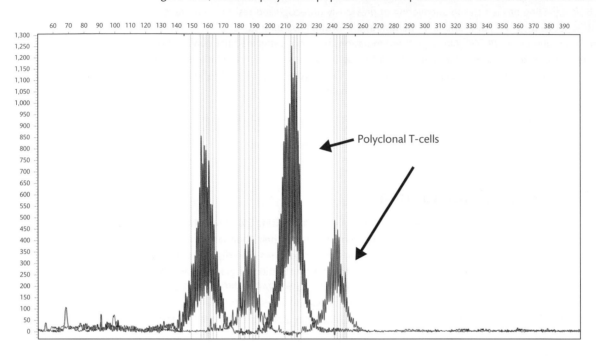

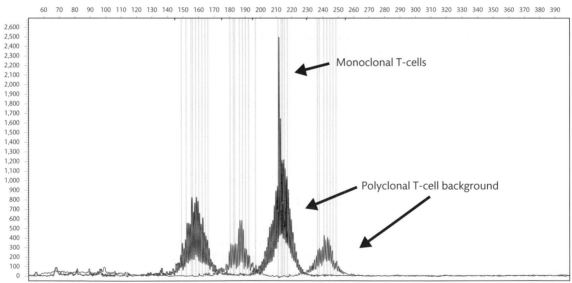

FIGURE 8.15

T-cell receptor clonality assessment in T-cell leukaemias. An example of *TRG* gene rearrangement analysis for a sample with a polyclonal (top) and monoclonal (bottom) T-cell population is shown. PCR using fluorescently labelled primers directed to the V and J regions of the *TRG* is analysed following size separation of the products by capillary electrophoresis. Polyclonal populations are represented as many different PCR products of different size, while monoclonal populations are represented by a prominent peak generated from a single PCR product.

multiple peaks. More recently TCR analysis has been performed by next-generation sequencing (NGS), which allows both length and sequence determination (Wu et al., 2012).

T-cell large granular lymphocytic leukaemia (T-LGL) is a clonal proliferation of circulating T lymphocytes characterized morphologically by lymphocytes with abundant cytoplasm often containing azurophilic granules. Up to 40% of T-LGL have activating mutations in the *signal transducer and activator of transcription 3* (*STAT3*; Jerez et al., 2012). In a normal cell, binding of cytokines to the IL-6 receptor initiates a signalling cascade, which converges on STAT protein dimerization, translocation to the nucleus, and transcriptional activation of target genes. Somatic *STAT3* mutations cluster within exon 12 encoding for the Src homology (SH2) phosphotyrosine-binding domain (often at residues Tyr640 or Asp661) and promote auto-dimerization in the absence of ligand. Rarely, mutations can be found outside of the SH2 domain of *STAT3*, or in *STAT5B*. Consequently, expression of *STAT*-responsive genes involved in apoptosis, proliferation, and immune response are upregulated in T-LGL. Treatment for T-LGL is reserved for patients with symptomatic disease as most cases have an indolent clinical behaviour.

T-cell prolymphocytic leukaemia (T-PLL) is an uncommon haematological malignancy characterized by a population of prolymphocytes in the peripheral blood and bone marrow. Although rare, T-PLL is an aggressive disease that is often resistant to conventional chemotherapy, owing in part to the finding of complex chromosomal abnormalities in malignant cells generally involving chromosomes 8, 11, 14, and X. Rearrangements of *TCL1A/B* and *MTCP1* on chromosomes 14q32 and Xq28 respectively with the T-cell receptor locus on chromosome 14 results in proto-oncogene activation (inv(14), t(14;14), t(X;14)). Additional changes include deletion or mutation of the *ATM* locus at 11q22 and trisomy 8q in three quarters of tumours. Frequent mutations are also seen in the interleukin 2 receptor-JAK-STAT pathway (illustrated in Figure 8.16) through mutually exclusive gain-of-function mutations in *IL2RG*, *JAK1*, *JAK3*, or *STAT5B*.

8.16 SELF-CHECK QUESTION

How is assessment of the T-cell receptor (TCR) used to determine clonality in T-cell leukaemia?

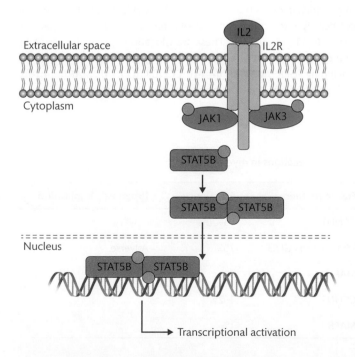

FIGURE 8.16

The interleukin-2-JAK-STAT pathway in T-PLL. IL-2 cytokine activation of the IL-2 receptor induces a conformational change of the cytoplasmic portion. This facilitates (i) JAK1/JAK3 recruitment, (ii) phosphorylation of the receptor, and (iii) STAT5B recruitment. STAT dimers translocate to the nucleus where they regulate transcription of genes involved in cell proliferation and survival. Activating mutations are found throughout this pathway in T-PLL.

8.6 **Multiple myeloma**

Multiple myeloma (MM) is a malignancy of plasma cells. This occurs through a stepwise process, whereby plasma cells acquire successive genomic changes over time that facilitates transformation from pre-malignant states to a malignant phenotype. The presence of excessive plasma cells within the bone marrow, and the associated production of cytokines and immunoglobulin (paraprotein) can result in progressive bone marrow failure, renal impairment, bony lesions, and hypercalcemia (a symptom of the breakdown of bone, which leads to the release of calcium into the blood). The most indolent phase is termed monoclonal gammopathy of uncertain significance (MGUS) and is characterized by a small population of malignant plasma cells in the bone marrow without clinical symptoms. Smouldering multiple myeloma (SMM) follows MGUS, which is also asymptomatic, but defined by a higher proportion of tumour cells. In both phases, the malignant cells maintain their normal function of producing immunoglobulin (either intact or light chain), which can be detected in the serum (known as a paraprotein) during the earliest phases of disease. A small proportion of patients with MGUS and 10% of patients with SMM will continue to acquire genetic lesions allowing the development of more aggressive clinical behaviour and the full malignant phenotype of MM. It is, therefore, helpful to understand the genomic changes observed in MM relative to this genomic hierarchy, i.e. primary genetic events contributing to plasma cell immortalization and secondary events contributing to disease progression.

8.6.1 **Primary genetic lesions**

Primary establishing lesions divide MM into two broad subtypes:

- *Hyperdiploid myeloma* display a characteristic hyperdiploid karyotype characterized by trisomies of the odd numbered chromosomes, commonly chromosomes 3, 5, 7, 9, 11, 15, 19, and 21. The generation of this karyotype is thought to occur at one single catastrophic mitotic event, rather than as successive chromosome gains over time.

- *IGH-translocated myeloma* which involves translocation of the IGH locus on chromosome 14q32, one of the most heavily transcribed genes in plasma cells, with various partner chromosomes including 4, 6, 11, 16, and 20. These translocations place the resulting fusion partner under the control of the IGH enhancer typically resulting in upregulated oncogene expression, including *FGFR3/MMSET* (t(4;14)), *CCND1* (t(11;14)), *MAF* (t(14;16)), *MAFB* (t(14;20)), and *CCND3* (t(6;14)). More information about these translocations is found in Table 8.6.

TABLE 8.6 Common IGH translocations in myeloma

Translocation	Fusion partner	Frequency	Prognostic implication
t(11;14)	*CCND1*	15–20%	
t(4;14)	*MMSET* and *FGFR3*	15%	Adverse
t(14;16)	*MAF*	5%	
t(6;14)	*CCND3*	1–2%	
t(14;20)	*MAFB*	1%	

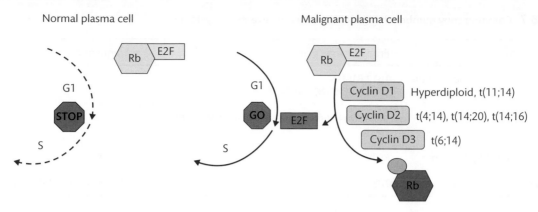

FIGURE 8.17

Dysregulation of the cell cycle is a key early molecular abnormality in multiple myeloma. In its unphosphorylated form Rb binds to and represses the E2F transcription factor preventing expression of E2F-regulated genes (which are involved in cell cycle progression). This inhibits cells from progressing through the G1/S checkpoint. The cyclin protein family, in complex with cyclin-dependent kinases, inhibit Rb by partial phosphorylation, allowing cell cycle progression. In myeloma the expression of cyclin D1 and cyclin D3 is increased as a result of *CCND1* or *CCND3* translocation to the IGH locus (t(11;14) and t(6;14)). Non-translocation-based upregulation of cyclin D can also occur. For example, the t(14;16) involving *MAF* upregulates *CCND2* by directly binding to its promoter.

A hyperdiploid karyotype and IGH translocations are detected in the MGUS phase of disease and, thus, contribute to the establishment of tumorigenesis, but are insufficient by themselves to result in the full malignant phenotype. As shown in Figure 8.17 both subtypes ultimately result in dysregulation of the cyclin D family of cell cycle control genes and result in transition through the G1/S checkpoint, either directly, for example, t(11;14) or t(6;14) involving *CCND1* and *CCND3*, or indirectly.

8.17 SELF-CHECK QUESTION

Describe the various pathways to cyclin D dysregulation in myeloma.

8.6.2 Secondary genetic lesions

In order for myeloma to develop a full malignant phenotype secondary genetic lesions are necessary. A common secondary driver is the acquisition of mutations involving the MAPK pathway which, as a group, are observed in approximately 40% of cases. These are typically activating mutations in *NRAS*, *KRAS* and *BRAF*. Other cellular pathways that are enriched for mutations are the NF-κB (*TRAF3*, *CYLD*, *LTB*, *IKBKB*, *BIRC2*, *BIRC3*, *CARD11*, and *TRAF3IPI*), DNA damage (*TP53*, *ATM*, and *ATR*), and genes involved in B-cell differentiation (*PRDM1*, *IRF4*, and *SP140*).

Copy number changes are also late genetic events and involve focal regions on chromosome 1, loss of chromosome 13, and heterozygous loss of the whole p arm of chromosome 17 as listed in Table 8.7. Genes frequently mutated in MM may also be involved in copy number changes, for example deletion of genes involved in the NF-κB pathway on chromosomes 11q (*BIRC2*, *BIRC3*), 14q (*TRAF3*), and 16q (*CYLD*).

MYC overexpression, consequent to chromosomal gain or translocation of the *MYC* locus (8q24), is also a frequent secondary event, observed in 50% of MM. *MYC* translocation partners include the immunoglobulin heavy, light, and kappa loci (t(8;14), t(8;22), and t(2;8), respectively) along with *FAM46C*, *FOXO3*, and *BMP6*. These translocations result in juxtapositioning of super-enhancers surrounding the partner gene to the *MYC* locus.

TABLE 8.7 Common copy number changes observed as late events in myeloma

CNV	Primary target gene	Frequency
1p monoallelic or biallelic deletion	1p12 *FAM46C* 1p32.3 *CDKN2C* and *FAF1*	30%
1q gain	Unknown but may involve *CSK1B, ANP32E, BCL9, PDZK1*	35–40%
Monosomy 13 or 13q deletion	*RB1*	45–50%
Monoallelic deletion of 17p	*TP53*	80% in advanced stage disease

8.6.3 Monitoring treatment response

Adding to the genomic complexity of myeloma is a relatively high level of clonal heterogeneity, which exists even in the early stages of disease. It is not unusual to observe multiple tumour subclones at diagnosis, each with a slightly different mutational profile, in addition to the predominant clonal population. Moreover, sampling of tumour from different sites can show different genetic lesions. Recognition of this intratumoural heterogeneity has important therapeutic relevance as minor clones will respond differently to different drugs and may harbour drug resistance mutations that can give rise to relapse.

Ultimately response to therapy and the amount of disease remaining following treatment is the most important determinant of outcome in MM. Sensitive PCR-based molecular methods that can detect low levels of remaining tumour cells following therapy can be used to monitor minimal residual disease in MM. These methods take advantage of the fact that each case of myeloma has a unique clonal immunoglobulin gene rearrangement comprised of a V, D, and J segment as discussed in Section 8.5.1. This clonal rearrangement can be identified at diagnosis by amplifying the region surrounding the rearranged allele using consensus primers followed by sequencing. After therapy, the monoclonal rearrangement can be identified among thousands of normal cells either by real-time quantitative allele-specific PCR using primers designed for each individual patient's unique sequence or by NGS using primers that amplify the IGH locus regardless of the VDJ rearrangement present. These methods have a sensitivity of one clonal cell in 10^5–10^6 normal cells.

8.18 SELF-CHECK QUESTION

Which molecular feature is used to monitor response in myeloma? Are there other leukaemias that could be monitored using the same method?

Key Point

Multiple myeloma is a very heterogeneous disease from a genetic point of view. Even within the same patient there can be multiple clones harbouring genetically distinct lesions. This has implications for targeted therapy because if a therapy is used which targets one particular lesion (e.g. *BRAF* Val600Glu) then the patient may have progressive disease from a clone that does not contain the *BRAF* mutation.

CASE STUDY 8.2 Multiple myeloma

Patient history

- A 55-year-old man presented with back pain and lethargy, and was found on investigation to have a normocytic normochromic anaemia and hypercalcemia.
- Serum electrophoresis detected a monoclonal IgG(κ) paraprotein of 34 g/L.
- A bone marrow biopsy was performed which showed 40% plasma cells present in the trephine with an abnormal immunophenotype being CD19–, CD56+, cyclin D1+, and kappa light chain restricted. Plasma cells were small and mature in appearance.

Results

Normocytic normochromic anaemia (i.e. anaemia in which the average size and haemoglobin content of red blood cells is normal) and hypercalcemia (elevated calcium in the blood) are common symptoms of multiple myeloma. The G-banded karyotype of the bone marrow aspirate showed a t(11;14), which was confirmed by FISH performed on CD138-positive plasma cells to be a translocation involving IGH and CCND1. A targeted next-generation sequencing panel revealed an activating KRAS mutation (Gly13Asp) and multiple mutations in CCND1.

Significance of results

The presence of a t(11;14) in myeloma has been associated with specific morphological features including small, mature plasma cells with a more 'lymphoplasmacytoid' type appearance. It has also been associated with aberrant CD20 expression. The cyclin D1 expression by immunocytochemistry is consistent with the observation of a t(11;14).

Chapter summary

- Leukaemias are generally classified by their cell of origin into myeloid and lymphoid leukaemias.

  Acute myeloid leukaemia is primarily subclassified on the basis of karyotype into various subgroups including acute promyelocytic leukaemia (t(15;17)), core binding factor AML (inv(16) and t(8;21)), normal karyotype AML, and complex karyotype AML.

  One of the most important decisions dictated by molecular/cytogenetic testing in AML is whether to perform an allogeneic stem cell transplant in first remission.

  Minimal residual disease testing is routinely used to monitor the response to therapy in t(15;17), t(8;21) and inv(16) AML.

- The defining genetic lesion in **chronic myeloid leukaemia** is t(9;22), also known as the Philadelphia chromosome, which creates the BCR-ABL1 fusion protein.

Different BCR-ABL1 transcript types are possible in CML and the most common ones are e13a2 and e14a2.

BCR-ABL1 monitoring by RT-PCR is used to monitor response to therapy and to detect resistance to therapy.

BCR-ABL1 mutations can occur that impart resistance to TKI therapy. An important one is the Thr318Ile which imparts resistance to almost all tyrosine kinase inhibitors apart from ponatinib.

■ **Acute lymphoblastic leukaemia** can be divided into those derived from B lymphocyte progenitors (B-ALL) or T lymphocyte progenitors (T-ALL).

B-ALL is genetically characterized by aneuploidies and recurrent translocations, for example t(12;21).

T-ALL can be subdivided into an early T-progenitor type, which have a mutational profile with some characteristics similar to myeloid malignancy and a cortical/mature type, which have NOTCH1 pathway dysregulation as a central theme.

Minimal residual disease testing is important in both B-ALL and T-ALL and can be performed by following the individual patient's rearranged IGH/TCR sequence.

■ Important cytogenetic lesions in **chronic lymphocytic leukaemia** are del 17p, del 11q, del 13q, and trisomy 12.

CLL mutations that impart an inferior prognosis are TP53, NOTCH1, SF3B1, and BIRC3.

Almost all cases of hairy cell leukaemia contain the BRAF Val600Glu mutation and this has been utilized for targeted therapy with mutation-specific inhibitors in relapsed/refractory disease.

Clonality in T-cell lymphoproliferative disorders can be established by looking for uniformity of length of PCR products of various TCR loci indicating the presence of a single clone.

■ The earliest lesions in **multiple myeloma** are hyperdiploidy and translocations involving the IGH loci on chromosome 14.

■ Mutations in multiple myeloma tend to recurrently involve the RAS/MAPK pathway.

 Further reading

● Blombery PA, Wall M, Seymour JF. The molecular pathogenesis of B-cell non-Hodgkin lymphoma. European Journal of Haematology 2015; 95: 280–293.

● Conter V, Bartram CR, Valsecchi MG, et al. Molecular response to treatment redefines all prognostic factors in children and adolescents with B-cell precursor acute lymphoblastic leukemia: results in 3184 patients of the AIEOP-BFM ALL 2000 study. Blood 2010; 115: 3206–3214.

● Holmfeldt L, Wei L, Diaz-Flores E, et al. The genomic landscape of hypodiploid acute lymphoblastic leukemia. Nature: Genetics 2013; 45: 242–252.

● Neumann M, Heesch S, Gokbuget N, et al. Clinical and molecular characterization of early T-cell precursor leukemia: a high-risk subgroup in adult T-ALL with a high frequency of FLT3 mutations. Blood Cancer Journal 2012; 2: e55.

● Papaemmanuil E, Gerstung M, Bullinger L, et al. Genomic classification and prognosis in acute myeloid leukemia. New England Journal of Medicine 2016; 374: 2209–2221.

- Walker BA, Boyle EM, Wardell CP, et al. Mutational spectrum, copy number changes, and outcome: results of a sequencing study of patients with newly diagnosed myeloma. Journal of Clinical Oncology 2015; 33: 3911–3920.

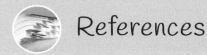 References

- Arber DA, Orazi A, Hasserjian R, et al. The 2016 revision to the World Health Organization classification of myeloid neoplasms and acute leukemia. Blood 2016; 127: 2391–2405.

- Baliakas P, Hadzidimitriou A, Sutton LA, et al. Recurrent mutations refine prognosis in chronic lymphocytic leukemia. Leukemia 2015; 29: 329–336.

- Cross NCP, White HE, Ernst T, et al. Development and evaluation of a secondary reference panel for BCR-ABL1 quantification on the International Scale. Leukemia 2016; 30: 1844–1852.

- Döhner H, Estey E, Grimwade D, et al. Diagnosis and management of AML in adults: 2017 ELN recommendations from an international expert panel. Blood 2017; 129: 424–447.

- Eichhorst B, Robak T, Montserrat E, et al. Chronic lymphocytic leukaemia: ESMO clinical practice guidelines for diagnosis, treatment and follow-up. Annals of Oncology 2015; 26(Suppl 5): v78–84.

- Grimwade D, Hills RK, Moorman AV, et al. Refinement of cytogenetic classification in acute myeloid leukemia: determination of prognostic significance of rare recurring chromosomal abnormalities among 5876 younger adult patients treated in the United Kingdom Medical Research Council trials. Blood 2010; 116: 354–365.

- Groschel S, Sanders MA, Hoogenboezem R, et al. Mutational spectrum of myeloid malignancies with inv(3)/t(3;3) reveals a predominant involvement of RAS/RTK signaling pathways. Blood 2015; 125: 133–139.

- Hughes T, Deininger M, Hochhaus A, et al. Monitoring CML patients responding to treatment with tyrosine kinase inhibitors: review and recommendations for harmonizing current methodology for detecting BCR-ABL transcripts and kinase domain mutations and for expressing results. Blood 2006; 108: 28–37.

- Jerez A, Clemente MJ, Makishima H, et al. STAT3 mutations unify the pathogenesis of chronic lymphoproliferative disorders of NK cells and T-cell large granular lymphocyte leukemia. Blood 2012; 120: 3048–3057.

- Landau DA, Tausch E, Taylor-Weiner AN, et al. Mutations driving CLL and their evolution in progression and relapse. Nature 2015; 526: 525–530.

- Mughal TI, Radich JP, Deininger MW, et al. Chronic myeloid leukemia: reminiscences and dreams. Haematologica 2016; 101: 541–558.

- Nowell P, Hungerford D. A minute chromosome in human granulocytic leukemia. Science 1960; 132: 1497–1501.

- Roberts KG, Li Y, Payne-Turner D, et al. Targetable kinase-activating lesions in Ph-like acute lymphoblastic leukemia. New England Journal of Medicine 2014; 371: 1005–1015.

● Rowley JD. Letter: a new consistent chromosomal abnormality in chronic myelogenous leukaemia identified by quinacrine fluorescence and Giemsa staining. Nature 1973; 243(5405): 290–293.

● Swerdlow SH, Campo E, Pileri SA, et al. The 2016 revision of the World Health Organization classification of lymphoid neoplasms. Blood 2016; 127: 2375–2390.

● Tiacci E, Trifonov V, Schiavoni G, et al. BRAF mutations in hairy-cell leukemia. New England Journal of Medicine 2011; 364: 2305–2315.

● White HE, Matejtschuk P, Rigsby P, et al. Establishment of the first World Health Organization International Genetic Reference Panel for quantitation of BCR-ABL mRNA. Blood 2010; 116: e111–e117.

● Wu D, Sherwood A, Fromm JR, et al. High-throughput sequencing detects minimal residual disease in acute T lymphoblastic leukemia. Science of Translational Medicine 2012; 4(134): 134ra63.

 Discussion questions

8.1 How is molecular assessment of the immunoglobulin heavy chain locus and T-cell receptor locus utilized in lymphoid malignancy?

8.2 Choose a subtype of leukaemia, and describe the techniques and utility of molecular methods used to monitor response to therapy.

8.3 Describe the broad categories of genomic abnormalities observed in leukaemia and their significance using specific examples.

Answers to the self-check questions and tips for responding to the discussion questions are provided on the book's accompanying website:

○ Visit www.oup.com/uk/warford

9

Haemopoietic Diseases 2: Lymphoproliferative Disorders

Mark Catherwood and Ken Mills

Learning objectives

After studying this chapter you should be able to:

- Understand the criteria used in the current WHO classification of lymphoid neoplasms.
- Describe the various laboratory techniques used to investigate lymphoid neoplasms.
- Discuss the role of immunoglobulin and T-cell receptor gene rearrangements in lymphoid neoplasms, and their role in clonality detection.
- Describe some of the common cytogenetic abnormalities associated with a range of lymphoid neoplasms, and outline the laboratory techniques available for their identification.

9.1 **Introduction**

The non-Hodgkin lymphomas (NHL) are a large group of clonal lymphoid tumours, with approximately 85% of B-cell and 15% of T- or NK (natural killer) cell origin. They are commonly referred to as lymphoproliferative disorders. Their aetiology is currently unknown, although infectious agents play a role in certain subtypes. Hodgkin lymphoma (HL) is a B-cell-derived malignancy that is characterized by the presence of Reed–Sternberg cells and its aetiology is currently unknown.

Lymphoproliferative disorders were originally categorized on morphological features and clinical behaviour alone. With the introduction of antibodies against cell surface markers this helped to place a given lymphoma into a diagnostic category when combined with appropriate morphology. However, it should be noted that, even within a distinct lymphoma category, considerable heterogeneity of clinical behaviour exists (Letai and Gribben, 2010).

The non-Hodgkin lymphomas (NHL) are a diverse group of haematological malignancies characterized by a broad range of morphological, immunophenotypic, and clinical features. A prominent example is the category of diffuse large B-cell lymphomas, in which approximately 50% are cured with chemotherapy, but the remainder die of disease, usually within a few years of diagnosis. In this case, as with many of the lymphomas, a prognostic index (the International Prognostic Index or IPI) based on a few pretreatment criteria is able to further subdivide the category and provide very useful prognostic information. However, even within IPI classes, significant clinical heterogeneity persists. Furthermore, it is likely that the IPI defines subclasses of lymphomas based on biological differences among these lymphomas. Accordingly, studies of genetic abnormalities are proving important tools for the improved classification and prognostication of disease. Additionally, a better understanding of the molecular pathophysiology of the disease will probably lead to improvements in treatment of lymphoma.

Lymphomagenesis in these disorders is driven by genetic alterations allowing escape from the normal physiological restrictions around growth, differentiation, proliferation, and cell death, resulting in a suite of genetic lesions, which differs both between various NHL subtypes, but also from case to case within each individual subtype. The underlying molecular heterogeneity of NHL is suggested by the highly variable clinical behaviour of different lymphoma subtypes, e.g. variable risk of involvement of the central nervous system and predilection for specific sites of extra-nodal involvement, and by a variation in the susceptibility to various molecularly targeted and novel therapies. The increasing availability of high-throughput sequencing techniques in both the research and diagnostic laboratory has significantly advanced the understanding of the mechanisms by which genetic lesions contribute to the initiation, maintenance, and progression of NHL (Rosenquist et al., 2016).

This chapter will explore the classification, immunoglobulin and T-cell receptor rearrangements, the application and detection of clonality, the role of translocations, and the use of FISH in the diagnosis of lymphoproliferative diseases.

9.2 Classification of non-Hodgkin lymphomas

The classification of NHL has been guided by the WHO classifications of haematopoietic and lymphoid tumours first published in 2001 (Jaffe et al., 2001), updated in 2008 (Swerdlow et al., 2008) and revised in 2016 (Swerdlow et al., 2016; Table 9.1). The classification has been adapted and developed from biological principles, and clinical relevance according to morphology, immunophenotypic, genetic, molecular, and clinical features. The 2001 WHO classification was based on the Revised European American Classification of Lymphoid Neoplasms (REAL) published in 1994 (Harris et al., 1994). In 2008 this publication established the guidelines for the diagnosis and classification of malignant lymphoproliferative diseases.

The understanding of the biology and genetics of lymphoid neoplasms has expanded greatly in recent years, which has resulted in improved insights into the mechanisms and management of lymphoid malignancies. The most recent 2016 revision (Swerdlow et al., 2016) has only limited alterations in the classification compared with 2008, but provides an update on the diagnostic categories related to biological and clinical correlates to facilitate improved patient care.

TABLE 9.1 **WHO classification of lymphomas**

Mature B-cell neoplasms	Chronic lymphocytic leukaemia/small lymphocytic lymphoma
	B-cell prolymphocytic leukaemia
	Splenic B-cell marginal zone lymphoma
	Hairy cell leukaemia
	Lymphoplasmacytic lymphoma
	Heavy chain diseases
	Plasma cell myeloma
	Solitary plasmacytoma of bone
	Extraosseous plasmacytoma
	Extranodal marginal zone lymphoma of mucosa-associated lymphoid tissue (MALT lymphoma)
	Nodal marginal zone lymphoma
	Follicular lymphoma
	Primary cutaneous follicle centre lymphoma
	Mantle cell lymphoma
	Diffuse large B-cell lymphoma (DLBCL), not otherwise specified (NOS)
	DLBCL associated with chronic inflammation
	Lymphoid granulomatosis
	Primary mediastinal (thymic) large B-cell lymphoma
	Intravascular large B-cell lymphoma
	Anaplastic (ALK) positive large B-cell lymphoma
	Plasmablastic lymphoma
	Large B-cell lymphomas arising in HHV8-associated multicentric Castleman disease
	Primary effusion lymphoma
	Burkitt lymphoma
Precursor lymphoid neoplasms	B-lymphoblastic leukaemia/lymphoma, (NOS)
	B-lymphoblastic leukaemia/lymphoma with recurrent genetic abnormalities
Mature T-cell and NK-cell neoplasms	T-cell prolymphocytic leukaemia
	T-cell large granular lymphocytic leukaemia
	Aggressive natural killer (NK)-cell leukaemia
	Systemic Epstein–Barr virus (EBV) positive T-cell lymphoproliferative disease of childhood
	Hydroa vacciniforme-like lymphoma
	Adult T-cell leukaemia/lymphoma
	Extranodal NK/T-cell lymphoma, nasal type
	Enteropathy-associated T-cell lymphoma
	Hepatosplenic T-cell lymphoma
	Subcutaneous panniculitis-like T-cell lymphoma
	Mycosis fungoides
	Sezary syndrome
	Primary cutaneous CD30-positive T-cell lymphoproliferative disorders
	Primary cutaneous gamma-delta T-cell lymphoma
	Peripheral T-cell lymphoma (NOS)
	Angioimmunoblastic T-cell lymphoma
	Anaplastic large cell lymphoma, (ALK positive)
Hodgkin lymphoma	Nodular lymphocyte predominant Hodgkin lymphoma
	Classical Hodgkin lymphoma

The WHO classification addresses the recognition of clonal expansions of lymphoid cells which correspond to early steps in lymphomagenesis. In some cases, it is not clear whether these lesions will ever progress to clinically significant disease, or whether they simply correspond to relatively stable 'benign lymphoid' clonal proliferations. The identification of these lesions opens new questions as to how to manage these patients.

Key Point

The WHO classification is now accepted as the reference lymphoma classification in clinical practice. The WHO classification was built upon the work of earlier systems, but took into account the role that molecular genetics played in refining disease categories. Forthcoming classifications will include additional genetic data derived from modern technologies.

9.1 SELF-CHECK QUESTION

Describe the features considered important by the WHO for the classification of lymphoid neoplasms.

9.3 Diagnostic haematopathology

Clonal B- and T-cell proliferations are identified by several different approaches including morphology, cell phenotyping, the detection of immunoglobulin, or T-cell receptor gene rearrangements, and by the detection of translocations by fluorescent *in situ* hybridization (FISH) and/or cytogenetics.

9.3.1 Immunophenotyping

Immunophenotyping involves the use of immunocytochemistry (ICC) and flow cytometry (FC). ICC is a method by which antibodies are used to detect cellular antigens in their morphological context. For lymphoma diagnosis ICC is normally applied to FFPE tissue sections. In contrast FC usually requires fresh blood, bone marrow, or tissue from which cell suspensions are made. The cells are then incubated with multiple fluorochrome-labelled antibodies and passed through a laser light beam in a flow cytometer (Craig and Foon, 2008). Multichannel instruments allow detection of several antigens simultaneously, as well as FS and SS measurements, which give an indication of the cell size and complexity, respectively. The stage of differentiation of lymphocytes may be recognized by their different patterns of surface antigen expression. Detailed information on the principles and practice of both techniques is provided in Chapter 3, Section 3.3 and Section 3.4.

The use of ICC and FC with various antibodies applied in 'panels' helps to identify the specific lineage and developmental stage of the lymphoma. For instance, B-cell markers, such as CD79a and PAX5, are expressed at the early stage of heavy chain gene rearrangements, whereas CD20 appears later with the light chain rearrangement. CD10 and bcl-6 are expressed by centrocytes and centroblasts and are positive in germinal centre-derived lymphomas such as follicular lymphoma. TdT, CD10 and CD34 are expressed by B and T-cell lymphoblasts and, therefore, are useful in recognizing lymphoblastic neoplasia. CD3, CD2, CD5, and CD7 recognize virtually all mature T-cells, and are useful in the diagnostic workup of T-cell lymphomas. Examples of the use of immunophenotyping panels are provided in Figure 9.1 for ICC and Figure 9.2 for FC.

Key Point

Immunocytochemistry and flow cytometry use antibodies in panels to assist in the accurate classification of lymphomas into B and T-cell subcategories.

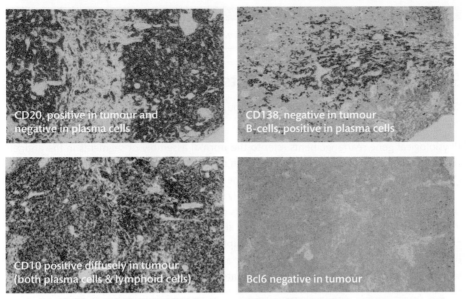

FIGURE 9.1

Immunocytochemistry from an orbital biopsy infiltrated by lymphoplasmacytic lymphoma. Tumour cells express CD20 and CD10 and are negative for CD138 and BCL6. FFPE tissue sections stained with an immunocytochemical method, terminating in the deposition of a brown chromogenic end-product for cells reactive with the antibodies. The sections have been counterstained with haematoxylin to allow for the simultaneous assessment of morphology.

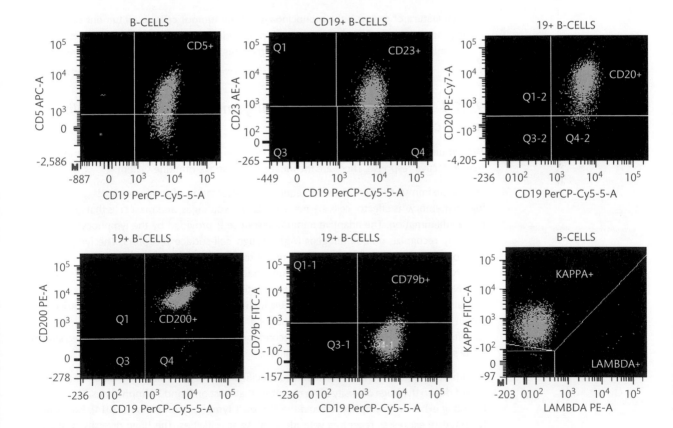

FIGURE 9.2

Flow cytometry of chronic lymphocytic leukaemia in peripheral blood. The chronic lymphocytic leukaemia (CLL) cells are positive for CD5, CD23, CD20, and CD200 (A–D), and negative for CD79b (E). The CLL cells demonstrated light chain restriction (clonal) positive for kappa and are negative for lambda light chains (F).

9.2 SELF-CHECK QUESTION

Describe the information that can be obtained with immunophenotyping techniques.

9.3.2 Gene rearrangements

Gene rearrangements can be conveniently grouped into two broad categories; physiological and pathological.

Physiological gene rearrangements refer to the normal assembly of segments in the antigen receptor genes (immunoglobulin, Ig) and T-cell receptor (TCR) in B- and T-cells, respectively. Pathological rearrangements refer mainly to the process of chromosomal translocations that lead to the movement of genes that are normally separated in nature. The diagnostic relevance of assessing physiological gene rearrangements compared with pathological rearrangements is important. For Ig/TCR gene rearrangement the major function is to determine if mono-clonality or polyclonality exists within the sample, whereas pathological rearrangements are usually qualitative, resulting in the presence or absence of a chromosomal translocation. The techniques behind these different approaches are examined further in the following sections.

> ## Key Point
>
> **A key feature of cancer is the monoclonality of the tumour cells, as all tumour cells are the progeny of a single malignantly transformed cell. This allows gene rearrangements to discriminate between polyclonal, reactive processes, and monoclonal malignant tumours.**

9.3.2.1 Physiological gene rearrangements of immunoglobulin and T-cell receptors

The immune system comprises two arms that function cooperatively to provide a comprehensive protective response. These are the innate and adaptive immune system.

The innate immune system does not require the presentation of antigen or leads to immunological memory. Its effector cells are neutrophils, macrophages, and mast cells that give rise to acute inflammation. The adaptive immune response is provided by the lymphocytes, which precisely recognize unique antigens (Ag) through cell-surface receptors. The lymphocytes develop in primary lymphoid tissue, bone marrow, and thymus, and circulate towards secondary lymphoid tissue that comprise lymph nodes (LN), spleen, and MALT. Antigens reaching the LNs are carried by lymphocytes or by dendritic cells via afferent lymphatics. Lymphocytes may also enter LNs from the bloodstream via high endothelial venules. Antigens are processed within the LNs by lymphocytes, macrophages and other immune cells in order to mount a specific immune response. In the other secondary lymphoid tissues, the response mechanism is the same, although the delivery of Ag differs.

The immune system has a unique ability to specifically recognize millions of different antigens in part due to the huge diversity of Ag receptors that exist on B and T lymphocytes. The surface bound Ig exists on B lymphocytes and the TCR on T lymphocytes (Parham, 2014). Each single lymphocyte expresses receptors with identical Ag specificities. The huge diversity of the Ag receptor is built on the process of downstream gene rearrangements of the Ig and TCR molecules. This is an essential process because, if the entire repertoire of Ig/TCR molecules were to be encoded by separate genes, they would occupy a major part of the human genome. However, due to the unique rearrangements and processing of Ag receptors only a limited number of genes are required to encode the receptor diversity (Parham, 2014).

The B lymphocytes developed in the bone marrow (B-cells) produce Ag-specific Ig that function as antibodies (Ab). Igs are proteins secreted by or present on the surface of B-cells assembled from identical couples of heavy (H) and light (L) chains (Figures 9.3 and 9.4). There are five classes of Ig: M, G, A, E, and D distinguished by different heavy chains (Klein and Dalla-Favera, 2008).

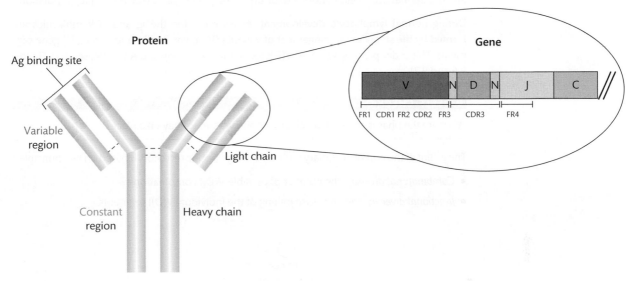

FIGURE 9.3

Structure of an immunoglobulin molecule.

Key: V, variable; D, diversity; J, joining; C, constant.

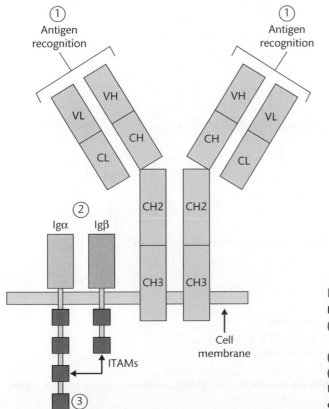

FIGURE 9.4

B-cell receptor complex. This comprises:

(1) An immunoglobulin molecule embedded in the cell membrane.

(2) Associated with it are Igα and Igβ chains that carry

(3) ITAMs to initiate signalling.

Key: ITAM, immunoreceptor tyrosine-based activation motif; C, constant region; V, variable region; H, heavy chain; L, light chain.

In contrast, T lymphocytes arise in the bone marrow and then migrate to the thymus, where they mature to express the Ag-binding TCRs on their cell membranes (Figure 9.5). The TCR is a dimer composed usually of αβ chains. After migrating to the secondary lymphoid organs, naïve T-cells are exposed to Ag that binds to the TCRs. TCR activation induces proliferation and differentiation. T-cells mature into distinct T-helper (Th) and T-cytotoxic (Tc) populations.

During normal lymphocyte development, genes encoding the Ig and TCR molecules are formed by the stepwise rearrangement of variable (V), diversity (D), and joining (J) gene segments. This order process is termed V(D)J recombination (Jung and Alt, 2004; see also Figures 9.6 and 9.7).

9.3 SELF-CHECK QUESTION

What are the phases of B-cell development and where do they take place?

The extent of the potential primary repertoire of Ag-specific receptors is based on two principles:

- *Combinatorial diversity*: the number of possible V(D)J combinations.
- *Junctional diversity*: the imprecise joining of the individual V(D)J segments.

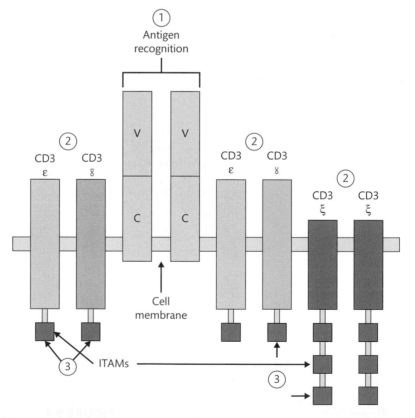

FIGURE 9.5

T-cell receptor complex. This comprises:
(1) A heteroduplex of either TCRα/β or TCR γ/δ chains.
(2) Associated with TCR is the CD3 complex that carry
(3) ITAMs to initiate signalling.
Key: ITAM, immunoreceptor tyrosine-based activation motif; C, constant region; V, variable region.

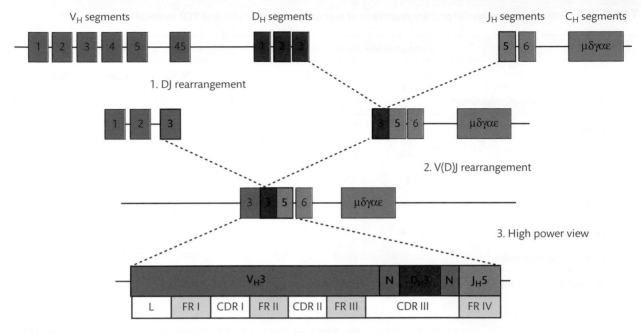

FIGURE 9.6

Gene rearrangement in the IgH locus. V(D)J rearrangement in the IgH locus is initiated by D–J rearrangement followed by V(D)J rearrangement. In between the rearranged V and D, and D and J are N (nucleotides) sequences, inserted by terminal deoxynucleotidyl transferase (TDT).

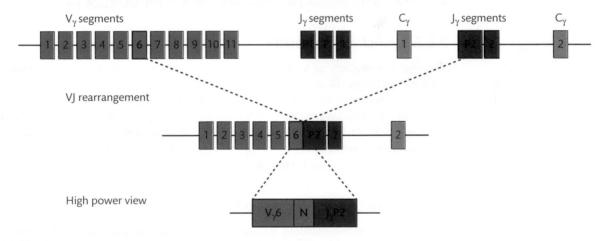

FIGURE 9.7

Gene rearrangement in the TCR gamma locus. V–J rearrangement in the TCR gamma (TCRG) locus is initiated by V–J rearrangement. The TCRG locus does not contain D segments. In between the rearranged V and J are N (nucleotides) sequences, inserted by terminal deoxynucleotidyl transferase (TDT).

9.3.2.2 Combinatorial diversity

The IGH locus is situated on chromosome 14q32.3 and, unlike the IGL locus, contains diversity segments in addition to V, J, and C segments. As the *IGH* gene complex contains approximately 40 V segments, 25 D segments, and 6 J segments recombination produces approximately 6000 V(D)J combinations. Combined with the estimated 175 and 115 V–J

TABLE 9.2 Estimation of potential primary repertoire of human immunoglobulin and TCR molecules

	Immunoglobulin molecules			TCRαβ molecules		TCRγδ molecules	
	IgH	Igκ	Igλ	TCRα	TCRβ	TCRγ	TCRδ
Number of functional gene segments*							
V gene segments	40-46	34-37	30-33	45	44-47	6	6
D gene segments	25	-	-	-	2	-	3
J gene segments	6	5	4	50	13	5	4
Combinatorial diversity	$>10^6$			$>10^6$		>5000	
Junctional diversity	++	±	±	+	++	++	++++
Estimation of total repertoire	$>10^{12}$			$>10^{12}$		$>10^{12}$	

*Numbers are based on the IMGT (ImMunoGeneTics) database (Lefranc, 2003).

combinations of the IGL that leads to a potential combinatorial diversity of more than 2×10^6 (Lefranc, 2003; Table 9.2).

Two different types of TCRs are known: the TCRαβ receptor, which consists of a TCRα and a TCRβ chain and a TCRγδ receptor composed of a TCRγ and a TCRδ chain. Similar diversity to the Ig chains can be obtained for TCRαβ molecules; see Lefranc (2003) and Table 9.2.

9.3.2.3 Junctional diversity

The junctional diversity of Ig/TCR molecules occurs due to the imprecise recombination involving V(D)J region DNA, when exonucleases cause deletions of nucleotides, as well as random insertions of nucleotides, which are mediated by the enzyme terminal deoxynucleotide transferase (TDT; Tuaillon and Capra, 2000). The junctional regions of the *Ig/TCR* genes encode the complementarity determining region 3 (CDR3), which is involved in Ag recognition. The random insertion and deletions of nucleotides significantly increases junctional diversity and provides a unique clonal marker for patients.

9.3.2.4 Somatic hypermutation

The repertoire of immunoglobulin molecules can be further increased through the process of somatic hypermutation (SHM; Odegard and Schatz, 2006). The primary function of SHM is to convert low affinity into high affinity Igs. SHM is a naturally occurring event, where random mutations are introduced into the *Ig* genes. The mutations are predominantly point mutations, although deletions and insertions also exist. This all takes place in a specialized microenvironment within LNs, known as the germinal centre, and results in B-cells that can optimally bind antigen. Those cells that are unable to optimally bind antigen are terminated by apoptosis.

9.4 SELF-CHECK QUESTION

What is meant by the term somatic hypermutation?

9.3.2.5 Class switch recombination

Class switch recombination (CSR) occurs in germinal centres when B-cells switch their Ig class. Typically, this involves a change from M heavy chain to G, while in MALT the switch is to A heavy chain. The result is enhanced effector functions for the antibody leaving the V(D)J region unaltered. Of interest, CSR and SHM do not occur in the DNA of T-cells, which could partly explain why B-cells are more prone to undergo malignant transformation than T-cells.

Rearrangements of *Ig* and *TCR* genes occur at the earliest stage of lymphoid differentiation and in a hierarchical order. During B-cell differentiation *IGH* genes rearrange first followed by *IGKappa* (*IGK*). If *IGK* rearrangements are non-functional then *IGLambda* (*IGL*) will start to rearrange. In T-cell differentiation, the *TCRD* genes rearrange first, followed by the *TCRG* genes. *TCRB* gene rearrangements take place prior to *TCRA* gene rearrangements (van Dongen et al., 1991a).

9.3.3 Application of clonality to diagnosis

As lymphoid malignancies are counterparts of normal lymphoid cells Ig/TCR rearrangements are present in almost all immature and mature lymphoid malignancies. As they are derived from a single transformed lymphoid cell they have their own unique Ig/TCR rearrangement, which can be readily employed for clonality assessment in lymphoid proliferations (van Dongen et al., 1991b). The variation that exists within Ig/TCR rearrangements will, therefore, identify the presence of polyclonal-activated cells as opposed to identically rearranged *Ig* or *TCR* genes reflecting the presence of a monoclonal and usually malignant lymphoid cell population.

Historically, Ig/TCR clonality testing was performed by means of Southern blot analysis. This is based on size detection of the restriction enzyme fragments that discriminate between the presence of germline smears in reactive polyclonal conditions and distinct bands, due to predominance of a single rearrangement in monoclonal cell proliferations (Beishuizen et al., 1993; van Krieken et al., 1991). This technique was both time-consuming and cumbersome, and was coupled with low sensitivity. Another major disadvantage was the need for comparatively large amounts of high molecular weight DNA, preventing the analysis of small biopsies and the routine use of FFPE material. This was due to the low amount of DNA available from biopsies and compromised DNA recovered from FFPE samples (see Chapter 1, Box 1.3). To compensate for this, PCR strategies were developed to assess clonality in Ig/TCR rearrangements (Brisco et al., 1990; McCarthy et al., 1990).

As most Ig/TCR studies in lymphomas are performed at the DNA level, PCR primers are used that are complementary to exons/intron sequences of V(D)J gene segments (Figures 9.6 and 9.7). In the first instance, consensus primers were designed that recognized particular gene segments. At the onset, as compared with Southern blot analysis, PCR-based Ig/TCR gene analysis had two major pitfalls:

- The risk of false negative results due to the inappropriate recognition of all Ig/TCR segments.
- The risk of false positive results due to the poor discrimination between polyclonal and monoclonal Ig/TCR rearrangements.

Given the numerous drawbacks with PCR Ig/TCR clonality, it was evident that a standardized approach was required to address these issues. In the late 1990s, a large European consortium of laboratories (BIOMED-2 Concerted Action BMH4-CT98-3936) was devised to establish a highly reproducible PCR-based clonality assessment tool (van Dongen et al., 2003). The issue of false negativity was addressed at a number of levels:

- Design of complete sets of primers to cover all possible V-J rearrangements of Ig/TCR loci.
- Inclusion of incomplete rearrangements in additional targets.
- Inclusion of multiple Ig and TCR targets to detect clonality.

The issue of false positivity was also addressed by the introduction of reliable methods for the evaluation of PCR products (Kneba et al., 1995; Langerak et al., 1997).

The BIOMED-2 clonality assay has been validated in many datasets and has now become a world standard for clonality assessment in lymphoproliferative disorders (Brüggemann et al., 2007; Evans et al., 2007; van Krieken et al., 2007).

Key Point

All forms of cancer are derived from a single cell that has acquired a number of generic abnormalities. During their normal development, T and B lymphocytes rearrange the genes coding for the T-cell receptor and immunoglobulin molecules. This mechanism allows the immune system to respond to a huge range of infections and foreign proteins and results in each individual cell having a unique genetic signature. In a lymphoid malignancy derived from a single T- or B-cell, all of the cells will share the same signature (a monoclonal population). Monoclonality can be determined phenotypically using flow cytometry and/or immunocytochemistry or geneotypically, usually using PCR methodologies.

9.3.4 Applications of PCR-based clonality testing

Although multiple applications of Ig/TCR clonality testing can be defined three broad scenarios exist (Table 9.3).

- Initial diagnosis.
- Assessment of minimal residual disease status.
- Prognostic subclassification of lymphoid malignancies.

9.3.4.1 Initial diagnosis

Normally in a diagnostic setting, the discrimination between neoplastic and benign lymphoid proliferations can be achieved through a combination of morphology and immunophenotype (see Section 9.3.1). However, in equivocal cases, Ig/TCR clonality provides a useful tool. The discrimination of tumour cells from reactive or normal lymphocytes is the most widely used approach with the current protocols being very reliable. Ig/TCR clonality testing is also very useful in establishing the clonal relationship between multiple lesions at the same time or in the case of recurrence of the disease (establishing relapse against a second malignancy).

Lineage determination (the assessment of either B-cell or T-cell origin of a tumour) may also be demonstrated by Ig/TCR assays, but this is hampered by the fact that the rearrangements are not exclusive for immature B and T lymphocytes, and can occur as cross-lineage rearrangements (Szczepanski et al., 1998).

TABLE 9.3 Applications of Ig/TCR clonality testing

Situation	Description
Diagnosis	
Neoplastic versus reactive	Atypical lymphoproliferations in which the lesion is either malignant or reactive by morphological analysis
Limited tissue	When small needle biopsies prevent the evaluation of the tissue architecture
Lineage determination	The assessment of either B-cell origin or T-cell origin of a tumour
Evaluating clonal relationships	Evaluation of clonal relationship between diagnosis and disease recurrence (relapse)
Minimal residual disease	Monitoring the effectiveness of treatment
Prognosis	Somatic hypermutation of the immunoglobulin heavy chain hypervariable region (IGHV)

9.3.4.2 Minimal residual disease (MRD)

Current treatment regimens induce complete remission in certain ALL and NHL patients, but many relapse with the same clone that was found at diagnosis. MRD describes the lowest level of disease detectable using available methods. When a patient is diagnosed, it is common for them to have approximately 10^{13}–10^{14} malignant cells. After successful treatment, a patient is described as being in remission, but malignant cells can still be present at a level, below the detection limit of conventional techniques. A well-recognized approach for MRD testing is in ALL, in which the ability to reduce the level of disease below a certain level is considered a favourable prognostic variable.

The junctional regions of the rearranged Ig/TCR genes are unique markers and can be used as tumour-specific targets for MRD-PCR analysis. The Ig/TCR tumour-specific clone is first recognized at diagnosis and the precise nucleotide junctional region (CDR3) is determined that acts as a molecular fingerprint. The current detection method of MRD using junctional regions of Ig/TCR is performed with real-time quantitative PCR (qPCR).

Cross reference

See Chapter 4 for more discussion of real-time PCR.

The quantitation of Ig/TCR clone-specific rearrangements by qPCR is now an internationally standardized technique. However, the advent of next-generation sequencing (NGS) technology has provided a proof of principle that its use for MRD assessment is feasible and, potentially, even more sensitive than the standard qPCR options (Faham et al., 2012).

Cross reference

See Chapter 5 for more discussion of NGS.

9.3.4.3 Prognostic subclassification of lymphoid malignancies

Chronic lymphocytic leukaemia/small lymphocytic lymphoma (CLL/SLL) is a clinically heterogeneous disorder, mainly showing an indolent disease course, but behaving more aggressively in a subset of patients. Various clinical and biological parameters have been associated with survival differences. However, the somatic hypermutation status of the variable region of the *IGH* gene (IGHV) has been determined as prognostically informative in CLL/SLL. CLL/SLL cases that show *IG* genes without somatic mutation in the *VH* gene segment (UNMUTATED) generally have a less favourable prognosis than cases that do show a *VH* mutation (MUTATED; Baliakas et al., 2014). The VH status in CLL/SLL cases is determined by amplification of the *Ig* gene followed by sequencing of the VH segment (Langerak et al., 2011).

9.5 SELF-CHECK QUESTION

When are clonality studies by PCR indicated?

9.3.5 Translocation detection

9.3.5.1 Chromosomal translocation

Many haematological malignancies are characterized by chromosomal translocations. This occurs when there is a reciprocal exchange of DNA between two chromosomes. In certain cases, this can lead to the splicing together of two genes (or parts of genes), which results in the production of a new and abnormal protein. Generally, this can lead to two types of gene activation:

- Formation of a chimeric gene.
- Production of a fusion protein or overexpression of a normal gene due to action of a translocated enhancer region.

Chromosomal translocations that result in the formation of a fusion gene are common in leukaemias and less so in NHL. Each one is identified according to a standard nomenclature system (see Box 9.1). A well-characterized example is that of the *NPM1-ALK* fusion/t(2;5), found in certain cases of anaplastic large cell lymphoma (Morris et al., 1994). The translocation produces a fusion or chimeric protein, and results in the normal wild-type function of the protein being abolished. The resulting chimeric protein acts as an aberrant transcription factor causing unregulated cell division.

Chromosomal translocations that result in the deregulation of a normal proto-oncogene usually involve the genes encoding for the *Ig* genes and, in particular, the heavy-chain gene due to somatic hypermutation and class switch recombination. Typically, the rearrangement places the proto-oncogene under the control of the IGH enhancer resulting in the constitutive activation thereby conferring a survival advantage to the B-cell. Many examples exist, such as the

BOX 9.1 *Translocation nomenclature*

When describing a translocation, a lower case 't' is used followed by two pairs of brackets. Within the first pair of brackets, the chromosome numbers involved in the translocation are presented in numerical order. Each chromosome is separated by a semicolon, e.g. t(14;18). Within the second set of brackets, breakpoints are included in an order that allows their relationship the chromosomes to be identified. Each breakpoint is also separated by a semicolon.

Example: A translocation between chromosomes 14 and 18 found in follicular lymphoma would be abbreviated as follows: t(14;18)(q32;q21). This particular translocation leads to the overexpression of the anti-apoptotic protein BCL-2.

overexpression of *BCL2* and *CCND1* as a consequence of being juxtaposed with the *Ig* gene in the t(14;18) and t(11;14) translocations, that are associated with follicular lymphoma and mantle cell lymphoma, respectively (Yunis et al., 1982; Tsujimoto et al., 1984). The causes for the generation of DNA strands-breaks in oncogenes in Ig-associated translocations have yet to be fully elucidated. However, certain genes undergo somatic hypermutation and, therefore, acquire DNA strand-breaks in regions where chromosomal translocations are located, e.g. Bcl-2-IgH in follicular lymphoma.

A further mechanism of class-switch recombination can also cause DNA strand-breaks and, as already discussed, it is of note that both of these processes do not occur in the DNA of T-cells, which may offer an explanation as to why B-cells are more prone to malignant transformation than T-cells (Lieber, 2016).

Molecular genetics has become increasingly used in the clinical practice of lymphoma diagnosis. These can be broadly divided into cytogenetic, FISH, and molecular genetic tests such as PCR or NGS.

Key Point

Reciprocal chromosomal translocations involving one of the Ig loci are a key feature of B-cell lymphoma. This causes the oncogene to come under the control of the Ig locus causing a deregulated expression of the former.

9.6 SELF-CHECK QUESTION

What are the two types of chromosomal translocations seen in lymphoid neoplasms?

9.3.5.2 Cytogenetic detection of translocations

Cytogenetic analysis (see Box 1.2) allows the evaluation of the state of chromosomes in the lymphoma cells to determine if they have any gross abnormalities. Several lymphomas have deletions or amplifications of parts of their chromosomes, although there are common balanced reciprocal translocations, which are characteristic of different types of lymphomas. Cytogenetic testing usually takes about 2–3 weeks because the lymphoma cells need to divide before their chromosomes are ready to be processed for viewing under the microscope. The detection of cytogenetic abnormalities in this way is infrequently requested due to the length of the procedure, the need for viable cells, and the high workload involved with the technique.

The detection of chromosomal translocations in lymphoma can be helpful as an additional clonality marker in those situations in which this cannot be firmly established by means of Ig/TCR clonality analysis. The translocations often involve the antigen receptor (*Ig/TCR*) genes with a partner oncogene. The oncogene is juxtaposed to the Ag receptor gene resulting in the inappropriate expression of the oncogene involved in cell cycle regulation (e.g. cyclin D1, *CCND1*) or apoptotic (e.g. BCL2) processes. However, a far more relevant application lies in the classification of the lymphoma. Correct classification is not only important with respect to prognosis, but also has implications for tailored therapy strategies. Specific translocations involving the IGH locus on chromosome 14 are associated with different types of lymphomas (Table 9.4) representing early B-cell (IGH-BCL2 or IGH-MALT1) or mature B-cell (IGH-MYC) development.

TABLE 9.4 Chromosomal abnormalities detected in lymphoproliferative disorders

Genetic abnormality	Oncogene	Lymphoma
t(8;14)(q24;q32)	MYC-IGH	Burkitt* DLBCL (10%)
t(8;22)(q24;q11)	MYC-IGK	Burkitt* DLBCL
t(2;8)(p12;q24)	MYC-IGL	Burkitt* DLBCL
t(14;18)(q32;q21)	IGH-BCL2	Follicular (90%) DLBCL (25%)
t(11;14)(q13;q32)	CCND1-IGH	Mantle cell (95%)
t(11;18)(q21;q21)	API2-MALT	MALT lymphoma
t(1;14)(p22;q32)	BCL10-IGH	MALT lymphoma
t(2;5)(p23;q35)	ALK-NPM	ALCL ALK+ (>25%)
t(1;2)(q21;p23)	TPM3-ALK	ALCL ALK+
3q27 translocations	BCL6	DLBCL (30-35%)

*>95% of Burkitts have one of these translocations.

DLBCL, diffuse large B-cell lymphoma; ALCL, anaplastic large cell lymphoma; MALT, mucosa-associated lymphoid tissue; ALK, activin receptor-like kinase.

9.3.5.3 *Fluorescent* in situ *hybridization (FISH)*

FISH utilizes nucleic acid probes labelled with fluorescent dyes that bind to specific genes or parts of chromosomes. FISH is very sensitive and can demonstrate most chromosome changes, particularly translocations, that can be detected in standard cytogenetic tests, as well as some deletions or insertions that are too small to be seen with normal cytogenetic testing. FISH can be used on blood, bone marrow, and FFPE tissue samples. It is very accurate and can usually provide results within a couple of days, which is why this test is now popular for diagnostic laboratories. In particular, it is used to demonstrate translocations involving *MYC* or *BCL2*, which are crucial to support, respectively, the diagnosis of Burkitt or follicular lymphoma in difficult cases. The disadvantage of FISH is that the probes used are specific for particular translocation breakpoints. This means that the overall pattern of cytogenetic disruption is not detected without conventional cytogenetic karyotyping.

In the classical FISH strategy, also referred to as fusion FISH, two differently labelled probes (mainly red and green fluorochromes) are selected for each of the genes involved in the translocation of interest. When a translocation occurs, co-localization of red and green signals result in the formation of a fused (yellow) signal (Figure 9.8).

FISH detection techniques have been thoroughly optimized and standardized. However, there are limitations to the technique, including a high false-positive rate due to co-localization of signals when viewing in two dimensions. Accordingly, attempts have been made to address these issues with the use of split-signal and break-apart FISH. In this approach, only probes that target one of the genes involved in the translocation are used.

(A)   (B)

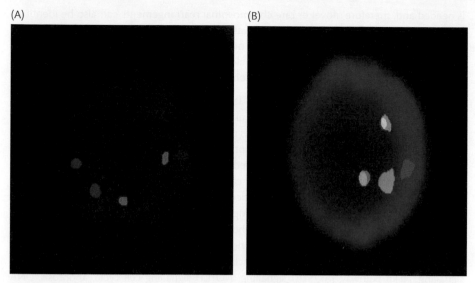

FIGURE 9.8

FISH testing for MALT Lymphoma-associated t(11;18)(q21;q21) translocation. FISH testing using probes for 11q22 and 18q21. Image A shows a normal nucleus with separated red and green genes. Image B shows the t(11;18) translocation as depicted by the yellow (red-green) fusion signal.

9.7 SELF-CHECK QUESTION

What are the advantages of FISH over conventional cytogenetics?

9.3.6 Next-generation sequencing (NGS)

NGS studies have led not only to major advances in better understanding of the lymphoproliferative neoplasms, but have contributed to the identification of mutations of diagnostic importance. For example, *BRAF* V600E mutations are found in almost all cases of HCL, but not in HCL-variant (HCL-v) or other small B-cell lymphoid neoplasms (Tiacci et al., 2011). More recently, mutations in *MAP2K1*, which encodes MEK1 downstream of BRAF, have been reported in almost half of HCL-v and in the majority of HCL that use IGHV4-34 and which, like HCL-v, lack *BRAF* V600E mutations (Waterfall et al., 2014).

Detection and characterization of clonal Ig/TCR rearrangements and translocations in lymphoproliferative neoplasms can be time-consuming as they include ascertaining the clonal nature of lymphoid proliferations, translocation detection, and MRD target identification. Until recently, collecting this information required a combination of different methodologies, including FISH and Sanger sequencing. The incorporation of NGS provides an opportunity for an integrated approach on sequence and structural variation in a single assay, including translocations and IG/TCR rearrangements (Wren et al., 2017).

Within the EuroClonality-NGS consortium, a capture-based protocol covering the coding *V*, *D*, and *J* genes of the IG/TCR loci has been designed. This approach allows the identification of D–J and V(D)J rearrangements, as well as chromosomal translocations involving *IG/TCR* genes by sequencing through the breakpoint regions in genomic DNA. An important advantage of these approaches lies in the fact that no prior knowledge of the translocation partner

Cross reference

Hairy cell leukaemia (HCL) is discussed in Chapter 8, Section 8.5.4.

is needed and, therefore, novel or rare chromosomal rearrangements can also be identified by this method, improving their diagnostic value. Sequencing of the V(D)J gene rearrangements in any of the IG/TCR loci can be used not only to assess clonality, and enable a more in-depth analysis of clonal relationships and clonal evolution, but also to identify targets for monitoring and analysis of the IG/TCR repertoire of diverse lymphoid populations.

NGS has also contributed to the mutational landscape of lymphoid neoplasms, identifying many genes that are mutated (Rosenquist et al., 2016). It has become possible to appreciate the vast spectrum of recurrently affected genes that contribute to disease pathogenesis and/or evolution in major lymphoma subtypes. Mounting evidence suggests that certain gene mutations have diagnostic, prognostic, and/or predictive impact. In few circumstances, a single recurrent mutation is identified in almost all cases of a given lymphoma and predominates by far in the genomic landscape of that particular tumour, e.g. the $MYD88^{L265P}$ mutation in Waldenström's macroglobulinemia (WM)/lymphoplasmacytic lymphoma (LPL; Treon et al., 2012). However, for the great majority of lymphomas studies NGS have revealed a quite diverse and complex mutation pattern, with a limited number of frequently mutated genes. Mutations are commonly found in cellular processes, including chromatin remodelling, B-cell receptor signalling, NFkB pathway, JAK/STAT signalling, NOTCH signalling cell cycle, apoptosis, DNA repair, and immune surveillance. NGS has also revealed that NHL harbour structural aberrations (Rosenquist et al., 2016; Wren et al., 2017), such as translocations, copy number variants, and point mutations. NGS has enormous potential to detect both known and unknown genetics changes that may play a critical role in understanding the pathogenesis of NHL.

NGS has the potential to provide the basis for precision medicine providing the identification of targetable genes or signalling pathways in individual patients. NGS has, for example, demonstrated that responders to ibrutinib in diffuse large B-cell lymphomas often carry mutations of both *MYD88* and *CD79A/B*, and a wild type *CARD11* (Wilson et al., 2015).

9.4 **Future directions**

Cross reference

See Chapter 7 for more details about liquid biopsy.

The concept of the 'liquid biopsy' has become accessible by using circulating tumour DNA (ctDNA) as primary material (Schwarzenbach et al., 2011). ctDNA is shed into the blood by tumour cells undergoing apoptosis and can be used as source of tumour DNA for the identification of mutations, clonal evolution, and genetic mechanisms of resistance.

NGS techniques enable the quantitation of ctDNA encoding the clonal rearranged V(D)J *Ig* receptor gene sequence of tumour cells (Faham et al., 2012). This has been successfully demonstrated in DLBCL, mantle cell lymphoma (MCL), chronic lymphocytic leukaemia (CLL), multiple myeloma and acute lymphoid leukaemia, and can be used for MRD detection (Logan et al., 2013; Martinez-Lopez et al., 2014; Roschewski et al., 2015). NGS results have also demonstrated that ctDNA genotyping is as accurate as genotyping of the diagnostic biopsy to detect clonally represented somatic tumour mutations. This method of testing represents a real-time and non-invasive approach to track clonal evolution and emergence of treatment resistance.

Signatures by quantitative multiplex polymerase chain reaction (PCR) of short fluorescent fragments (QMPSF) or qPCR of selected genes are promising as prognostic tools, for example, in diffuse large B-cell lymphoma, and may soon be incorporated into clinical practice (Malumbres et al., 2008; Jardin et al., 2010). Standardization of such techniques on frozen and paraffin-embedded tissue has, nevertheless, to be addressed.

CASE STUDY 9.1 A 53-year-old female presents with a painless right breast lump noted on self-examination

A 53-year-old woman presented to a symptomatic breast clinic with a 1-month history of a painless right breast lump noted on self-examination (Windrum et al., 2001). No nipple bleeding or discharge was present, and there was no family history of breast carcinoma. On examination, she had a hard 3-cm diameter mass above the right nipple. No lymphadenopathy or organomegaly was detected.

Full blood count revealed a mild lymphocytosis for her age (total leucocyte count, 11.83 $\times 10^9$/L; lymphocyte count, 4.81 $\times 10^9$/L; neutrophil count, 6.04 $\times 10^9$/L). The erythrocyte sedimentation rate was 19 mm/hour. Biochemical screening tests and immunoglobulin levels were within normal limits. Mammography showed an asymmetric density above the right nipple, with distortion of the surrounding breast tissue, but no evidence of calcification.

A fine needle aspirate was performed and the resultant cytology was suspicious of malignancy. A core biopsy was then taken and the patient was referred for haematology assessment. A computed tomography scan of chest, abdomen, and pelvis did not reveal further evidence of disease.

The core biopsy of the breast lesion showed a B-cell NHL, probably of large cell type and of high grade. Morphological and immunophenotypic analysis of peripheral blood and bone marrow samples was suggestive of mantle cell lymphoma. This was confirmed by the detection of a t(11;14)(q13;q32) in the bone marrow aspirate and breast tissue by FISH.

The patient received standard chemotherapy, but due to residual disease the patient proceeded to high dose chemotherapy with autologous peripheral blood stem cell transplantation.

CASE STUDY 9.2 A 69-year-old male presented to his GP with a cough and left groin pain

A 69-year-old male presented to his GP with a cough and left groin pain. Left axillary and inguinal lymph nodes showed non-tender lymphadenopathy. Full blood count revealed a marked lymphocytosis (total leucocyte count, 45.53 $\times 10^9$/L; lymphocyte count, 43.81 $\times 10^9$/L; neutrophil count, 6.04 $\times 10^9$/L). Biochemical screening tests were within normal limits.

Immunophenotyping was performed on the blood sample and demonstrated a population of B lymphocytes positive for surface immunoglobulin, CD10, CD19, and CD20.

The patient underwent computed tomography (CT) of the chest with contrast due to his coughing and axillary lymphadenopathy, and a CT of the pelvis due to his inguinal lymphadenopathy. The thorax CT with contrast revealed an increase in the number of mediastinal and axillary lymph nodes.

An inguinal node biopsy showed BCL-2, Ki-67, CD20, CD3, and CD10 positive monoclonal kappa B-cell population consistent with a B-NHL and suggestive of follicular lymphoma. This was confirmed by the detection of a t(14;18)(q32;q21) in the blood and inguinal biopsy by FISH.

The patient received standard chemotherapy and remained well, with no evidence of local or systemic disease.

Chapter summary

This chapter has:

- Provided an overview of the criteria used in the current WHO classification of lymphoid neoplasms and how this has evolved to its current state.
- Described the role of immunoglobulin and T-cell receptor rearrangements in lymphoid neoplasms and their role in clonality detection.
- Discussed a variety of laboratory techniques that are used in the diagnosis and monitoring of lymphoid neoplasms.
- Detailed some of the common cytogenetic abnormalities associated with a range of lymphoid neoplasms.
- Provided an insight into the potential role of NGS for the diagnosis and monitoring of lymphoid neoplasms.

Further reading

Books

- Swerdlow, S.H., Campo, E., Harris, N.L., Jaffe, E.S., Pileri, S.A., Stein, H., Thiele, J., Vardiman, J.W (Eds). *WHO Classification of Tumours of Haematopoietic and Lymphoid Tissue*. Geneva, IARC, WHO Classification of Tumours, 2008.

Journal articles

- Bollard CM, Gottschalk S, Leen AM, et al. Complete responses of relapsed lymphoma following genetic modification of tumour-antigen presenting cells and T-lymphocyte transfer. Blood 2007; 110: 2838–2845.

- Jevremovic D, Viswanatha DS. Molecular diagnosis of hematopoietic and lymphoid neoplasms. Hematology & Oncology Clinics of North America 2009; 4: 903–933.

- Keen-Kim D, Nooraie F, Rao PN. Cytogenetic biomarkers for human cancer. Frontiers in Biosciences 2008; 13: 5928–5949.

- Rowley JD. Chromosomal translocations: revisited yet again. Blood 2008; 112: 2183–2189.

- Tonegawa S. Somatic generation of antibody diversity. Nature 1983; 302: 575–581.

Useful websites

- *EuroClonality*: http://www.euroclonality.org/.
- *Haematological Malignancy Diagnostic Service*: http://hmds.info/.
- *Leukaemia and Lymphoma Society*: http://www.lls.org/.

References

- Baliakas P, Hadzidimitriou A, Sutton LA, et al. Clinical effect of stereotyped B-cell receptor immunoglobulins in chronic lymphocytic leukaemia: a retrospective multicentre study. Lancet Haematology 2014; 1: e74–e84.

- Beishuizen A, Verhoeven MA, Mol EJ, et al. Detection of immunoglobulin heavy-chain gene rearrangements by Southern blot analysis: recommendations for optimal results. Leukemia 1993; 7: 2045–2053.

- Brisco MJ, Tan LW, Orsborn AM, et al. Development of a highly sensitive assay, based on the polymerase chain reaction, for rare B-lymphocyte clones in a polyclonal population. British Journal of Haematology 1990; 75: 163–167.

- Brüggemann M, White H, Gaulard P, et al. Powerful strategy for polymerase chain reaction-based clonality assessment in T-cell malignancies. Report of the BIOMED-2 Concerted Action BHM4 CT98-3936. Leukemia 2007; 21: 215–221.

- Craig FE, Foon KA. Flow cytometric immunophenotyping for haematological neoplasms. Blood 2008; 111: 3941–3967.

- Evans PA, Pott CH, Groenen PJ, et al. Significantly improved PCR-based clonality testing in B-cell malignancies by use of multiple immunoglobulin gene targets. Report of the BIOMED-2 Concerted Action BHM4-CT98-3936. Leukemia 2007; 21: 207–214.

- Faham M, Zheng J, Moorhead M, et al. Deep-sequencing approach for minimal residual disease detection in acute lymphoblastic leukemia. Blood 2012; 120: 5173–5180.

- Harris NL, Jaffe, ES, Stein H, et al. A revised European–American classification of lymphoid neoplasms: a proposal from the International Lymphoma Study Group. Blood 1994; 84: 1361–1392.

- Jaffe ES. Harris NL. Stein H. Vardiman J.W. *Pathology and Genetics of Tumours of Haematopoietic and Lymphoid Tissues: 2001 WHO/IARC Classification of Tumours.* Lyon, IARC Publications, 2001.

- Jardin F, Jais JP, Molina TJ, et al. Diffuse large B-cell lymphomas with CDKN2A deletion have a distinct gene expression signature and a poor prognosis under R-CHOP treatment: a GELA study. Blood 2010; 116: 1092–1104.

- Jung D, Alt FW. Unraveling V(D)J Recombination: insights into gene regulation. Cell 2004; 116: 299–311.

- Klein U, Dalla-Favera R. Germinal centres: role in B-cell physiology and malignancy. Nature Reviews Immunology 2008; 8: 22–33.

- Kneba M, Bolz I, Linke B, et al. Analysis of rearranged T-cell receptor beta-chain genes by polymerase chain reaction (PCR) DNA sequencing and automated high resolution PCR fragment analysis. Blood 1995; 86: 3930–3937.

- Langerak AW, Szczepański T, van der Burg M, et al. Heteroduplex PCR analysis of rearranged T cell receptor genes for clonality assessment in suspect T cell proliferations. Leukemia 1997; 11: 2192–2199.

- Langerak AW, Davi F, Ghia P, et al. Immunoglobulin sequence analysis and prognostication in CLL: guidelines from the ERIC review board for reliable interpretation of problematic cases. Leukemia 2011; 25: 979–984.

- Lefranc MP. IMGT databases, web resources and tools for immunoglobulin and T cell receptor sequence analysis, http://imgt.cines.fr. Leukemia 2003; 17: 260–266.

- Letai AG, Gribben JG. In: Provan D and Gribben JG (Eds). *Molecular Haematology*, 3rd edn. Oxford: Wiley-Blackwell, 2010, pp. 117–126.

- Lieber MR. Mechanisms of human lymphoid chromosomal translocations. Nature Reviews: Cancer 2016; 16: 387–398.

- Logan AC, Zhang B, Narasimhan B, et al. Minimal residual disease quantification using consensus primers and high-throughput IGH sequencing predicts post-transplant relapse in chronic lymphocytic leukemia. Leukemia 2013; 27: 1659–1665.

- Martinez-Lopez J, Lahuerta JJ, Pepin F, et al. Prognostic value of deep sequencing method for minimal residual disease detection in multiple myeloma. Blood 2014; 123: 3073–3079.

- Malumbres R, Chen J, Tibshirani R, et al. Paraffin-based 6-gene model predicts outcome in diffuse large B-cell lymphoma patients treated with R-CHOP. Blood 2008; 111: 5509–5514.

- McCarthy KP, Sloane JP, Wiedemann LM. Rapid method for distinguishing clonal from polyclonal B cell populations in surgical biopsy specimens. Journal of Clinical Pathology 1990; 43: 429–432.

- Morris SW, Kirstein MN, Valentine MB, et al. Fusion of a kinase gene, ALK, to a nucleolar protein gene, NPM, in non-Hodgkin's lymphoma. Science 1994; 263: 1281–1284.

- Odegard VH, Schatz DG. Targeting of somatic hypermutation. Nature Reviews: Immunology 2006; 6: 573–583.

- Parham P. *The immune system*, 4th edn. New York, Garland Science Publishing, 2014.

- Roschewski M, Dunleavy K, Pittaluga S, et al. Circulating tumour DNA and CT monitoring in patients with untreated diffuse large B-cell lymphoma: a correlative biomarker study. Lancet: Oncology 2015; 16: 541–549.

- Rosenquist R, Rosenwald A, Du MQ, et al. Clinical impact of recurrently mutated genes on lymphoma diagnostics: state-of-the-art and beyond. Haematologica 2016; 101: 1002–1009.

- Schwarzenbach H, Hoon DS, Pantel K. Cell-free nucleic acids as biomarkers in cancer patients. Nature Reviews: Cancer 2011; 11: 426–437.

- Swerdlow SH, Campo E, Harris NL, et al. WHO *Classification of Tumours of Haematopoietic and Lymphoid Tissues: 2008 WHO/IARC Classification of Tumours*, 4th edn. Lyon, IARC, 2008.

- Swerdlow SH, Campo E, Pileri SA, et al. The 2016 revision of the World Health Organization classification of lymphoid neoplasms. Blood 2016; 127: 2375–2390.

- Szczepański T, Langerak AW, van Dongen JJ, et al. Lymphoma with multi-gene rearrangement on the level of immunoglobulin heavy chain, light chain, and T-cell receptor beta chain. American Journal of Hematology 1998; 59: 99–100.

- Tiacci E, Trifonov, Schiavoni G, et al. BRAF mutations in hairy-cell leukemia. New England Journal of Medicine 2011; 364: 2305–2315.

- Treon SP, Xu L, Yang G, et al. MYD88 L265P somatic mutation in Waldenström's macroglobulinemia. New England Journal of Medicine 2012; 367: 826–833.

- Tsujimoto Y, Finger LR, Yunis J, et al. Cloning of the chromosome breakpoint of neoplastic B cells with the t(14;18) chromosome translocation. Science 1984; 226: 1097–1099.

- Tuaillon N, Capra JD. Evidence that terminal deoxynucleotidyltransferase expression plays a role in Ig heavy chain gene segment utilization. Journal of Immunology 2000; 64: 6387–6397.

- van Dongen JJ, Wolvers-Tettero IL. Analysis of immunoglobulin and T cell receptor genes. Part I: Basic and technical aspects. Clinica Chimica Acta 1991a; 198: 1–91.

- van Dongen JJ, Wolvers-Tettero IL. Analysis of immunoglobulin and T cell receptor genes. Part II: Possibilities and limitations in the diagnosis and management of lymphoproliferative diseases and related disorders. Clinica Chimica Acta 1991b; 198: 93–174.

- van Dongen JJ, Langerak AW, Brüggemann M, et al. Design and standardization of PCR primers and protocols for detection of clonal immunoglobulin and T-cell receptor gene recombinations in suspect lymphoproliferations: report of the BIOMED-2 Concerted Action BMH4-CT98-3936. Leukemia 2003; 17: 2257–2317.

- van Krieken JH, Elwood L, Andrade RE, et al. Rearrangement of the T-cell receptor delta chain gene in T-cell lymphomas with a mature phenotype. American Journal of Pathology 1991; 139: 161–168.

- van Krieken JH, Langerak AW, Macintyre EA, et al. Improved reliability of lymphoma diagnostics via PCR-based clonality testing: report of the BIOMED-2 Concerted Action BHM4-CT98-3936. Leukemia 2007; 21: 201–206.

- Waterfall JJ, Arons E, Walker RL, et al. High prevalence of MAP2K1 mutations in variant and IGHV4-34-expressing hairy-cell leukemias. Nature: Genetics 2014; 46: 8–10.

- Wilson WH, Young RM, Schmitz R, et al. Targeting B cell receptor signaling with ibrutinib in diffuse large B cell lymphoma. Nature: Medicine 2015; 21: 922–926.

- Windrum P, Morris TC, Catherwood M.A, et al. Mantle cell lymphoma presenting as a breast mass. Journal of Clinical Pathology 2001; 54: 883–886.

- Wren D, Walker BA, Brüggemann M, et al. Comprehensive translocation and clonality detection in lymphoproliferative disorders by next-generation sequencing. Haematologica 2017; 102: e57–e60.

- Yunis JJ, Oken MM, Kaplan ME, et al. Distinctive chromosomal abnormalities in histologic subtypes of non-Hodgkin's lymphoma. New England Journal of Medicine 1982; 307: 1231–1236.

 # Discussion questions

9.1 Discuss the role of clonality testing in the diagnosis and monitoring of lymphoproliferative disorders.

9.2 Discuss the role of cytogenetics in the investigation of a lymphoid malignancy providing an example in your answer.

Answers to the self-check questions and tips for responding to the discussion questions are provided on the book's accompanying website:

Visit: www.oup.com/uk/warford

10

Breast Cancer

Mary Falzon, Elaine Borg, and Alexandra Saetta

Learning objectives

After studying this chapter, you should be able to:

- Understand the changing landscape in breast cancer diagnosis and treatment options.
- Link the expression of hormone and HER2 receptors with targeted therapy and increased survival.
- Recognize the role of molecular profiling in directing the use of chemotherapy.
- Recognize the potential of using gene expression profiling to provide prognostic and predictive information.
- Explain why immunotherapy may prove beneficial to those patients with triple negative breast cancers that are hard to treat with conventional therapy.
- Appreciate the importance of using molecular profiling to subtype breast cancer so that informed treatment decisions can be made.

10.1 Introduction

Breast cancer is the most commonly diagnosed cancer in women in the United Kingdom (UK). Each year 53,000 women are diagnosed with the cancer and 5500 additional women are also diagnosed annually with an earlier (non-invasive) form of breast cancer, called *in situ* breast ductal carcinoma (DCIS). Breast cancer is second only to lung cancer as the cause of cancer death in women in the UK with 11,470 women dying of the condition in 2013. Set against these figures, it is important to note that, since peaking in the mid-1980s, female breast cancer death rates have fallen by 40% in the UK. This can be attributed to the introduction of national screening for the cancer in 1988 and targeted therapies, such as tamoxifen drug treatment for hormone receptor-positive breast cancers (http://scienceblog.cancer-researchuk.org/2012/10/15/high-impact-science-tamoxifen-the-start-of-something-big/). Indeed, in the last 10 years female death rates for breast cancer in the UK have fallen by around a fifth.

For many years, patient age, axillary lymph node status, tumour size, histological features (especially histological grade and lymphovascular invasion), hormone receptor status, and *HER2* status have been the major factors used to categorize patients with breast cancer in order to assess prognosis and determine the appropriate therapy. More recently, various molecular techniques, in particular gene expression profiling, have been increasingly used to help refine breast cancer classification, and to assess prognosis and response to therapy. Immunotherapy is also currently being explored in triple negative breast cancers (hormone receptors and *HER2*- negative) and metaplastic breast cancers, and may have a role to play in these poor prognosis tumours.

This chapter seeks to highlight the impact of these advances on the molecular classification of breast cancer, and their link to targeted treatment and management of this disease.

10.2 Histopathological classification

There are several types of breast cancers, but they are all adenocarcinomas arising from the terminal duct lobular units (see Figure 10.1). Breast cancers predominantly fall into two main categories, the non-invasive and invasive. Infiltrating ductal carcinoma is the most common histologically diagnosed breast cancer in women and men accounting for about 80% of all breast cancers. Invasive lobular carcinoma is the second most common and accounts for 10% of breast cancers.

10.2.1 Non-invasive breast cancer

Ductal carcinoma *in situ* (DCIS) is the most common type of non-invasive breast cancer, accounting for about 15% of all new breast cancer cases. DCIS refers to an uncontrolled proliferation of cells that are confined to the breast duct with the basement membrane of the breast duct remaining intact (Figure 10.1). Eventually, the cells outstrip their blood supply and become centrally necrotic. This debris can calcify and be detected mammographically. DCIS is considered to be a precancerous condition and is classified into five histological subtypes associated with varying prognostic implications.

Recognized patterns of DCIS include solid, papillary, cribriform, micropapillary, and comedo (Allred, 2010). Most lesions contain a combination of at least two of these subtypes. Comedo carcinoma is considered high grade and predictive of recurrence. In patients with DCIS the invasive cancer usually occurs within the same breast. However, women with DCIS are also at a higher risk of developing cancer in the contralateral breast. For a regularly updated listing of DCIS publications, the following link is available: https://www.uptodate.com/contents/ductal-carcinoma-in-situ-treatment-and-prognosis.

Lobular carcinoma *in situ* (LCIS) is demonstrated by a benign-appearing proliferation of terminal ducts and ductules, and these are often multifocal and bilateral. Their growth continues in a lobular pattern, and they rarely develop central necrosis or become calcified. When these are seen, then the LCIS is classified as high grade. LCIS is much less common than DCIS and is associated with a lower risk of the development of invasive cancer than DCIS. LCIS is considered a marker that identifies women at increased risk of invasive breast cancer; this can be ductal or lobular. The risk of developing an invasive cancer remains elevated beyond two decades from initial diagnosis, and most subsequent breast cancers are ductal, rather than lobular.

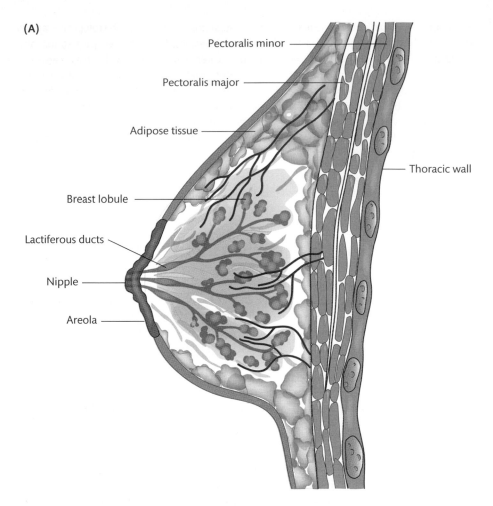

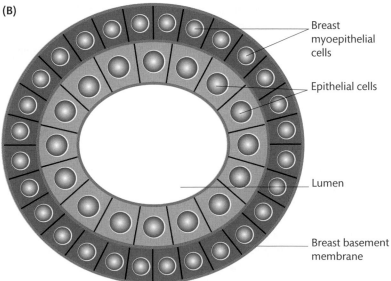

FIGURE 10.1
Diagram of normal breast structure.
Patrick J. Lynch, Medical Illustrator/CC-BY.

10.2.2 Invasive breast cancer

Invasive ductal carcinoma (IDC) is the most common type of breast cancer with about 80% of invasive breast cancers classified as IDC. Here, the tumour cells lose their myoepithelial layer, penetrate the ductal basement membrane and infiltrate into the surrounding breast parenchyma. Invasion can be demonstrated by immunocytochemistry (ICC) for smooth muscle actin, which unlike in DCIS, is lost. The invasive tumour is characterized by tumour cells arranged in cords, islands, and glands embedded within a dense fibrous stroma. The invasive tumour cells have the potential to metastasize to other parts of the body through the bloodstream or lymphatic system. Most IDC have no specific histological characteristics and are classified as no special type (NST).

Invasive lobular carcinoma (ILC) originates in the lobules, where it infiltrates into the surrounding breast parenchyma. It is much less common than IDC, accounting for about 10% of invasive breast cancers. This tumour has a tendency to be multifocal and multicentric. Tumour cells are often arranged in single files/strands, which is sometimes referred to as Indian filing, but at times ILC may be difficult to distinguish from ductal carcinomas. ICC staining for E-cadherin is helpful in differentiating ILC from IDC as in the former there is loss of E-cadherin expression.

Other types of invasive breast cancer that have a low incidence are summarized in Table 10.1). Among these, it is noteworthy that 70–80% of metaplastic breast carcinomas overexpress epidermal growth factor receptor (Reis-Filho et al., 2005). Furthermore, basal type cancers, first identified in 2000 (Perou et al., 2000), are associated with genetic changes. The *TP53* gene is either mutated or lost, and the tumour cells make large amounts of cytokeratin 5/6. They are often triple negative, with no receptors for oestrogen, progesterone or *HER2* (a member of the epidermal growth factor receptors family) present and can be *BRCA1* positive (breast cancer gene 1) linked to hereditary breast cancer due to a germline mutation (Foulkes et al., 2003). Hereditary breast cancers are also more likely than non-*BRCA1/2*-related breast cancers to express a basal epithelial phenotype. In Paget's disease of the nipple, the tumour cells within the epidermis contain intracytoplasmic mucin and also stain positive for HER2.

Key Point

Breast tumours can be non-invasive and invasive. The former can progress to the latter, which happens when the tumour cells lose the surrounding myoepithelial cells and invade the surrounding stroma. This can be detected on histological sections and by demonstrating loss of smooth muscle actin by immunocytochemical staining.

10.1 SELF-CHECK QUESTION

Which subtypes of non-invasive and invasive breast cancers are most common, and what are their frequencies?

TABLE 10.1 Characteristics and prognosis of infrequent types of invasive breast cancer

Type	Incidence	Characteristics	Prognosis compared with ductal carcinoma of no special type
Tubular carcinoma	2%	Highly differentiated; cells are regular and arranged in well-defined tubules. Diagnosis requires the presence of tubular formation in at least 75% of the specimen	Better
Medullary carcinoma	5–7%	Tumours have well-defined pushing, rather than infiltrative margins. Histologically, the tumour is characterized by larger than average cancer cells, and with a lymphoplasmacytic inflammatory response present at the periphery of the tumour	Relatively favourable
Mucinous carcinoma	2%	Large amounts of extracellular mucin production	Relatively favourable
Paget's disease of the nipple	1%	Paget cells are large cells surrounded by a clear halo-like area containing mucin, invade the epidermis. An underlying ductal carcinoma is almost always present, and the cancer must be differentiated from eczema, contact dermatitis, and basal cell carcinoma.	Relatively Favourable
Metaplastic breast carcinoma (MBC)	<1%	Various combinations of adenocarcinoma, mesenchymal, and other epithelial components present. Five subgroups of MBC, have been described.	Variable
Invasive cribriform carcinoma	5%	Well-differentiated cancer. It shares some features with tubular carcinoma.	Better
Invasive papillary carcinoma	<1–2%	Formation of numerous, irregular, finger-like projections of fibrous stroma that is covered with a surface layer of neo-plastic epithelial cells.	Relatively poor
Invasive micropapil-lary carcinoma	<3%	Cohesive tumour cell clusters within prominent spaces resembling dilated lymphatics or vessels. There is frequent skin invasion and extensive nodal involvement.	Poor
Inflammatory	1–5%	Lymphatic involvement of skin by the underlying carci-noma, causes red, swollen, hot skin resembling an inflam-matory process (peau d'orange).	Poor
Adenoid cystic carcinoma	<1%	Have both glandular (adenoid) and cylinder-like (cystic) features. They rarely spread to the lymph nodes or distant areas.	Good
Basal type breast cancer	15–20% of ductal carcinoma	High grade histological features with triple negative immu-nophenotype. They usually exhibit necrosis within the tumour mass.	Poor
Angiosarcoma	<1%.	A type of breast sarcoma. It starts in the cells that line the blood or lymphatic vessels. These cancers are more com-mon in premenopausal women in their 30s and 40s. The lump is usually at least 4 cm in size, and the skin over it may turn a bluish colour.	Poor
Phyllodes tumours	0.3–1%	Can be either benign, borderline or malignant. Malignant tumours are very rare. Phyllodes tumours are biphasic meaning that they are composed of both benign epithelial elements and cellular connective tissue stroma. The stroma dictates whether the tumour will be benign, borderline or malignant depending on mitotic activity, pleomorphism of nuclei, and an infiltrative margin.	Dependent on histo-logical grading

10.3 **Grading of tumours**

The grade and histological subtyping of breast cancer offers prognostic information on how the tumour may behave clinically. The tumour grade and type are two of the most important intrinsic characteristics that can be determined by histological assessment of the tumour. The morphological features are indicators of the degree of differentiation and the proliferative activity, which can predict the tumour's aggressiveness. The histological grade and the tumour subtypes have, therefore, been incorporated into various algorithms to further assess the prognosis of patients with breast cancer.

The Bloom–Richardson grading system was put forward in 1957 (Bloom and Richardson, 1957) and refers to a semi-quantitative method of the breast cancer classification system to grade breast cancers, and was the precursor of the present criteria, the modified Bloom–Richardson, also called the Nottingham system (Haybittle et al., 1982; Elston and Ellis, 1991). This grading system is dependent on light microscopic histological assessment of the tumour. It assesses three parameters: tubule formation, nuclear pleomorphism, and mitotic activity giving each a score from 1 to 3 (see Table 10.2 and Figure 10.2). The higher the score the worse the grade of the tumour.

TABLE 10.2 Modified Bloom–Richardson/Nottingham system for the histological grading of invasive breast cancer

Score	Tubule formation	Nuclear pleomorphism	Mitotic activity
1	>75%	Mild	Dependent on field diameter of the microscope (Score 1–3)
2	75–10%	Moderate	
3	<10%	Severe	

The higher the score the worse the grade:

Grade 1	Score 3–5
Grade 2	Score 6–7
Grade 3	Score 8–9

(A)

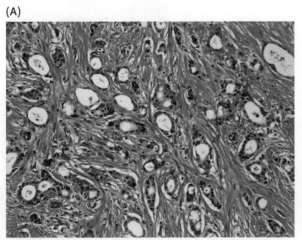

(B)

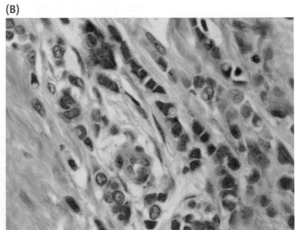

FIGURE 10.2
Examples of application of the modified Bloom–Richardson/Nottingham system for the histological grading of invasive breast cancer. (A) Invasive ductal carcinoma, Grade I (H & E, ×10). (B) Invasive lobular carcinoma, Grade II (H & E, ×20).

10.4 **Sentinel lymph node assessment**

The sentinel node is the first lymph node in the axilla to receive metastasis from the **ipsilateral** breast cancer (Figure 10.3). If the sentinel node is free of tumour, the probability of non-sentinel lymph nodes harbouring metastases is <0.1%. Sentinel lymph node assessment has revolutionized how the axilla is managed in women with breast cancer. This has decreased the frequency of unnecessary axillary lymph node dissection in women with negative sentinel lymph node.

The technique used for intraoperative assessment of sentinel node involvement usually employs touch imprint cytology. For this, the lymph node is divided longitudinally in half. The cut surface of both halves is then pressed onto a coated glass. The slides are air dried and stained by a rapid Giemsa technique, and examined microscopically for the presence of metastatic cells. The accuracy of this technique is dependent on the size of metastases, histological type of the primary tumour, and the experience of the pathologist.

Post-operative histological assessment is performed by cutting the lymph node at no more than 2 mm thickness and after FFPE by taking at least three step-sections through the tissue blocks (Figure 10.4). This will ensure that all macrometastases, defined as metastases that are 2 mm or more in diameter, are identified. This method is also suitable for identification of micrometastases, which are defined as 0.2–2 mm diameter in size. Submicrometastases and individual tumour cells are often identified by the use of ICC using a pan-cytokeratin marker such as AE1/AE3.

Sentinel Lymph Node

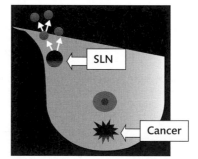

FIGURE 10.3
Axillary sentinel lymph node.
SLN, sentinel lymph node, first lymph node to receive lymphatic metastases.

FIGURE 10.4
Metastatic carcinoma into sentinel lymph node. The majority of the metastatic cancer cells are present in the left and centre of the image, with the normal lymphoid component on the right (Haematoxylin and eosin, ×20).

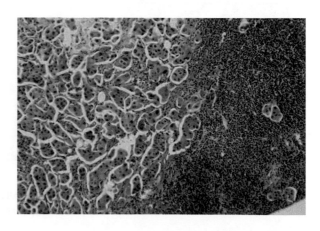

Sysmex has developed an automated intraoperative method for the analysis of sentinel lymph nodes called OSNA (one-step nucleic acid amplification). The OSNA method amplifies cytokeratin 19 (CK19) mRNA by a specific and sensitive isothermal procedure called RT-LAMP (reverse transcription loop-mediated isothermal amplification. The progress of the amplification is monitored in real time. The fresh intraoperative lymph node is serially sliced into four slices of 1 or 2 mm width, with two alternate slices being used, or the OSNA assay and the others being used for multilevel histological investigation. The OSNA process takes around 30 minutes and the assay quantitatively measures the amount of CK19 mRNA, which is directly related to the size of metastatic foci. Discrimination of macrometastases, micrometastases, and negative lymph nodes is provided. Results are displayed as ++ for macrometastases, + for micrometastases, and – for metastases negative. In one study, the sensitivity for OSNA was calculated as 100%, specificity as 90.47% in comparison with frozen section assessment (Szychta et al., 2016). Additional references to the sensitivity and specificity of OSNA for the detection of breast cancer metastatic deposits in SLNs can be accessed using the following link: http://www.sysmex.se/products/oncology/sentinel-lymph-node-analysis-osna/clinical-evidence-osna.html.

Cross reference
See Chapter 4, Section 4.2.1.1.2 for information about variations around standard PCR.

Key Point

Axillary sentinel lymph node status using intraoperative and post-operative techniques have proven to be reliable indicators of axillary lymph node status. This has changed the landscape of axillary surgery. When the sentinel lymph node is negative, no further axillary surgery is indicated.

10.2 SELF-CHECK QUESTION

Why is it important to identify metastatic carcinoma in the sentinel lymph nodes and how does this influence the future management of the patient?

10.5 Receptors

Receptors are proteins present within cells or on cell membranes, which can attach to certain substances in the blood. Normal breast cells have receptors that attach to the hormones oestrogen and progesterone, and depend on these hormones to grow. Breast cancer cells may have one, both, or none of these receptors. The ER was first identified in the 1960s and with the PR it became recognized as a 'predictive' marker for women with breast cancer who would respond to hormone treatment. ICC is used most often to find out if cancer cells have ERs and PRs, and to guide hormone inhibitor therapy in patients who are positive for the receptors. There are a number of different hormone therapies that work in different ways to block the effect of oestrogen on cancer cells such as tamoxifen and letrozole.

The HER2 coded by the gene *ERBB2* has been at the forefront of the therapeutic management of breast cancer since 1998, following the FDA approval of trastuzumab (Herceptin) for therapeutic intervention. It is now an established prognostic and predictive factor in invasive breast cancer. Positive confirmation of *HER2* status, in conjunction with the wider clinicopathological characteristics of the patient and their disease, now determines eligibility for treatment with a range of HER2-targeted therapies. These include trastuzumab (Herceptin), lapatinib (Tykerb), and pertuzamab (Perjeta).

In current clinical practice, NICE has mandated that ER, PR, and HER2 testing is evaluated in all newly diagnosed invasive breast cancers irrespective of subtype. Accurate results are critical in determining the use of targeted therapy in individual patients. ICC is the method of choice to detect the nuclear steroid receptors in tumour cells and cell membrane staining for HER2. Approximately 75% of all breast cancers are oestrogen hormone receptor positive, while 15–25% are found to be *HER2* positive.

HER2 is also assessed in cases of high grade DCIS. While there is no sufficient current data to support the use of HER2 targeted therapy for DCIS; *HER2* positive DCIS has been shown to have a lower risk of recurrent invasive breast cancer (Borgquist et al., 2015).

The hormone receptors oestrogen and progesterone are assessed using the Allred or Quick score methods (see Table 6.2), where numerical values are given to the percentage of nuclei staining for the hormone and for the intensity of staining. According to the current international guideline, when 1% of the nuclei of the cancer cells are either ER- or PR-stained, then a positive result is reported for hormone status (Hammond et al., 2010). This does not imply that hormonal inhibitor therapy will be recommended for the patient by the attending clinical oncologist. Indeed, the 'Ontario' guidelines suggest that a score of 10% cells is more predictive of a positive response to hormonal inhibitor therapy (Nofech-Mozes et al., 2012). Images of positive and negative hormone receptor ICC are provided in Figure 10.5.

The ICC HER2 score (Box 6.1) is dependent on the percentage of cancer cells staining, and the intensity and completeness of the cell membrane staining of these cells. The score is reported as negative, 1+, 2+, and 3+ (see Figure 10.6). A score of 3+ is considered positive and patients would be suitable for Herceptin or equivalent therapy. A score of 2+ is borderline and the tumour must be tested for gene amplification using ISH. Standard laboratory practice for ISH testing includes assessing the HER2 and control centrometric probe for chromosome 17 (CEP17) signals in 20–60 intact tumour nuclei to determine the average HER2 and CEP17 copy numbers per cell, followed by calculating the HER2/CEP17 gene ratio. When this dual probe ISH procedure is employed a HER2 positive result is reported when the HER2/CEP17 ratio is 2.0 with an average HER2 copy number of 4.0 signals per counted cell (Figure 10.7).

Cross reference

See Chapter 6 for more details of analysing and interpreting molecular data.

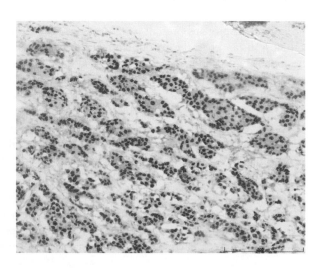

FIGURE 10.5

Immunocytochemical staining for oestrogen receptor in an invasive breast cancer.

Note the moderate to strong brown nuclear staining of the majority of the tumour cells.

(A)

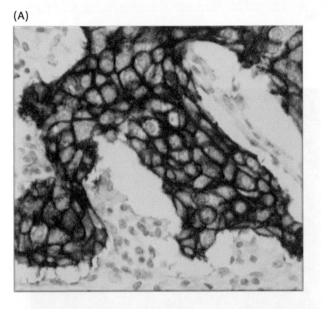

(B)

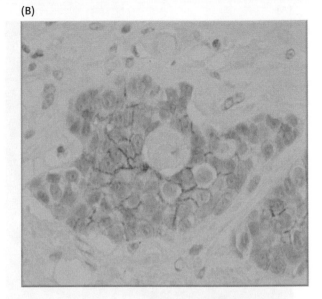

(C)

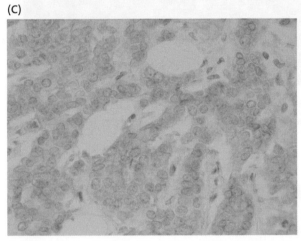

FIGURE 10.6
HER2 Immunocytochemical staining. (A) 3+ (positive). Note presence of strong complete membrane staining in all cancer cells. (B) 2+ (equivocal). Note weak to moderate staining membrane staining that in most cancer cells is incomplete. (C) 1+ (negative). Note lack of any membrane staining.

Alternatively, ISH can be performed using the HER2 probe alone. In this situation, a positive HER2 result is reported when the average HER2 signal per counted cell is 6.0 (Wolff et al., 2014; Rakha et at., 2015).

Tumours that express ER or and PR have been associated with a much more favourable prognosis, while the overexpression of HER2 has been associated with poorer outcomes. Around 15–20% of all breast cancers do not have hormone or HER2 receptors. Sometimes referred to as 'triple negative' breast cancer, it is typically more aggressive and difficult to treat.

Key Point

Receptor status for ER and PR is determined by ICC. *HER2* status is initially tested using ICC with the adjunct use of ISH in borderline cases. Determining receptor status is now part of the accepted standard of care. Receptor status is important as both a prognostic indicator, as well as guiding various treatment modalities available to oncologists.

(A)

(B)

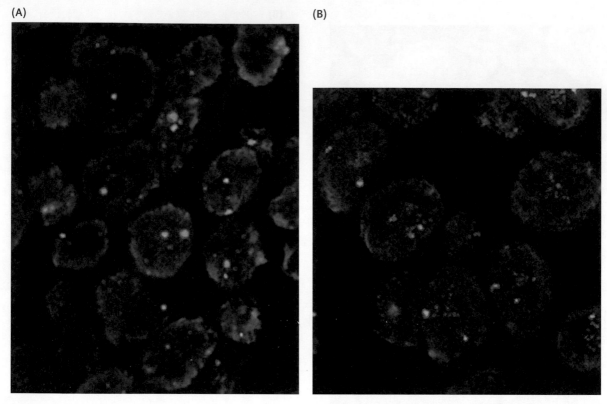

FIGURE 10.7

HER2 fluorescent *in situ* hybridization. HER2 hybridization signal is orange; CEP17 hybridization signal is green. (A) HER2, non-amplified; ratio 1.03. (B) HER2, amplified; ratio 4.73.

10.3 SELF-CHECK QUESTION

What are the hormones that are regularly tested for in breast cancer and which laboratory techniques are used for these tests?

10.6 Molecular classification

The traditional clinical approach to treat *in situ* and invasive breast cancer would usually involve a combination of surgery, together with chemical- and radiation-based therapies. The surgical options might include wide local excision of the tumour, quadrantectomy and/or mastectomy. The surgical procedure can be preceded or followed by hormonal therapy and/or chemotherapy. Chemotherapy is invariably associated with a degree of toxicity, which can affect the quality of life. Conventionally, the traditional histological staging, focusing mainly on the tumour size and lymph node status, has been of great importance; however, the patients' outcomes can be very dissimilar and new molecular techniques have highlighted that tumours of the same stage might have a different biology. Taken together, the insights into the molecular diversity of breast cancer offer the hope of better diagnostic and treatment pathways.

In the last decade, a new conceptual approach to the biology of breast cancer has emerged and provided new hope to better understanding the biology of the disease and guiding of therapy (Kittaneh et al., 2013). This approach was initially based on gene expression arrays, and has later been translated to quantitative real-time polymerase chain reaction (qRT-PCR) and other molecular methods as discussed in Chapters 4 and 5.

At the molecular level, breast cancer is not one disease, but many different diseases. Even when tumours are classified together, based on their morphology, they can act differently because of distinctive genetic characteristics. The advent of genomics has revolutionized the understanding of breast cancer as several different biologically and molecularly distinct diseases.

Breast cancer arises from the terminal duct units of the breast lobules (see Figure 10.1). Luminal tumour cells look the most like cells of breast cancers that start in the inner (luminal) cells lining the mammary ducts. Using receptor-based results, tumours can be classified as luminal A, B, HER2, or basal (see Table 10.3; Alizart et al., 2012). Of the four subtypes, luminal A tumours tend to have the best prognosis, with fairly high survival rates and fairly low recurrence rates.

Molecular techniques have also been used to understand the taxonomy of breast cancers better. For example, using the combination of ICC and gene expression profiling has shown that some cancers represent discrete entities (for example, invasive micropapillary carcinoma), but that others are very similar at the transcriptome level despite morphological differences (for example, tubular carcinomas and lobular carcinomas). Furthermore, when classified by expression profiling, lobular carcinomas (similar to ductal carcinomas) are composed of tumours in all molecular categories (luminal, HER2, and basal), whereas most other cancers belong to only one molecular category (for example, tubular, mucinous, endocrine, and micropapillary carcinomas are luminal; medullary, adenoid cystic, and metaplastic carcinomas are basal).

It has been shown that ER-positive breast cancer have various genetic changes, which include 16q deletions and 1q gains, while ER negative cancers have more than 53 mutations, including *HER2* amplifications, *BRCA1* mutation, as well as higher genomic instability.

TABLE 10.3 Molecular subtypes of breast cancer

Subtype	ER status	HER2 status	Proliferation index	Histological grade	Additional features
Luminal A	Positive	Negative	Low	1 or 2	Luminal CK positive, E-cadherin variable
Luminal B	Positive	Variable	High	2 or 3	Luminal CK positive, TP53 mutations
HER2	Negative	Positive	High	2 or 3	TP53 mutations
Basal like	Negative	Negative	High	3	Basal CK positive, TP53 mutations, DNA repair loss

CK = cytokeratin
Adapted from Alizart et al., 2012.

The data collected also suggests that ER-positive cancers may progress from low to higher grade tumours, and that more than 50% of the high and low grade ER-positive cancers have 16q and 1q alterations, which would be indicative of a common pathway (Allison, 2012). The morphologically higher grade tumours are also noted to have higher proliferation rates and more genomic instability. Furthermore, it has been reported that *in situ* and invasive ER-positive ductal and lobular tumours have similar genetic alterations, again involving 16q and 1q changes with additional E-cadherin loss in the lobular tumours (Allison, 2012). Multiple sets of significantly mutated genes, that are likely to be the primary drivers of breast cancer, have been identified and these include *PIK3CA, TP53, GATA3, MAP3K1, AKT1*, and *CBFB* (Pereira et al., 2016).

Furthermore, markers of increased risk of breast cancer, such as columnar cell change with atypia in the ducts, atypical ductal epithelial hyperplasia, and lobular intraepithelial neoplasia have been noted to show similar pathways, which also supports the fact that they are often seen together in the same patient. It has also been observed that, generally, basal-like breast cancers do not usually have an associated *in situ* component. High grade solid DCIS especially the comedo type has been consecutively associated with *HER2* positive invasive disease.

Molecular studies of breast cancer have shown that the less common tumours known as tumours of special type, e.g. medullary, cribriform, mucinous carcinomas, are usually more homogeneous in comparison with the more common invasive ductal carcinoma of no special type or invasive lobular carcinoma. Various studies that compared metaplastic carcinoma with basal-like invasive ductal carcinoma, the pathways and genes related to DNA repair, and chemotherapy response such as *BRCA1, PTEN*, and *TOP2A* were significantly downregulated in metaplastic cancers, which would account for decreased response to chemotherapy (Weigelt and Reis-Filho, 2009).

The awareness of this new molecular information, as well as the heterogeneity observed, suggests that different subpopulations of tumour cells have different acquired mutations and, hence, may lead to different progress and metastatic potential. At the time of writing (2018) the intrinsic subtype classification introduced above, e.g. luminal A and B, and the integrative cluster molecular classifications for breast cancer (Dawson et al., 2013) are not used in diagnostic breast pathology. However, the intrinsic subtype has been used in clinical trials to classify tumours. For further information see further reading at the end of the chapter.

Key Point

Breast cancer is very heterogeneous at the molecular level with various driver mutations being discovered. Further work is in progress into further establishing the different mutations involved.

10.4 SELF-CHECK QUESTION

Which are the commonest driver mutations in breast cancer?

10.7 Gene expression profiling of breast cancer

For more than a decade, high-throughput microarray-based gene expression methods have been extensively employed to the study of breast cancer in attempts to determine the molecular features that highlight metastatic propensity and histological grade, with associated signatures determining prognosis and response to therapy.

Several RNA-based tools are commercially available to give this additional information. Most of these focus on gene expression microarrays or quantitative reverse transcription analysis. Although RNA is degraded in FFPE tissues, RT-PCR is well suited to amplify short RNA fragments. These microarrays measure RNA expression levels, rather than gene copy number or protein expression and distribution. Unlike RT-PCR, microarrays are high throughput and can assess the expression levels of many genes at once.

Expression array analysis may be of value in refining the grading of invasive breast cancers. Two studies have suggested that, although histological grade I and grade III tumours have distinct expression signatures, there is no characteristic expression signature for histological grade II tumours. Furthermore, in these studies applying genomic grading to histological grade II tumours was able to separate that group into those which had a favourable prognosis (similar to that of grade I tumours) and those in which the prognosis was poor (similar to that of grade III tumours).

Some of gene expression-based assays include Theros H/I and MGI, Oncotype DX, Prosigna (Pam 50), Mammaprint, and Endopredict. All tests can provide an overall risk assessment of breast cancer recurrence, but are distinctive in the genes they assess. These assays were designed to be more specific than conventional clinico-pathologic parameters in the selection of patients for **(neo-) adjuvant treatment** and in effect help to avoid unnecessary cytotoxic treatment.

Theros H/I measures the ratio of HOXB13:IL17BR gene expression as a predictor of clinical outcome for breast cancer patients treated with tamoxifen. A high level of expression of the two-gene ratio has been associated with the tumour aggressiveness and failure to respond to the drug. *Theros MGI* is an additional test that uses a five-gene expression index to stratify ER-positive breast cancer patients into high or low risk of recurrence by re-classifying the Grade 2 into higher (Grade 3) or lower grade (Grade 1) outcomes. Both assays rely on the recovery of mRNA from FFPE samples that is used to quantitatively measure gene expression via qRT PCR. In combination, these two tests provide much more information apart from re-classifying the tumours, but the assays have not been objectively proven to be superior to the methods which have been conventionally used.

Oncotype DX was approved by NICE for use in the UK for NHS cancer patients in 2013. It is a 21-gene expression assay that uses qRT-PCR and microarray technologies on FFPE breast cancer material to identify patients who may be successfully treated with chemotherapy and to estimate the risk of the invasive tumour to recur after treatment (Table 10.4). This test is usually used in patients with early stage, lymph node negative or up to three positive lymph nodes, ER positive, *HER2* negative invasive breast cancer. It indicates the probability of the cancer to recur within 10 years from the original diagnosis and sub-categorizes the patient's risk into low, intermediate, or high. The Oncotype DX test is both a prognostic test and a predictive test, since it predicts the likelihood of benefit from chemotherapy or radiation therapy treatment. The result is reported as a recurrence score:

- *Recurrence Score lower than 18:* the cancer has a low risk of recurrence. The benefit of chemotherapy is likely to be small and will not outweigh the risks of side effects.

Neoadjuvant therapy
Treatment given as a first step to shrink a tumour before the main treatment, which is usually surgery, is undertaken.

- *Recurrence Score of 18 up to and including 30:* the cancer has an intermediate risk of recurrence. It's unclear whether the benefits of chemotherapy outweigh the risks of side effects.

- *Recurrence Score greater than or equal to 31:* The cancer has a high risk of recurrence and the benefits of chemotherapy are likely to be greater than the risks of side effects.

MammaPrint is another molecular diagnostic tool to assess the risk of cancer recurrence. It has not yet been approved by NICE for the UK NHS. It uses a 70-gene expression profile, which was initially developed from whole-genome expression (25,000 genes) arrays. MammaPrint has been shown to be a prognostic marker, independent of clinicopathological risks assessment in patients with lymph node negative breast cancer. The test requires a fresh sample to be put in a stabilizing solution and sent to an experienced laboratory to isolate RNA. The RNA is isolated, amplified, and co-hybridized with a standard reference to the MammaPrint microarray to obtain the profile. This has been shown to have very high prognostic correlation. This test is indicated for patients with Stage 1 and 2 disease, younger than 61 years, and lymph node negative, with invasive tumours measuring less than 50 mm in diameter. The test requires a substantial amount of material making simultaneous histological assessment of the tumour, at times, not possible.

EndoPredict is a molecular test that is approved for use in Europe, but not in the USA or by NICE for the NHS in the UK. It is a test that analyses 12 genes within a tumour sample to predict the probability that the cancer may recur. The information provided by EndoPredict can help decide whether chemotherapy may be of benefit. The test is not suitable for all types of breast cancer and is only useful for women with hormone receptor positive, *HER2*-negative breast cancer where up to three lymph nodes are involved. If there is a low risk of recurrence, chemotherapy may not be necessary and hormone-based therapy may be sufficient on its own.

Prosigna Breast Cancer Prognostic Gene Signature Assay (formerly called the PAM50 test) Thus far, it has not been approved by NICE for the UK NHS use. The assay uses a qRT-PCR to assess 50 genes using FFPE material. The assay is able to identify luminal A, luminal B, *HER2* enriched, and basal-like cancers, while also being an independent predictor of survival, irrespective of the clinicopathological data such as nodal status, receptor status, tumour grade, and stage. It is able to predict the 5-year risk of tumour recurrence, and indicate whether chemotherapy or hormonal therapy might be helpful. This assay is predominantly helpful in post-menopausal women with hormone-positive breast cancers, with either node negative or up to three lymph nodes with metastatic carcinoma.

TABLE 10.4 Genes identified by Oncotype DX RNA expression assay

Genes	Assessing
ER, PR, BCL2, SCUBE2	Oestrogen receptor
Ki67, STK15, Survivin, Cyclin B1, MYBL2	Proliferation
HER2, GRB7	HER2
GSTM1, CD68, BAG1	Other
b actin, GAPDH, RPLPO, GUS, TFRC	Sample and assay

Adapted from: http://www.oncotypeiq.com/en-GB/breast-cancer/healthcare-professionals/oncotype-dx-breast-recurrence-score/about-the-test.

Key Point

Various commercially available gene expression-based molecular tests can be used to predict the behaviour, prognosis, and the utility of chemotherapy benefit for patients.

10.5 SELF-CHECK QUESTION

Why is molecular testing important in some patients with breast cancer and how can it help in the management of their disease?

10.8 **Future directions**

Molecular technologies are providing for a better understanding of breast cancer. Some, or perhaps all, of these may provide a significant step change in the treatment of breast cancer.

NGS is considered to be the new breakthrough in molecular diagnostic technology, as it allows the simultaneous sequencing of a huge number of genes in a timely and cost-effective manner. Several breast cancers have been sequenced with this technique over the past few years and, although no new driver mutations have been identified, novel mutations have been identified in different cancers. Unlike the known cancer genes, *TP53* and *PIK3CA*, that are mutated in >30% of breast cancer patients in ER-negative and ER-positive tumours, respectively, most of the newly identified cancer genes are present in less than 10% of patients. NGS has highlighted the large genetic diversity in mutations amongst the different breast tumours (Verigos and Magklara, 2015) and has allowed further exploration of intra-tumour heterogeneity, in addition to revealing that, although subclonal mutations are present in all tumours, at least 50% of the tumours have shown to have a dominant clone (Hagemann, 2016).

PD-1 is a checkpoint receptor, expressed primarily by activated T cells, and it limits T-cell effector functions. PD-L1, a T-cell inhibitory molecule, is expressed on cancer cells, tumour-infiltrating inflammatory cells, and immune cells. The binding of PD-L1 to PD-1 on T-cells is a major mechanism of tumour immune evasion. While breast cancer has not been considered as an immunogenic tumour recent studies have shown that between 19–26% of triple negative breast cancers (TNBC) have an elevated expression of PD-L1 (Mittendorf et al., 2014). This suggests that it is reasonable to investigate therapies that target PD-1/PD-L1 in TNBC.

The poly (ADP-ribose) polymerase (PARP) inhibitors have recently come to the forefront as promising treatments for BRCA (hereditary)-associated cancers and TNBC. PARP proteins are important for repairing single-strand breaks in the DNA. Drugs that inhibit the action of PARP cause multiple breaks to form in the DNA and in tumours with BRCA mutations these double-strand breaks cannot be repaired, leading to cell death. There are at least five PARP inhibitors that are currently in clinical trials. The trials are for women who have developed metastatic breast cancer. Currently, Pharma has drugs that appear to improve progression-free survival in these women, see https://www.medscape.org/viewarticle/711369. Combining radiation therapy with PARP inhibitors also offers promise, since these inhibitors would cause formation of double-strand breaks from the single-strand breaks generated by radiotherapy in tumour tissue with *BRCA* mutation. This combination could lead to a decrease in radiation dose being administered to patients.

Cross reference
See Chapter 5 for more details of NGS.

When patients experience metastatic disease from breast cancer, the clinicians are posed with new challenges. The current standard of assessment of metastatic disease by morphological and functional imaging does not provide adequate information on tumour biology and the presence of micrometastasis, limiting the possibility of predicting tumour metastatic potential. Recent studies on circulating tumour cells (Giuliano et al., 2014; De Luca et al., 2016) have investigated estimating the risk of metastatic relapse and progression in patients with breast cancer. Deep amplicon sequencing has also enabled the rapid assessment of somatic mutations and structural changes in multiple cancer genes in DNA isolated from tumour and circulating cell-free DNA (ctDNA). This has the potential to combine assessment of metastasis, as well as prognosis, and may be able to help in treatment planning.

In human breast cancers, of all the malignant cells, only a small minority, possibly as few as one in 100, appear to be capable of forming new malignant tumours. These stem cells make copies of themselves in a process called self-renewal and produce all the other kinds of cells associated with the tumour, such as stromal cells. The tumour microenvironment is a critical regulator of cancer stem cell driven metastasis and the stromal cells or extracellular matrix may regulate the dormancy at the metastatic site. In addition to this, complex cellular programmes may allow the tumour cells to modify their microenvironment through various secretions of autocrine/paracrine molecules that may increase the potential for the development of a metastatic deposit. To achieve any real cures in metastatic breast cancer, it will be absolutely necessary to eradicate these cells (Velasco-Velázquez et al., 2011). Both hereditary and sporadic breast cancers may develop through dysregulation of stem-cell self-renewal pathways, which are normally tightly regulated. Breast cancer cells that overexpress HER2 are thought to have four or five times as many cancer stem cells compared with HER2-negative breast cancer cells (Shah and Osipo, 2016). Novel approaches that might target breast cancer stem cell microenvironment could be the basis for the generation of even more effective and clinically useful therapies to decrease recurrence and metastatic potential.

CASE STUDY 10.1

A 51-year-old female presented to her GP with a breast lump. After examination, she was referred on a 2-week wait referral to the breast clinic. She was seen at the one-stop breast clinic by a consultant breast surgeon. The woman was sent for an ultrasound and a mammogram, and was seen again by the surgeon. The ultrasound showed a 30-mm hypoechoic mass graded U5. The mammogram was also graded M5. The woman had a FNA cytology, which showed malignant cells and this was followed by a core biopsy.

The biopsy was sent to the histology laboratory for processing and for examination by a pathologist. The histology showed a grade 2 invasive ductal carcinoma of no special type. There were areas of DCIS. The hormone receptor status for oestrogen and progesterone was investigated using ICC and both were determined as positive using Quickscore. The HER2 score was 2+ (Borderline) by ICC. HER2 ISH showed a HER2/CEP ratio of 1.6, indicating that the gene was non-amplified.

The case was discussed at the multidisciplinary team meeting, where it was agreed that the patient would undergo a wide local excision and a sentinel LN biopsy.

The post-surgical histology showed a 30-mm grade 2 invasive ductal carcinoma with admixed high grade DCIS. The sentinel lymph node was negative.

The case was rediscussed at the multidisciplinary team meeting, where the oncologists requested the specimen to be tested for Oncotype DX to determine whether the patient would benefit from chemotherapy in addition to tamoxifen. The Oncotype DX result showed a recurrence score of 12 and it was decided that adding chemotherapy would be of negligible benefit to the patient.

Chapter summary

- Breast cancer is the commonest cancer in women. The most common invasive tumour is ductal carcinoma.

- Classification of breast cancer is determined not only by the histological features and stage, but also by the receptor profile.

- Invasive breast cancers are tested for ER and PR, and for *HER2* status.

- The use of hormone receptor inhibitors and Herceptin in *HER2* positive tumours has revolutionized neo-adjuvant and adjuvant treatment of breast cancer.

- Molecular tests for gene expression are increasingly being used to determine the recurrence risk and benefit of chemotherapy in both lymph node negative and low positive nodal status women.

- Molecular tests, especially gene expression profiling, have great potential to refine breast cancer classification, enhance the understanding of the tumour biopsy, and improve patient management.

- The role of molecular testing is still evolving and the tumour type, grade, staging and receptor status is still of paramount importance in everyday diagnostic practice.

Further reading

- Dai X, Li T, Bai Z, et al. Breast cancer intrinsic subtype classification, clinical use and future trends. American Journal of Cancer Research 2015; 5: 2929–2943.

- Rakha ER, Green AR. Molecular classification of breast cancer: what the pathologist needs to know. Pathology 2017; 49: 111–1119.

- Russnes HG, Lingjaerde OC, Borresen-Dale AL, Caldas C. Breast cancer molecular stratification: from intrinsic subtypes to integrative clusters. American Journal of Pathology 2017; 187: 2152–2162.

- Orchard G, Nation B. Breast tissue. In: *Cell Structure & Function*, pp. 478–481. Oxford, Oxford University Press.

References

- Alizart M, Saunus J, Cummings M, Lakhani SR. Molecular classification of breast carcinoma. Diagnostic Histopathology 2012; 18: 97–103.

- Allison KH. Molecular pathology of breast cancer: what a pathologist needs to know. American Journal of Clinical Pathology 2012; 138: 770–780.

- Allred DC. Ductal carcinoma *in situ*: terminology, classification, and natural history. Journal of the National Cancer Institute Monographs 2010; 41: 134–138.

- Bloom H, Richardson WW. Histological grading and prognosis in breast cancer; a study of 1409 cases of which 359 have been followed for 15 years. British Journal of Cancer 1957; 11: 359–377.

- Borgquist S, Zhou W, Jirström K, et al. The prognostic role of HER2 expression in ductal breast carcinoma in situ (DCIS); a population-based cohort study. BMC Cancer 2015; 15: 468.

- Dawson S-J, Rueda OM, Caldas C. A new genome-driven integrated classification of breast cancer and its implications. EMBO Journal 2013; 32: 617–628.

- De Luca F, Rotunno G, Salvianti F, et al. Mutational analysis of single circulating tumor cells by next generation sequencing in metastatic breast cancer. Oncotarget 2016; 18: 26107–26119.

- Elston CW, Ellis IO. Pathologic prognostic factors in breast cancer. I. The value of histological grades in breast cancer. Experience from a large study with long-term follow-up. Histopathology 1991; 19: 403–410. Republished: Histopathology 2002; 41: 154–161.

- Foulkes WD, Stefansson IM, Chappuis PO, et al. Germline BRCA1 mutations and a basal epithelial phenotype in breast cancer. Journal of the National Cancer Institute 2003; 95: 1482–1485.

- Giuliano M, Giordano A, Jackson S, et al. Circulating tumour cells as early predictors of metastatic spread in breast cancer patients with limited metastatic dissemination. Breast Cancer Research 2014; 16: 440.

- Hagemann IS. Molecular testing in breast cancer: a guide to current practices. Archives of Pathology & Laboratory Medicine 2016; 140: 815–824.

- Hammond ME, Hayes DF, Dowsett M, et al. American Society of Clinical Oncology/ College of American Pathologists guideline recommendations for immunohisto-chemical testing of estrogen and progesterone receptors in breast cancer. Archives of Pathology & Laboratory Medicine 2010; 134: 48–72.

- Haybittle JL, Blamey RW, Elston CW, et al. A prognostic index in primary breast cancer. British Journal of Cancer 1982; 45: 361–366.

- Kittaneh M, Montero AJ, Glück S. Molecular profiling for breast cancer: a comprehensive review. Biomarkers in Cancer 2013; 5: 61–70.

- Mittendorf EA, Philips AV, Meric-Bernstam F, et al. PD-L1 Expression in triple-negative breast cancer. Cancer Immunology Research 2014; 2: 361–370.

- Nofech-Mozes S, Vella ET, Dhesy-Thind S, et al. Cancer care Ontario guideline recommendations for hormone receptor testing in breast cancer. Clinical Oncology 2012; 24: 684–696.

- Pereira B, Chin SF, Rueda OM, et al. The somatic mutation profiles of 2,433 breast cancers refines their genomic and transcriptomic landscapes. Nature: Communication 2016; 7: 11479.

- Perou CM, Sorlie T, Eisen MB, et al. Molecular portraits of human breast tumours. Nature 2000; 406: 747–752.

- Rakha EA, Pinder SE, Bartlett JM, et al. National Coordinating Committee for Breast Pathology. Updated UK Recommendations for HER2 assessment in breast cancer. Journal of Clinical Pathology 2015; 68: 93–99.

- Reis-Filho JS, Milanezi F, Silvia Carvalho, et al. Metaplastic breast carcinomas exhibit EGFR, but not HER2, gene amplification and overexpression: immunohistochemical and chromogenic in situ hybridization analysis. Breast Cancer Research 2005; 7: R1028–R1035.

- Shah D, Osipo C. Cancer stem cells and HER2 positive breast cancer: The story so far. Genes & Diseases 2016; 3: 114e123.

- Sotiriou C, Neo SY, McShane LM, et al. Breast cancer classification and prognosis based on gene expression profiles from a population-based study. Proceedings of the National Academy of Sciences, USA 2003; 100: 10393–10398.

- Szychta P, Westfal B, Maciejczyk R, et al. Intraoperative diagnosis of sentinel lymph node metastases in breast cancer treatment with one-step nucleic acid amplification assay (OSNA). Archives of Medical Sciences 2016; 12: 1239–1246.

- Velasco-Velázquez MA, Popov VM, Lisanti MP, Pestell RG. The role of breast cancer stem cells in metastasis and therapeutic implications. American Journal of Pathology 2011; 179: 2–11.

- Verigos J, Magklara A. Revealing the complexity of breast cancer by next generation sequencing. Cancers (Basel) 2015; 4: 2183–2200.

- Weigelt B, Reis-Filho JS. Histological and molecular types of breast cancer: is there a unifying taxonomy? Nature Reviews: Clinical Oncology 2009; 6: 718–730.

- Wolff AC, Hammond ME, Hicks DG, et al. American Society of Clinical Oncology; College of American Pathologists. Recommendations for human epidermal growth factor receptor 2 testing in breast cancer: American Society of Clinical Oncology/College of American Pathologists clinical practice guideline update. Archives of Pathology & Laboratory Medicine 2014; 138: 241–256.

Useful websites

General information on breast cancer

- *American Cancer Society*: https://www.cancer.org/cancer/breast-cancer.html.

- *Breast Cancer Now*: http://breastcancernow.org/about-breast-cancer/want-to-know-about-breast-cancer/breast-cancer-facts?gclid=CL6pxPzL8dkCFcu_7QodQcADMA.

- *Imaginis*: http://www.imaginis.com/breast-cancer-resource-center.

Specific information as provided in chapter

- *Ductal carcinoma* in situ *updated publication listing curated by Collins LC, Laronga C, Wong JS*: https://www.uptodate.com/contents/ductal-carcinoma-in-situ-treatment-and-prognosis.

- *Oncotype DX gene expression panel*: http://www.oncotypeiq.com/en-GB/breast-cancer/healthcare-professionals/oncotype-dx-breast-recurrence-score/about-the-test.

- *PARP inhibitors*: https://www.medscape.org/viewarticle/711369.

- *Sentinel lymph node using OSNA*: http://www.sysmex.se/products/oncology/sentinel-lymph-node-analysis-osna/clinical-evidence-osna.html.

- *Tamoxifen treatment for hormone receptor positive breast cancer*: http://scienceblog.cancerresearchuk.org/2012/10/15/high-impact-science-tamoxifen-the-start-of-something-big/.

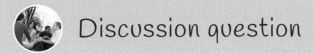 Discussion question

10.1 How do molecular assessments help in the management of breast cancer?

Answers to the self-check questions and tips for responding to the discussion question are provided on the book's accompanying website:

⊙ Visit: www.oup.com/uk/warford.

11

Epithelial Tumours and Melanoma

Manuel Rodriguez-Justo

Learning objectives

After reading this chapter you should be able to:

- Understand the epidemiology and tumour types of colorectal cancer, lung cancer, gastro-oesophageal cancer, and melanoma.
- Understand signalling pathways in epithelial carcinogenesis.
- Give examples of the most common molecular abnormalities found in epithelial cancers and melanoma, and the therapeutic implications.
- Consider the potential impact of immuno-checkpoint inhibition in the treatment of epithelial tumours and melanoma.

11.1 Introduction

Carcinomas are epithelial tumours that arise from the lining layer of internal organs (gastrointestinal digestive tract, lung, urinary tract, ...) or skin. They are the most common type of cancers accounting for 85% of all cancers. There are several types depending on the organ involved and tissue of origin. Melanomas arise from melanocytes, specialized cells found in the epidermis and hair follicles in the skin, but also in other tissues, such as the eye and brain.

Carcinomas involving the lung, and upper and lower gastrointestinal tracts are among the most frequent type of cancers in Western countries. Numerous advances in the understanding of how these cancers develop have been made in the last decade, leading to the introduction of targeted therapies, which have improved the overall survival of these patients, even in the presence of metastatic disease.

This chapter will cover epidemiological features of lung, colorectal, and gastro-oesophageal cancers and melanomas (incidence and mortality), signalling pathways, and most frequently mutated genes, together with therapeutic options. In addition, the role of immune-checkpoint inhibitors will be discussed and how their use has dramatically changed the prognosis of cancer and has become standard of care for some types of cancers.

11.2 Colorectal cancer

11.2.1 Epidemiology and aetiology

Colorectal cancer (CRC) is the third most common cancer and the second leading cause of death from cancer among adults in Western countries (Arnold et al., 2017). In the USA, it is estimated that approximately 135,000 new cases of colorectal cancer were diagnosed in 2017 and more than 50,000 Americans died of CRC (https://seer.cancer.gov/statfacts/). In the UK in 2015, there were 41,804 new cases of large bowel cancers and 15,903 persons died from the disease (http://www.cancerresearchuk.org/health-professional/cancer-statistics/). Worldwide, an estimated 1.36 million new cases of colorectal cancer were diagnosed in 2012 with a slight male predominance (Ferlay et al., 2015).

The rates of colorectal cancer vary by ethnicity, with Caucasians and northern Europeans having highest incidence (Ferlay et al., 2013). Age is a major risk factor for sporadic CRC and more than 90% of CRCs are diagnosed in people aged over 50. The lifetime risk for men of being diagnosed with colorectal cancer in the UK is estimated to be 1 in 16 for men and for women 1 in 20.

Polygenic inheritance
A trait produced from the cumulative effects of many genes.

The aetiology of bowel cancer is incompletely understood, but includes a combination of genetic and environmental factors. Family history plays an important role, mainly due to **polygenic inheritance**, but well-characterized cancer syndromes, such as Lynch syndrome, and familial adenomatosis polyposis (FAP) are known to significantly increase the risk of developing bowel cancer. Obesity, diet (high in fat and red meat and low in fibre), and a high intake of alcohol are all associated with an increased risk. Finally, inflammatory conditions of the gastrointestinal tract, such as inflammatory bowel disease, also play a role.

11.2.2 Common tumour types

More than 95% of colorectal cancers are gland-forming adenocarcinomas, which invade through the muscularis mucosae into the submucosa or deeply into the colonic wall. Based on architectural and cytological features (the preservation of normal glandular architecture, nuclear polarity, cytological anaplasia), adenocarcinomas are classified as well, moderately, and poorly differentiated adenocarcinomas. Eighty-five per cent are well to moderately differentiated, while the remainder are poorly differentiated (Qizilbash, 1982).

In addition to adenocarcinomas, the World Health Organization (Bosman et al., 2010) recognizes other histological types and variants, which are clinically important as their behaviour and prognosis are different.

11.2.2.1 Signet ring carcinoma

The incidence of primary colorectal signet-ring (SR) cell carcinoma in Western countries varies between 1% and 2.5% of all colorectal cancer (Secco et al., 1994). A poorer survival compared with ordinary adenocarcinomas has been reported and SR is recognized as a stage-independent prognostic factor for adverse outcome in CRC. The molecular and genetic characteristics of SR are largely unknown, but errors in the DNA-replication system (microsatellite instability) appear to play a role in the carcinogenesis of this type of tumour (Kawabata et al., 1999).

11.2.2.2 Mucinous adenocarcinoma

Carcinomas composed of at least 50% of mucin are classified as mucinous carcinomas (MC) (Figure 11.1). This variant accounts for less than 10% of all colon cancers and 33% of rectal cancers (Bosman et al., 2010). A strikingly higher frequency of mucinous carcinomas in young adult patients has been reported suggesting that the pathogenesis of this tumour variant might be different that non-mucinous tumours. High frequency of microsatellite instability, a higher *KRAS* mutation rate and downregulation of *TP53* are involved in the genesis of these tumours (Garcia-Aguilar et al., 2011). As with signet-ring cell carcinomas, MC have also been described frequently in patients with inflammatory bowel disease-associated cancer.

11.2.2.3 Adenosquamous adenocarcinoma

Adenosquamous (Ad-SCC) carcinoma of the colon and rectum are very rare, and it has been estimated that its incidence varies between 0.025% and 0.85% of all colorectal malignancies (Cagir et al., 1999). Ad-SCC contains both malignant glandular and squamous histological components with metastatic potential. They tend to present in young patients and the prognosis is worse compared with ordinary adenocarcinomas, particularly in those cases with lymph node involvement or metastatic disease at presentation (Frizelle et al., 2001). The pathogenesis of Ad-SCC is poorly understood. The role of human papilloma virus (HPV) has been extensively investigated and the results are contradictory. Some authors have detected HPV type 16 in these tumours (Kong et al., 2007), while others have found no evidence of HPV infection (Frizelle et al., 2001). Similarly, the role of p16 is yet to be defined, as high levels of p16 have been detected in Ad-SCC, even in the absence of HPV infection (Dong et al., 2009).

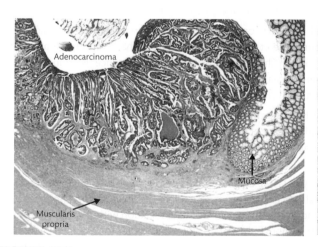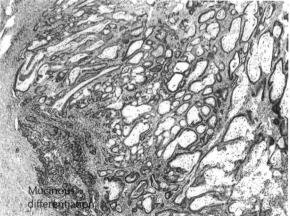

FIGURE 11.1

Mucinous differentiation in a colonic adenocarcinoma (Haematoxylin and eosin stain). (Left) Histological section of colonic wall showing an adenocarcinoma extending into the muscularis propria. (Right) Higher magnification showing areas of mucinous differentiation (mucinous lakes and neoplastic glands).

11.2.4 Signalling pathways

Several signalling pathways play a pivotal role in controlling and maintaining growth and proliferation in colorectal cancer. Key pathways in colorectal tumorigenesis include:

- Chromosomal instability pathway, also known as adenoma–carcinoma pathway.
- Serrated pathway.
- DNA repair defect pathway/microsatellite instability (MSI).

Alterations in these pathways may lead to activation of the epidermal growth-factor receptor, and downstream transcription factors and tyrosine-kinases in the RAS/RAF/MAPK and the phosphoinositide 3-kinase (PI3K)/AKT cascades.

Other pathways such as Notch (Katoh and Katoh, 2007) and Hedgehog (Hh; Taipale and Beachy, 2001) have attracted research interest in the investigation of the development of colorectal cancer. In addition, over the past decade, advances in the knowledge of epigenetics mechanisms have drawn renewed attention to the cytosine polyguanine (CpG) island methylator pathway (CIMP) in MSI sporadic colorectal cancers (Bae et al., 2013).

11.2.4.1 Chromosomal instability (CIN) pathway

Vogelstein (Vogelstein et al., 1988) described the first model of colorectal cancer, occurring in a stepwise sequence from normal tissue. CIN refers to chromosomal gains and losses (parts or whole chromosomes) resulting in alterations in the karyotype and imbalances in the number of chromosomes. The CIN pathway combines abnormalities in the chromosomes and mutations in tumour suppressor genes and oncogenes (see later) and is responsible for 70% of colorectal cancers.

Adenoma

A benign growth that may progress to a malignant lesion.

1. The first step in the adenoma-carcinoma sequence is the development of a conventional **adenoma** with low-grade dysplasia, arising from normal tissue due to mutations in various genes, of which *APC* (adenomatous polyposis coli) is the most relevant. *APC*, a tumour suppressor gene, is involved in the Wnt-signalling pathway that regulates β-catenin concentrations in the cytoplasm. β-catenin is then translocated to the nucleus leading to the activation of a wide range of proliferation and anti-apoptotic pathways. Loss-of-function mutations in the *APC* gene are found in over 90% of CRCs, and are one of the first steps in the adenoma–carcinoma sequence.

2. A mutation in one of the *Ras* (rat sarcoma) genes results in enlargement of the adenoma. *KRAS* mutations are present in up to 40–50% of sporadic CRCs and *NRAS* in 5%. RAS proteins are mediators in the mitogen-activated protein kinase (MAPK) pathway, which regulates cell proliferation in response to external growth factors, such as epidermal growth factor (EGF).

3. LOH at chromosome 18q, the most common large-scale chromosomal event in CRC, is the next step and is usually accompanied by the loss of expression of genes in the same region (e.g. *SMAD4*). These alterations lead to the development of high-grade dysplasia in adenomas.

4. LOH also results in a loss of the *TP53* gene (located on chromosome 17). The gene product p53 is a well-known tumour suppressor involved in apoptotic and cell-arrest pathways, which are initiated by cell stress and DNA damage. It is believed that loss of p53 function seems responsible for the transition between high-grade dysplasia and invasive carcinoma.

11.2.4.2 Microsatellite instability

The MSI pathway occurs in both hereditary (e.g. Lynch syndrome; 2–5% of all CRCs) and sporadic colorectal cancer (15%). Mismatch repair proteins are involved in the repair of point mutations and insertion/deletions (indels), which occur during DNA replication. Mismatch repair (MMR) mechanisms maintain genomic stability by eliminating indels loops of short repeated nucleated sequences. In eukaryotes interaction of MMR proteins, usually heterodimers (MLH1-PMS2, MSH3-MSH6, MSH2-MSH6) is required to correct these indels loops. Mutations in genes encoding for MMRs lead to MSI, where failure to clear damaged cells during replication leads to accumulation of mutations.

MSI can be routinely detected by examining the DNA sequence and the degree of MSI can be graded as MSI-H (high) or MSI-L (low), depending upon the degree of instability. ICC detection of MLH1, MSH2, PMS2, and MSH6 proteins is also available for assessing the function of MMR repair mechanism, and it is routinely performed in most pathology diagnostic laboratories (NICE guidelines 2017; Figure 11.2).

Alterations in the Wnt signalling are also initial events in the MSI pathway, but in contrast with the CIN pathway, MSI tumours rarely have mutations in *RAS* genes or show large-scale chromosomal abnormalities. Most commonly, MSI carcinomas are associated with mutations in *BRAF* (an early event), *CDC4*, and *BAX*. Like Ras proteins (KRAS and NRAS), BRAF is also a mediator in the MAPK pathway. *RAS* and *BRAF* mutations are usually almost mutually exclusive in colorectal tumours. The detection of MSI/dMMR (deficient mismatch repair proteins tumours) and *BRAF* mutations is useful in clinical practice for prognosis and therapeutic approach. Tumours with MSI or dMMR by ICC phenotype have better prognosis, do not respond to 5-fluorouracil (5-FU)-based chemotherapy and potentially respond to immunotherapy (Hutchins et al., 2011; Jover et al., 2011). Recently, alterations in gene coding for DNA polymerases δ and ε (*POLD* and *POLE*) have been described as associated with predisposition to colorectal cancer (Spier et al., 2015). These polymerases have a proofreading function alongside their role in DNA replication. Germline mutations in *POLD* and *POLE* result in a

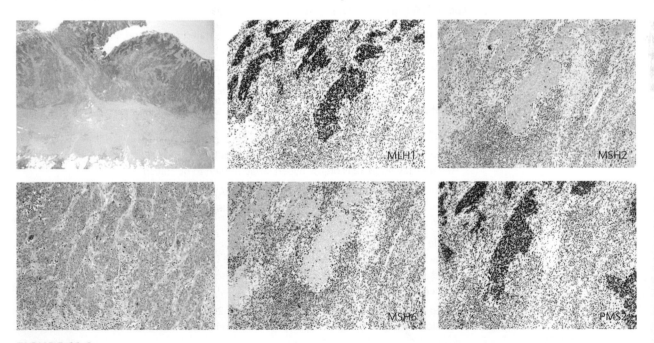

FIGURE 11.2

Mismatch repair proteins: immunocytochemical detection. Poorly differentiated adenocarcinoma with small number of intratumour lymphocytes (H&E, upper and lower left pictures). The tumour cells show normal nuclear expression of MLH1 and PMS2, but there is loss of MSH2 and MHS6 expression (scattered intratumour lymphocytes are positive).

very high rate of mutations (the hyper/ultramutator phenotype) with phenotypically similar tumours to MSI cancers, high immunogenicity, and favourable prognosis.

Finally, MSI has been detected in colorectal cancers associated with inflammatory bowel disease (IBD), but the CIN pathway is more common in these settings. IBD-associated CRC patients have somatic mutations in classical CIN genes such as *APC*, *KRAS*, and *TP53*, as well as a number of inflammation pathways such as *COX-2*. However, in contrast to non-IBD CRCs, these tumours appear to develop more rapidly and p53 mutations are initiating mutations in the majority of the lesions (Leedham et al., 2009).

11.2.4.3 Serrated pathway

Our understanding of serrated **polyps**/lesions, defined by the characteristic 'saw-toothed' crypts, have evolved from the classical hyperplastic polyps in the distal colon to include several other entities, some of which carry an increased risk of developing colorectal cancer. Classifications of serrated polyps have been proposed (East et al., 2017) and nowadays serrated lesions of the large bowel include:

1. Hyperplastic polyps.

2. Serrated adenomas.

3. Mixed hyperplastic/adenomatous polyps.

4. Sessile serrated polyps/lesions.

5. Hyperplastic polyposis.

6. Serrated carcinomas.

A high level of inter and intra-observer variability in classifying serrated polyps has been reported (Wong et al., 2009).

- Hyperplastic polyps (HP) account for 75% of all serrated polyps and it has been estimated that the prevalence of these in the adult population in Western countries is around 10% (Yano et al., 2005). These have been further subclassified in three subgroups:

  - microvesicular serrated polyps, the most frequent subtype;
  - goblet cell hyperplastic polyps (very small size and little serration in crypts);
  - mucin poor type.

  Whether these morphological variants have a different behaviour and whether sporadic HPs are truly neoplasias or just 'hyperplasias' remain to be solved.

- Serrated adenomas (SA) comprise 5% of serrated polyps and usually are present in the left colon. By definition, SA have dysplasia, and frequently have an abundant eosinophilic cytoplasm and pencillated nuclei.

- Sessile serrated polyps/lesions (SSP) account for 15–20% of all serrated polyps. SSPs are characterized by hyperserration throughout the crypt and horizontal orientation of deep crypts. Up to 14% of SSP show dysplasia and these findings appears to be more frequent in women than in men (Lash et al., 2010).

- Mixed polyps account for 0.7–1.7% (Carr et al., 2009) of serrated polyps, and represent a collision between hyperplastic polyps and adenomas.

Diagnostic criteria for hyperplastic polyposis, as recommended by the WHO classification (Burt and Jass, 2000), are:

- at least five HPs proximal to the sigmoid colon, of which two are >10 mm in diameter; or

- any number of HP proximally to the sigmoid colon in an individual who has a first-degree relative with HPS; or

- >30 HP of any size distributed throughout the colon.

A high incidence of colorectal cancer in hyperplastic polyposis syndrome has been reported (Rubio et al., 2006).

Serrated adenocarcinomas comprise 7–12% of all colorectal cancers (Bae et al., 2016) and might arise in traditional serrated adenomas or in 15–20% of cases in sessile serrated polyps/lesions. They are more frequent in females and, although they can occur in the rectum, they predominate in the right colon.

Two alterations have been described to play a role in the development of colorectal cancers arising in serrated lesions: CpG island methylator phenotype and mutations in *BRAF* (v-Raf murine sarcoma viral oncogene homolog B1) oncogene.

1. *Somatic mutation in BRAF* (single point mutations in V600E) is the most common alteration in all serrated polyps and is less frequent in adenomatous polyps (Rosty et al., 2013).

2. Approximately 50% of all protein-encoding genes contain CG (cytosine-guanine)-rich regions in their promoters, known as *CpG islands*. Although DNA methylation plays a role in physiologically normal development, hypermethylation of these islands might also result in tumour suppressor gene inactivation (type C methylation). Excess of methylation of CpG islands, known as CIMP, occur in up to 50% of CRCs, but it is also well known that age-related methylation (type A methylation) also occurs and methylation of the oestrogen receptor gene in ageing normal colonic epithelium has been described (Issa, 2000). In contrast with determination of MSI status, there is no agreement on the optimal panel of CpG for CIMP determination and several panels have been proposed, varying from the 'classical' panel of five markers (*CDKN2A, MINT1, MINT2, MINT31, MLH1*) (Park et al., 2003) to panels, including 807 cancer-related genes (Ang et al., 2010). Comprehensive reviews on detection of CIMP can be found elsewhere in the literature (Curtin et al., 2011). Clinicopathologically, CIMP-positive tumours are usually right-sided, present at an older age at diagnosis, have a female predominance, are poorly differentiated, microsatellite unstable, carry *BRAF* mutations, and are *TP53* wildtype (Ogino et al., 2009b).

It is now believed that an activating mutation in *BRAF* is the first step in the canonic serrated pathway (Leggett and Whitehall, 2010). Gene inactivation through methylation of CpG islands is then required to reverse the senescence induced by BRAF in initial serrated lesions. Among the genes inactivated are *p16INK4* and particularly insulin-like growth factor-binding protein 7 (*IGFBP7*). It has been shown that inactivation of IGFBP7 enables CIMP tumours to escape from p53-induced senescence (Suzuki et al., 2010). The senescence–neoplasia sequence that also plays a key role in the development of melanomas from naevi and, similarly, oncogene-induced senescence drives tumorigenesis in serrated lesions (Minoo and Jass, 2006). If *MLH1* is also methylated, the tumours are microsatellite unstable and, therefore, more prone to acquire more mutations as described in MSI-associated tumours (see MSI pathway).

All the above models are not exclusive and overlapping among different mechanisms and alternative pathways do exist. Recently, an international expert consortium (Guinney et al., 2015) have proposed four molecular subtypes (CMS) based on analysis of 18 different *CRC* gene expression datasets, including data from *The Cancer Genome Atlas TCGA* (The Cancer Genome Atlas Network, 2012; see Figure 11.3).

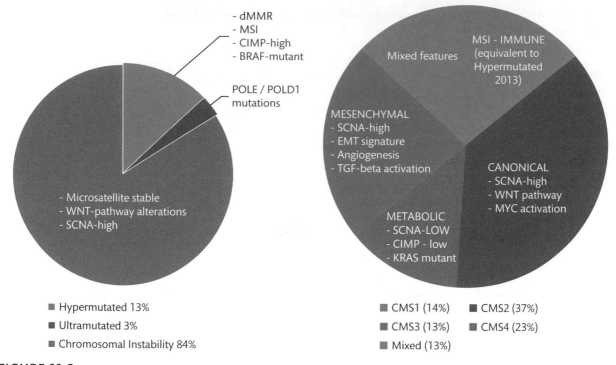

- dMMR
- MSI
- CIMP-high
- BRAF-mutant

POLE / POLD1 mutations

- Microsatellite stable
- WNT-pathway alterations
- SCNA-high

Mixed features

MSI - IMMUNE (equivalent to Hypermutated 2013)

MESENCHYMAL
- SCNA-high
- EMT signature
- Angiogenesis
- TGF-beta activation

CANONICAL
- SCNA-high
- WNT pathway
- MYC activation

METABOLIC
- SCNA-LOW
- CIMP - low
- KRAS mutant

■ Hypermutated 13%
■ Ultramutated 3%
■ Chromosomal Instability 84%

■ CMS1 (14%)   ■ CMS2 (37%)
■ CMS3 (13%)   ■ CMS4 (23%)
■ Mixed (13%)

FIGURE 11.3

Molecular classification of colorectal cancer based on TCGA (2013) and international consortium (Guinney et al., 2015).

Key Point

Several signalling pathways control and maintain growth and proliferation in colorectal cancer. These include the chromosomal instability, serrated, and DNA repair defect/MSI pathways.

11.1 SELF-CHECK QUESTION

How do changes in signalling pathways influence the change from benign adenomas to malignant colorectal tumour?

11.2.5 Specific somatic mutations

11.2.5.1 RAS mutations

The signalling cascade RAS–RAF–MEK–ERK defines the MAPK pathway. Alterations in this pathway occur in >50% of CRC, with activating mutations in the *KRAS* gene the most frequent abnormality (40% of all CRCs). Initially, mutations affecting exclusively codons 12 and 13 of exon 2 were described as significant in prognosis and predictive of response to treatment in CRC. However, in the last few years, additional mutations in exons 3 (codons 59 and 61) and 4 (codons 117 and 146) have also been reported with significant implications in terms of response to monoclonal antibody targeted therapy. Therefore, wider RAS testing panels (e.g. targeted NGS (cancer panel), real-time PCR (Cobas®), Idylla™, Biocartis) are recommended before treatment decision (Sorich et al., 2015).

Activating mutations of the *KRAS* gene lead to aberrant activation of EGFR signalling. *Cetuximab* and *panitumumab* (Figure 11.4) are antibodies that bind to the extracellular domain of EGFR, blocking the EGFR cascade. In the presence of *KRAS* mutations, there is resistance to anti-EGFR monoclonal antibodies. It has been suggested that some mutations, e.g. G13D, were associated with sensitivity to anti-EGFR, but recent meta-analysis found no significant differences between KRAS G13D and other mutations in terms of benefit from anti-EGFR therapy (Rowland et al., 2016).

It has to be noted that absence of *KRAS* mutation does not accurately predict response to anti-EGFR monoclonal antibody therapy. *Amphiregulin* (AREG) and *epiregulin* (EREG) are ligands to EGFR, and it has been reported that high mRNA levels of these ligands predict response to anti-EGFR and are associated with favourable prognosis in KRAS wild-type metastatic colorectal cancer (Razis et al., 2014).

11.2.5.2 BRAF mutations

BRAF is a proto-oncogene located on chromosome 7q34 and is mutated in about 7% of all human cancers, in 8–10% of CRC and 50% of melanomas (Sclafani et al., 2013). *KRAS* and *BRAF* mutations are usually mutually exclusive although co-existence of both mutations have been described (De Roock, et al., 2010). *BRAF* mutant (*mBRAF*) in CRCs are more common in older patients and females, and the tumours are more likely to have MSI, be of higher grade, have more lymph node involvement, and thus of advanced disease.

mBRAF tumours have strong negative prognostic factor in the metastatic setting regardless of treatment. Whether the presence of *BRAF* mutation predicts lack of response to anti-EGFR monoclonal antibody is still controversial (Rowland et al., 2016). Targeting *mBRAF* cancers with inhibitors such as vemurafenib has failed to improve survival in CRC patients, as *BRAF* inhibitors induce aberrant upstream signalling of the pathway through activation of PTEN-PI3K-AKT pathway. In these settings different approaches targeting BRAF, MEK, and EGFR are currently being considered in phase I and II clinical trials.

Recently, it has been recognized that non-V600 *BRAF* (e.g. codons 594 and 596) mutations make up nearly 5% of all *BRAF* mutations found in metastatic colorectal cancer (Cremolini et al., 2015). As a whole, non-V600 mutant CRCs appear to behave favourably to V600 *BRAF* mutant CRC (Jones et al., 2017).

11.2.5.3 PIK3CA and PTEN

Activation of this pathway has also been associated with resistance to anti-EGFR therapy. The activation could be secondary to mutations in *PIK3CA* (10–18% of CRCs) or loss of expression of PTEN (30% CRCs). *KRAS* and *PIK3CA* mutations frequently co-exist in CRC (Janku et al., 2011).

A recent meta-analysis has shown that *PIK3CA* mutation has a neutral prognostic effect on overall survival and progression-free survival in CRC (Mei et al., 2016). Currently, there is insufficient evidence to recommend PIK3CA mutational analysis or PTEN analysis (expression by ICC or deletion by *in situ hybridization* (ISH)) in colorectal carcinoma tissue for patients who are being considered for therapy selection outside clinical trials (Figure 11.5).

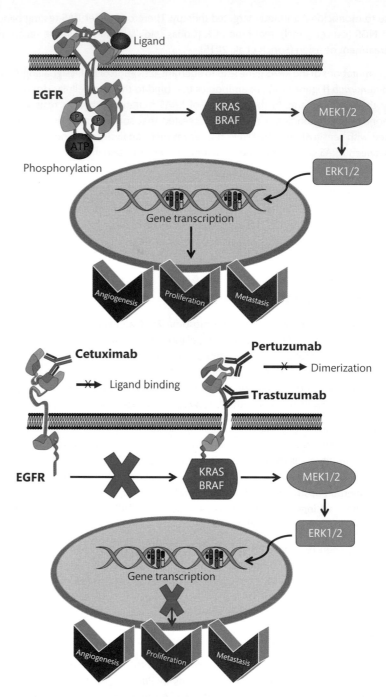

FIGURE 11.4

EGFR-KRAS/MEK pathway and pharmaceutical targets. Mitogen-activated protein kinase (MAPK) pathway initiated by activation of EGFR (left). Monoclonal antibodies, cetuximab that blocks ligand binding, or pertuzumab/transtuzumab that block receptor dimerization, lead to deactivation of the pathway. In the presence of *KRAS* mutation the pathway is activated downstream and anti-EGFR monoclonal antibodies have no effect in the pathway.

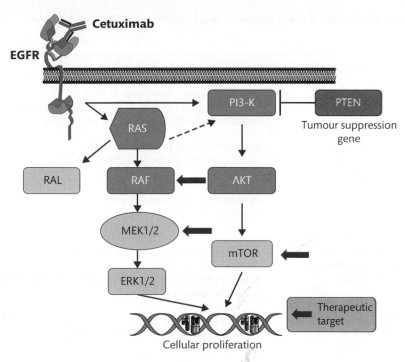

FIGURE 11.5

EGFR-MAPK pathway in colorectal cancer and agents targeting signalling proteins. Signalling pathways involved in colorectal cancer progression. Several agents are being evaluated in the context of CRC resistant to anti-EGFR therapy: (1) BRAF inhibitor: dabrafenib; (2) MEK inhibitor: trametinib; (3) mTOR inhibitor: everolimus; and (4) SAR245409 for Pi3K.

11.2.5.4 HER2

Overexpression and amplification of *HER2* has been identified as a potential therapeutic target in colorectal cancers. *HER2* amplification occurs in approximately 3.5% of all patients with colorectal cancer, and in approximately 5–6% of patients with colorectal cancer who have wild-type *RAS* and *BRAF* tumours (Pectasides and Bass, 2015). 3–4% of patients with colorectal cancer have a potentially activating mutation in *HER-2*, but the prognostic significance of HER-2 alterations is still unknown in CRC.

The HERACLES study (HER-2 Amplification for Colo-recta cancer Enhanced Stratification, phase II study) has shown that dual attack on *HER2* with trastuzumab and lapatinib in cetuximab-resistant metastatic colorectal cancer is an effective therapy in up to 40% of *KRAS* wild type metastatic colorectal cancer patients (Sartore-Bianchi et al., 2016).

The mutations that are presently known to be associated with CRC are summarized in Table 11.1.

Key Point

The MAPK pathway is the most studied signalling pathway in colorectal cancer. Mutations in genes in the pathway have prognostic implications and predict response to targeted therapies.

TABLE 11.1 Summary of most common genetic alterations in colorectal cancer, pathway involved, and clinical significance

Gene	Lesion	Mechanisms	Significance
APC	Deleted or mutated in 90% of CRCs	Wnt signalling dys-regulation and β-catenin accumulation	Little prognostic significance Not routinely tested outside hereditary cancers (germline mutations in FAP)
KRAS	Point mutation in 30–40% of CRC	Over-activation of EGF/MAPK signalling and proliferation	Predicts response to anti-EGFR mAbs Routinely tested in the metastatic setting
NRAS	Point mutation in 5% of CRC	Over-activation of EGF/MAPK signalling and proliferation	Predicts response to anti-EGFR mAbs Routinely tested in the metastatic setting
BRAF	Point mutation in 8–10% of CRC	Over-activation of EGF/MAPK signalling and proliferation	Predicts prognosis and probably response to some treatments Routinely tested in the metastatic setting and in dMMR colorectal cancers
PI3KCA	Somatic mutation in up to 20%	Over-activation of EGF/MAPK signalling and proliferation	Controversial data in predicting response to anti-EGFR Routinely tested in the metastatic setting
MLH1	Hypermethylation (80%) or point muta-tion in 10% to 15% of CRC	MMR deficiency and MSI	Predicts prognosis and response to some treatments Used in Lynch syndrome screening Routinely assessed by ICC in clinical practice
TP53	Mutation or deletion due to LOH 17p in 50% of CRCs	Increased proliferation and progression to cancer	Little prognostic significance, not routinely tested

Key: mAbs, monoclonal antibodies.
Adapted from Colling (2017).

11.2 SELF-CHECK QUESTION

Up to 40–60% of wild-type KRAS colorectal tumours do not respond to anti-EGFR therapy. What other mechanisms can explain the resistance to anti-EGFR in these tumours?

11.3 Gastric and gastro-oesophageal adenocarcinoma

11.3.1 Epidemiology and tumour types

Gastric cancer is the fourth most common cancer worldwide and the second leading cause of cancer mortality. There is significant geographic variability, with the highest incidence in East Asia, Eastern Europe, and parts of South and Central America. In the UK, stomach cancer is the sixteenth most common cancer (http://www.cancerresearchuk.org/health-professional/cancer-statistics/), thirteenth most common in males, and the eighteenth most common in females. Its incidence has decreased significantly in the last decade worldwide, but mortality remains still high (<15% of patients survive >10 years) mainly as a result of most cancers being diagnosed at advanced stage.

Oesophageal cancers in the UK account for almost 2% of all the new cases of cancer with 8900 diagnosed in 2014. Oesophageal cancer is the fourteenth most common cancer in the UK (http://www.cancerresearchuk.org/health-professional/cancer-statistics/). Oesophageal cancers usually occur in the lower third of the oesophagus and comprise both adenocarcinomas (OAC) and squamous cells carcinomas (SCC). The risk factors for each type of cancer are different, but smoking has been linked to both types. Patients with Barrett's oesophagus, a complication of gastrointestinal reflux in which the squamous lining of the oesophagus is replaced by columnar epithelium, have increased risk of developing OAC. In Western countries, gastro-oesophageal reflux, obesity, and Barrett's oesophagus have contributed to a significant rise in the incidence of OAC.

Oesophageal SCC is a multistep process through the sequence chronic oesophagitis, dysplasia (low- and high-grade) and invasive carcinoma. As in the colorectal carcinogenesis the pathway is characterized by progressive acquisition and accumulation of molecular abnormalities involving the inactivation of tumour suppressor genes and the activation of oncogenes. In this multistep process the main molecular abnormalities are: *TP53* mutations, LOH of **tylosis with oesophageal cancer (TOC)** and amplification of cyclin D1, EGFR, or cMYC. *TP53* mutations occur at an early stage (oesophagitis/low grade dysplasia), and amplification of cyclin-D1 and cMYC are more commonly detected in the high-grade dysplasia/invasive carcinoma stage. Several variants of SCC have been described, some with prognostic implications—undifferentiated, verrucous carcinoma, sarcomatoid, and basaloid carcinoma.

> **Tylosis with oesophageal cancer**
>
> A genetic disorder characterized by thickening of the palms and soles, oral leukoplakia, and a very high risk of squamous cell carcinoma of the oesophagus.

Gastric adenocarcinomas have traditionally been classified in two types (Lauren, 1965) with different histological findings and clinical behaviour (Table 11.2).

Helicobacter pylori infection has been linked to the development of gastric cancer, particularly in endemic areas of *Helicobacter* infection. However, *Helicobacter* infection by itself does not lead to gastric cancer as only 5% of those infected develop cancer. Specific *H. pylori* strains, such as *cagA*, are associated with severe inflammatory response and higher risk (Huang et al., 2003).

11.3.2 Molecular pathways and common somatic alterations

The TCGA consortium (The Cancer Genome Atlas Research Network, 2014), as for colorectal cancer, have defined four molecular subtypes of gastric cancer. Cancer driver mutations include *TP53*, *PIK3CA*, and *RHOA* (the gene encodes the GTPase RhoA and this alteration is

TABLE 11.2 Main types of gastric adenocarcinoma

	Intestinal	Diffuse
Site	Antrum	Gastric body
Gross appearance	Polypoid, fungating	'Linitis plastica'
Histology	Forming glands, cohesive	Discohesive, signet ring type cells
Precursor lesions	Atrophic gastritis	Superficial gastritis
Prognosis	Better survival than diffuse type	Poor

Key: Linitis plastica; adenocarcinoma that affects the whole of the stomach invading the external muscle layers and hindering normal contractions. It is also known as 'leather bottle stomach'.

more frequent in diffuse gastric cancers), and several genes are frequently amplified (*ERBB2, FGFR2, MET, KRAS*). Recent studies have shown that Epstein–Barr virus subtype is associated with the best prognosis, while the genomic stable type carries the worst one. In addition, patients with MSI subtype get the greatest benefit from chemotherapy (Sohn et al., 2017; Table 11.3).

11.3.2.1 Epidermal growth factor family

EGFR (ErbB1) is overexpressed in up to 65% of gastric cancers, but the addition of anti-EGFR monoclonal antibodies, such as panitumumab or cetuximab to standard therapy, has failed to produce any clinical benefits in terms of overall survival in gastric cancer patients (Lordick et al., 2013).

The role of *HER2 (ErbB2)* is clearer in gastro-oesophageal and gastric cancers. The Trastuzumab for Gastric Cancer (ToGA) trial (Bang et al., 2010) showed a clear benefit of addition of trastuzumab to standard chemotherapy. Depending on the histological type and site of the tumour, HER2 overexpression varies between 6% and 35%, with the higher expression in intestinal type adenocarcinomas at the gastro-oesophageal junction.

HER3 (ErbB3) forms a heterodimer with HER2 and it has been shown to play a vital role in breast cancer. Although the data in gastric cancer is still scanty, a recent meta-analysis has shown that HER3 overexpression is associated with adverse clinical and pathological features in gastric cancer (Cao et al., 2016).

11.3.2.2 Fibroblast growth factor receptor (FGFR)

Fibroblast growth factor receptors (FGFRs) are a family of four transmembrane receptor tyrosine kinases (FGFR1–4), whose ligand fibroblast growth factor (FGF) has 23 subtypes. Its activation triggers the downstream MAP kinase pathway (PIK3CA/pAKT/mTOR). Deregulation in the pathway is linked to cancer development, angiogenesis, tumour progression, and chemotherapy resistance in multiple tumour types. The pathway might be dysfunctional due to gene amplification, gene fusions, or activating mutations. Amplification of *FGFR2* has been described in 5–10% of gastric cancers and *FGFR2* mutations in 4% of these tumours (Su et al.,

TABLE 11.3 Gastric cancer types according to molecular changes

Subtype (Frequency)	Main alteration
CIN subtype (50%)	*TP53* mutation Tyrosine-kinase receptor and RAS activation
MSI subtype (22%)	Hypermethylation MLH1 silencing
Genomic stable	Cadherin (*CDH1*)/*RHOA* mutations Fusions in *CLDN18* family
EBV subtype (8%)	Immune cell signalling *PIK3CA* mutation High hypermethylation Overexpression PD-L1/2

Cancer Genome Atlas Research Network. Comprehensive molecular characterization of gastric adenocarcinoma. Nature 2014; 513: 202–209.

TABLE 11.4 Summary of most common genetic alterations in gastric cancer

Gene	Frequency	Mechanism
TP53	>40%	Mutation
PIK3CA	25%	Mutation and amplification
MYC	>10%	Amplification
CDKN2A / p16	>10%	Deletion
ErbB2 (HER2)	15%	Amplification and mutation
ErbB3 (HER3)	10%	Mutation and amplification
KRAS	15%	Mutation and amplification
ARID1A	20%	Mutation

2014). Amplifications are most commonly seen in the diffuse-type gastric adenocarcinoma. There are several ongoing clinical trials assessing the use of highly selective pan-FGFR inhibitors in *FGFR2*-amplified gastric cancers.

11.3.2.3 cMET/HGF

Proto-oncogene *c-MET* (*MET*), also known as hepatocyte growth factor (HGF), belongs to the receptor-tyrosine kinase family. Its phosphorylated form also activates the PI3K/AKT and ERK/MAPK pathways. cMET has been shown by immunocytochemistry to be over-expressed in 75–90% of gastric cancers and amplified at gene level in 1.5–20% of the cases (Pyo et al., 2016). Monoclonal antibodies against HGF (rilotumumab) or MET (onartuzumab) has been developed and tested in GC patients with conflicting results (Shah et al., 2016).

A summary of the most common genetic alterations in gastric cancer is provided in Table 11.4.

Key Point

The molecular classification of gastro-oesophageal cancer has replaced classic classification of diffuse versus intestinal type cancer. The different subtypes are characterized by different driver mutations and pathways with prognostic significance.

11.3 SELF-CHECK QUESTION

Which infective agents have been associated with gastric cancer?

11.4 Lung cancer
11.4.1 Epidemiology and histological subtypes

Lung cancer is the leading cause of cancer-related deaths among men and women in the USA and Western countries. According to the Surveillance, Epidemiology, and End Results (SEER) programme data, lung and bronchial cancers account for 57.3 new cases of cancer per 100,000 individuals and 46.0 deaths per 100,000 in the USA each year (2009–2013). The estimated 5-year survival rate is < 20% (https://seer.cancer.gov/statfacts/). In the UK, lung cancer is the third most common cancer, accounting for 13% of all new cases. It is the second most common cancer in both males (14% of all male cases) and females (12% of all female cases) (http://www.cancerresearchuk.org/health-professional/cancer-statistics/).

There are two major types of lung carcinomas:

1. *Non-small cell lung cancer (NSCLC)* (>80%)

    (a) *Adenocarcinoma* (40%), show glandular formation/differentiation and mucin production by tumour cells. By ICC the tumour cells express: CK7, TTF1 (thyroid transcription factor 1) and Napsin-A proteins.

    (b) *Squamous cell carcinoma* (20%) is histologically characterized by keratin formation and/or the presence of intercellular bridges. By ICC the tumour cells are p63, p40, and CK5 positive, but negative for CK7 and TTF1.

    (c) *Large cell carcinoma*, undifferentiated non-small cell carcinoma account for the rest of the NSCLC.

2. *Small cell carcinomas* (15%) are poorly differentiated tumours composed of cells with scanty cytoplasm and ill-defined cell borders. The cells have granular chromatin and, usually, there is extensive necrosis and high mitotic activity. These tumours have a neuroendocrine origin. By ICC these tumours are CD56, CK7, and TTF1 positive, but negative with p63 and CK5.

Up to 86% of lung and bronchial carcinomas are linked to smoking as the main risk factor and smoking accounts for at least 30% of all cancer deaths and 87% of lung cancer deaths (USA Department of Health & Human Services, 2012).

11.4.2 Molecular pathways and common somatic alterations

Molecular abnormalities in squamous cell carcinomas involve alterations in *TP53* and *CDKN2A*, more frequently, while mutations in *retinoblastoma* (*RB*) and *TP53* genes are most prevalent in small cell carcinomas. The most frequent genetic alterations in adenocarcinomas are described below (Figure 11.6).

11.4.2.1 *EGFR mutations*

EGFR mutations are present in 10–15% of lung cancers, 80% of them are in adenocarcinomas, and they are usually more frequent in females and non-smokers. Somatic mutations are most commonly seen in exons 18–21, the most frequent events are deletion in exon 19 (45–60%) and L858R mutation in exon 21 (37–45%). Other uncommon mutations are: exon 20 insertions (~4%), exon 18 substitutions (~3%), exon 18 deletions or insertions, and exon 20 substitutions (Sharma et al., 2007). The assessment of EGFR mutation status relies of the use of sequencing methods, as described in Chapter 5, can be used to detect L858R point mutations and E746–A750 deletions.

Both deletion in exon 19 and L858R mutation are associated with good response to *EGFR* kinase inhibitors (gefitinib/erlotinib). However, not infrequently, patients develop resistance to these drugs, mainly as a result of a second *EGFR* mutation. This has been reported in up to 50–70% of the cases (Stewart et al., 2015). Recently, T790M in *EGFR* has been described in patients with cancer progression and failure to tyrosin-kinase inhibitors (Takahama et al., 2016). This mutation can be detected in circulating tumour DNA (ctDNA) from plasma in patients with NSCLC. The National Institute for Health and Clinical Excellence (NICE) in the UK has approved the use of osimertinib (3rd generation EGFR TKI) in locally advanced or metastatic *EGFR* T790M mutation-positive NSCLC (NICE guideline TA416, 2016).

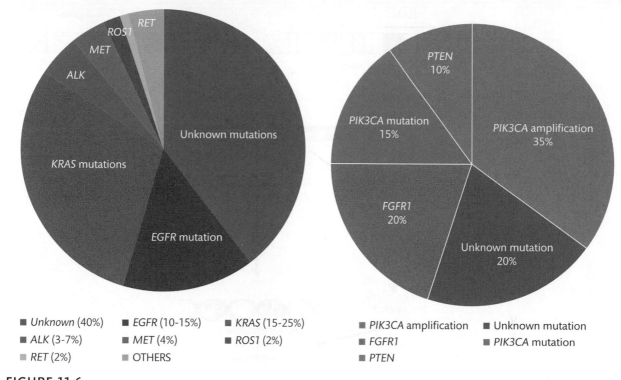

Cross reference

See Chapter 6, Section 6.3.1 for discussion of the analysis of genes in interphase nuclei.

FIGURE 11.6

Molecular abnormalities in non-small cell lung cancer. Left, adenocarcinoma; right, squamous cell carcinoma.

11.2.4.2 ALK rearrangement

ALK fusions are detected in 3–7% of lung adenocarcinomas. The most common rearrangement is due to chromosome 2p inversion usually with *EML4*, although up to 13 different *ALK* fusions have been reported (Shackelford et al., 2014); Figure 11.7). The fusion protein results in the activation of several pathways leading to cell proliferation, decrease in apoptosis, and invasiveness. *ALK* rearrangements can be detected by ICC or using the break-apart FISH assay. ICC correlates well with the presence of *ALK* fusions (Yi et al., 2011) and might be a first approach to screen lung carcinomas.

Patients with *ALK* rearrangements show excellent response rates to the kinase inhibitor crizotinib (also indicated in tumours with *MET* or *ROS1* amplification). As with patients treated for EGFR inhibitors drug resistance may develop due to different *ALK* rearrangements arising or activation of alternative pathways.

Other genetic/molecular alterations seen in lung adenocarcinomas are summarized in Table 11.5.

Key Point

Squamous cell carcinomas and adenocarcinomas of the lung have different pathogenic mechanisms and molecular aberrations. Although there have been significant advances in understanding the molecular pathways involved in lung adenocarcinoma, in 40% of cases there are no known mutations, precluding use of targeted therapies.

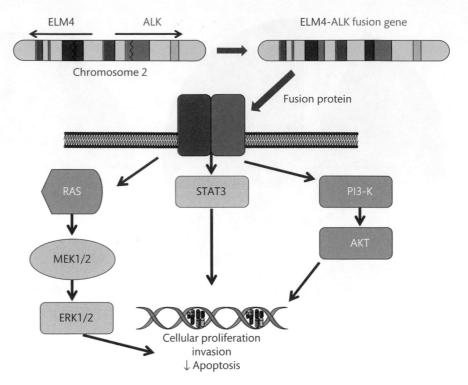

FIGURE 11.7

ALK/EML **fusion-driven adenocarcinoma of lung.** ALK fusions are the result of rearrangement in chromosome with *EML4*, resulting in a fusion protein responsible for the activation of several oncogenic pathways.

TABLE 11.5 Genetic and molecular alterations in MET / KRAS / RET and ROS1 genes in adenocarcinoma of lung

Gene	Frequency	Comments
MET	3–4% NSLCL, mainly adenocarcinomas	MET amplification present in 4–7% of anti-EGFR resistant lung carcinomas
KRAS	Mutations in 15–25%	Conflicting data on prognosis and predictive significance of *KRAS* mutation ASCO guidelines do not recommend *KRAS* mutation testing as a sole determinant of EGFR TKI therapy
RET	1–2% adenocarcinomas, fusions *RET-KIF5B*	Younger patients, poorly differentiated adenocarcinomas
ROS1	1–2% adenocarcinomas, several fusion partners (*SLC34A2, SDC4, LRIG3...*)	More likely to be never-smokers with more advanced disease

Cancer progression and resistance to anti-*EGFR* mutations has been reported in up to 50–70% of lung adenocarcinoma patients. Please describe any mechanism leading to anti-EGFR resistance. Is there any role in ctDNA to monitor response to monoclonal antibody targeted therapy?

11.5 Melanoma

11.5.1 Epidemiology and tumour types

Melanomas are malignant tumours that originate from melanocytes that are a normal cellular component of the epidermis of skin. However, melanocytes are also present in other organs and melanomas can also be seen in mucosa and central nervous system (e.g. uveal melanomas). Basal cell carcinomas and squamous cell carcinomas of the skin are far more frequent, but malignant melanomas (MM) are more aggressive. In the UK MM is the fifth most common cancer with over 15,000 cases diagnosed in 2014 with a similar incidence among women and men (http://www.cancerresearchuk.org/health-professional/cancer-statistics/statistics-by-cancer-type/skin-cancer/incidence). In the USA more than 87,000 cases were diagnosed in 2017 (https://www.cancer.org/cancer/melanoma-skin-cancer/about/key-statistics.html). Unfortunately, the incidence of MM has increased more than 100% in the last two decades. If the tumour is confined to the skin (early stage), the 5-year survival rate is 85–90%, but this drops to 4–5% in cases with distant metastases.

There is strong evidence that UV exposure/radiation is a mutagenic factor as people living in geographic regions with increased UV exposure are at a greater risk of developing melanoma. Genomics studies have shown that melanomas affecting chronically sun-damaged skin carried a high burden of mutations. It has to be noted that some melanomas might arise from benign melanocytic naevi, but most primary melanomas are not associated with precursor naevi.

The WHO recognizes at least four different subtypes of melanoma, based on the growth pattern and anatomical site (de Vries et al., 2006):

- superficial spreading melanoma (SSM);
- lentigo malignant melanoma (LMM);
- nodular melanoma (NM);
- acral lentiginous melanoma (ALM).

The majority of melanomas start as flat lesions which grow in a radial fashion (known as radial growth phase) in which there is no invasion into the dermis. Later they progress to a vertical growth phase with dermal invasion and potential metastatic spread. Depth of invasion (Breslow thickness), presence of ulceration, and high mitotic rate are considered adverse prognostic factors. On the other hand, it is recognized that patients with melanoma tumours infiltrated with T-cells have better long-term survival. Monoclonal antibody therapy using ipilimumab (targeting cytotoxic T lymphocyte-associated protein 4/*CTLA-4*) or nivolumab or pembrolizumab (targeting the programmed cell death protein1/PD-1) have the capacity of inducing tumour immunity and have been shown to dramatically improve durable survival rates in melanoma patients in tumours that express these cell membrane proteins (see Section 11.6).

11.5.2 Molecular pathways and common somatic alterations

The main molecular pathways dysregulated in malignant melanomas and its precursors include the mitogen associated protein kinase (MAPK); the phosphatidylinositol 3 kinase (PI3K)/phosphatase and tensin homologue deleted on chromosome 10 (PTEN)/AKT/mammalian target of rapamycin (mTOR) pathways and MET (hepatocyte growth factor). *BRAF, NRAS, GNAQ, GNA11,* and *KIT* mutations are also frequent in malignant melanomas. Germline defects in DNA repair mechanism, mainly involving cyclin-dependent kinases (*CDKN2A* and *CDK4*), might be a contributing factor particularly in combination with UV radiation.

11.5.2.1 RAS/MAPK pathway

BRAF and *NRAS* represent the most studied genes in melanomas. The mutation frequency varies depending on the type and site of the melanoma, but roughly 60–80% of MM carry a *BRAF* mutation and 20% a *NRAS* mutation (Ascierto et al., 2012). Interestingly, discrepancies in the rate of mutations have been described between primary and metastatic sites in up to 29% of cases; due to considerable tumour heterogeneity (Bradish et al., 2015).

The $BRAF_{V600E}$ mutation is most frequent. In this, a single point mutation in codon 600(CTG) of exon15 replaces valine with glutamine [*p.(Val600Glu)*]. This mutation accounts for 70–75% of all *BRAF* mutations in melanomas. $BRAF_{V600D}$, $BRAF_{V600R}$, and mutations in codon 11 are much less frequent (Richtig et al., 2016). Also of interest is the fact that $BRAF_{V600E}$ is also reported in benign naevi, raising the potential role of *BRAF* mutations in the transformation of 'benign' naevi into malignant melanomas (Wu et al., 2007). However, the presence of a *BRAF* mutation alone is not sufficient for malignant transformation.

With regard to *NRAS*, mutations at codon 61 are, by far, the most frequent alterations in primary sporadic melanomas, although mutations at codon 18 have also been described (Jakob et al., 2012). Although activating mutations in *KRAS* and *NRAS* also occur, these are thought to play a weak oncogenic role in melanoma development (Milagre et al., 2010).

Malignant melanomas with *BRAF600* mutations respond to *BRAF* inhibitors, such as vemurafenib and dafrabenib (Klein et al., 2013). The presence of these mutations, rather than the presence of NRAS mutation, is also predictive of response to *MEK* and *BRAF* inhibitors.

11.5.2.2 PTEN (phosphatase and tensin homolog deleted on chromosome 10)

PTEN is a tumour suppressor gene located at 10q23.3, which has been shown to be deleted or mutated in several malignant tumours. Loss of chromosome 10q has been reported in 30–50% of malignant melanomas and some authors have suggested an association with poor clinical outcome (Bucheit et al., 2014).

11.5.2.3 c-KIT

Cross reference
See Chapters 8 and 12 for more discussion of leukaemias and mesenchymal tumours.

c-KIT is a transmembrane receptor tyrosine kinase, the activation of which has been implicated in gastrointestinal stromal tumours and haematological malignancies. C-KIT activates other oncogenic pathways such as MAPK and JAK (janus kinase)/STAT (signal transducers and activators of transcription) pathways. The role of KIT in the genesis of melanoma is still controversial. It was believed that loss of c-KIT expression was associated with progression,

TABLE 11.6 Summary of most common genetic alterations in melanoma

Gene	Frequency	Mechanism
BRAF	50–70%	Mutation
NRAS	15–30%	Mutation and amplification
Cyclin-D1	6–44%	Amplification
CDK4	? (acral and mucosal)	Amplification
CKIT	30% (sun-damaged skin)	Mutation and amplification
CDKN2A (p16)	30–70%	Mutation or deletion
PTEN	5–20%	Mutation or deletion

but amplification of *c-KIT* or activating mutations have been described in mucosal (39%), acral (36%), and up to 30% of melanomas arising inchronically sun-damaged skin (Curtin et al., 2006). A summary of the most common genetic alterations in melanoma is provided in Table 11.6.

Key Point

Malignant melanomas are aggressive skin cancers. Their incidence has increased significantly in the last two or three decades. The MAPK/PIK3/mTOR pathway is the main signalling cascade altered in these tumours with *BRAF* mutation being the commonest somatic mutation found in melanoma patients.

11.5 SELF-CHECK QUESTION

Immune checkpoint blockade has provided durable responses across several tumour types, particularly successful in melanoma. Can you think of any mechanism in the pathogenesis of melanoma, which can explain the response to immuno-checkpoint inhibitors?

11.6 Immuno-checkpoints in epithelial tumours and melanoma

The importance of tumour-infiltrating lymphocytes (TILs) as reflection of primary host immune response against solid tumours is now recognized (Schatton et al., 2014). There is strong evidence that tumour infiltration by activated CD8+ cytotoxic T lymphocytes correlates with better survival of CRC patients. Programmed cell death ligand 1 (PD-L1) is a key immunoregulatory molecule that, upon interacting with its receptor, PD-1, suppresses the CD8+ cytotoxic immune response in both physiologic and pathologic conditions. In the tumour micro-environment TILs or tumour-specific T-cells can potentially recognize tumour cells. However, tumour cells can express PD-L1 (or PD-L2) and the interaction between PD-1 and its ligands leads to downregulation of antigen-stimulated lymphocyte proliferation and cytokine production. This results in what is called 'tumour tolerance' or 'anergy' with lymphocyte depletion and immunological tolerance. By blocking inhibitory molecules or, alternatively, activating stimulatory molecules, immunotherapy is designed to unleash or enhance pre-existing anti-cancer immune responses (Figure 11.8).

PD-L1 overexpression, by ICC detection, in tumour cells has been validated as a predictive marker for tumour response to anti-PD-1 or PD-L1 immunotherapy in different malignancies

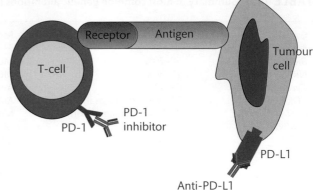

FIGURE 11.8
PD-1 Programmed cell death receptor 1.
PD-1 inhibitors pembrolizumab and
nivolumab, PDL-1 inhibitors atezolizumab
and durvalumab.

such as lung, head and neck, urothelial, and upper gastrointestinal cancers (Wang et al., 2017). However, the prognostic value of PD-1 and PD-L1 expression in colorectal cancers is still controversial.

Pathological and molecular determinants that define a subset of tumours more likely to respond to PD-1 or PD-L1 blockade include PD-L1 overexpression (ICC), RNA expression signatures, lymphocytic infiltrate, and mutational burden. It is also known that mismatch repair-protein deficient tumours (dMMR or MSI-high) have a very high mutational rate leading to mutation-associated neoantigens that are potentially recognized by the immune system.

Several immune-checkpoint inhibitors have been developed amongst which are nivolumab, an anti PD-1 antibody and pembrolizumab, a humanized IgG4 monoclonal antibody that blocks the interaction of PD-1 with its ligands PD-L1 and PD-L2. A summary of clinical trials and cancers involving the use of immune-checkpoint inhibitors is provided in Tables 11.7 and 11.8.

TABLE 11.7   Summary of most relevant clinical trials using immunocheckpoint inhibitors

Drug	Target	Tumour type	Clinical trials
Pembrolizumab	PD-1 inhibitor	NSCLC	*KEYNOTE-024*: NSCLC patients with no previous chemotherapy and no EGFR or ALK alterations. Progression-free survival (PFS) and overall survival (OS) superior in patients receiving PD-1 inhibitors vs standard chemotherapy
Pembrolizumab	PD-1 inhibitor	Gastric/ gastro-oesophageal	*KEYNOTE-059*: Patients with advanced gastric or gastroesophageal junction cancer who progressed on at least two prior lines of chemotherapy
Nivolumab	PD-1 inhibitor	Melanoma	Approved in advanced melanoma independently of PD-L1 expression
Nivolumab	PD-1 inhibitor	Non-squamous NSCLC	*CheckMate057*: overall survival was longer with nivolumab than with docetaxel in patients who had progressed during or after platinum-based chemotherapy
Atezolizumab	PD-L1 inhibitor	Urothelial carcinoma	*IMVigor 210 and 211*: Locally advanced or metastatic urothelial carcinoma after failure of prior platinum-based chemotherapy
Durvalumab	PD-L1 inhibitor	Urothelial carcinoma	*NCT02516241*: Ongoing phase III study of durvalumab with or without tremelimumab versus standard of care chemotherapy (cisplatin) in urothelial cancer

Key: NSCLC, non-small cell lung cancer.

TABLE 11.8 Immuno-checkpoint inhibitors approved (FDA and/or EMA) or in clinical trials and tumour type (solid tumours)

Tumour type	Drug	
Melanoma	Pembrolizumab	1st line treatment, including previously untreated advanced melanoma regardless of *BRAF* mutation status
	Nivolumab	In combination with ipilimumab for unresectable or metastatic melanoma irrespective of *BRAF* mutation status
Non-small cell lung cancer	Pembrolizumab	In combination with carboplatin as first line treatment of metastatic NSCLC irrespective of PD-L1 expression
	Nivolumab	In metastatic squamous NSCLC
	Atezolizumab	In previously-treated NSCLC
	Durvalumab	Granted FDA breakthrough designation in adjuvant treatment of locally advanced NSCLC
Urothelial cancer	Pembrolizumab	Locally advanced or metastatic cancers who have disease progression after platinum-based chemotherapy
	Nivolumab	Locally advanced or metastatic urothelial cancer
	Atezolizumab	First PD-L1 inhibitor approved for locally advanced and metastatic urothelial cancers
	Durvalumab	In platinum-refractory locally advanced or metastatic UC
Renal cell carcinoma	Nivolumab	Treatment-refractory clear cell renal cell carcinoma
Head and neck cancers	Pembrolizumab	Metastatic head and neck SCC with disease progression
	Nivolumab	In patients with recurred or progressed disease after platinum-containing chemotherapy
Gastric cancer	Pembrolizumab	Advanced gastro-oesophageal cancer PD-L1 positive and refractory to previous chemotherapy
Colorectal	Nivolumab	In MSH-H or dMMR metastatic colorectal cancer refractory to chemotherapy
	Pembrolizumab	MSH-H or dMMR colorectal cancer patients with unresectable or metastatic tumours
Hepatocellular carcinoma	Nivolumab	In advanced HCC refractory to sorafenib

Several studies and clinical trials have demonstrated that colorectal cancer patients with metastatic disease and mismatch-repair deficient tumours treated with these agents may achieve complete response and prolonged overall survival (Le et al., 2015). The FDA and European Medicines Agency (EMA) have granted a priority review to pembrolizumab for MSI-H metastatic colorectal cancer.

Long-term survival for patients treated with immuno-checkpoint inhibitors has been reported, for example, in malignant melanoma patients continued response (in terms of disease control) has been reported 5–10 years after starting anti-CTLA-4 therapy (Ugurel et al., 2016). However, the optimal duration of treatment with these drugs is yet to be defined and immunotherapies are also associated with immune-related adverse events such as colitis, hepatitis, and endocrinopathies. Early diagnosis and correct management of these side effects are currently areas of further work.

Key Point

Immunotherapy with immune checkpoint inhibitors has profoundly changed cancer treatment. The number of available monoclonal antibodies and their indications in solid and blood cancers has grown significantly in the last few years. PD-L1-positive tumours may indicate immune active tumours that can be a response to anti-PD-1 and/ or PD-L1 therapy.

11.6 SELF-CHECK QUESTION

Tumours with high mutational burden/microsatellite instability tend to respond better to immuno-checkpoint inhibitors. Please explain the possible mechanisms underlying this response.

11.7 **Future directions**

Although very significant advances have been made in the last few years in the understanding of solid epithelial tumours and melanomas, numerous challenges remain. Perhaps one of the most important is tumour heterogeneity with critical implications for the treatment of cancer patients (Fisher et al., 2013). Despite having similar histology the cells within the same tumour have diverse genomes and interact differently with the micro-environment leading to intratumour heterogeneity. The process is dynamic and 'driver mutations' and 'passenger mutations' (concepts yet to be defined) play a role in clonal and subclonal evolution of tumours. Different studies are looking at the evolution of cancer genomic clones and quantifying the dynamics of subclonal variation in epithelial tumours. In addition, there is also significant variation between primary tumours and metastatic deposits, and even between metastatic deposits at different sites. Clonal evolution seems to follow the Darwinian model, in which the 'fittest' clones are selected during the metastatic process. Understanding this process, through the combination of bioinformatics and mathematical modelling, will allow better design of therapeutic strategies for individual patients and their tumours.

Drug resistance is a common and difficult challenge to address in the treatment of epithelial cancers and melanomas, usually as a result of activation of collateral signalling pathways or secondary mutations. Combination of biological agents (monoclonal antibodies) can be effective in overcoming drug resistance and promising results, in terms of partial response in colorectal cancer, have been shown combining BRAF and EGFR inhibitors, BRAF and MEK inhibitors and even triple regimens with BRAF/EGFR/MEK inhibitors. In this area of monitoring cancer patients' disease the introduction of 'liquid biopsy' may overcome challenges associated with analysis of metastatic deposits and tumour heterogeneity. Blood samples can be analysed for circulating tumour cells (CTCs), circulating tumour DNA (ctDNA), and exosomes (Merker et al, 2018). Identification of CTCs might prove to be useful in patients with minimal residual disease who might develop recurrence in a short period of time. As mentioned

Cross reference

Details about liquid biopsy can be found in Chapter 7.

previously in this chapter, the detection of *EGFR* gene mutation (*TT90M*) in plasma samples has been approved as a companion diagnostic for osimertinib in metastatic *EGFR* mutation positive non-small lung cell cancer. Despite the extraordinary advances achieved in the last few years, liquid biopsy analyses is still far from being implemented in routine clinical practice.

In addition to the new immuno-checkpoint inhibitors new targets are being investigated, e.g. lymphocyte activation gene 3 protein (LAG-3) and combination therapies, such as ipilimumab and nivolumab in melanoma, show better response rates, although frequently associated with an increase in toxicity. Identification of predictive biomarkers is an unmet need to identify which tumour are 'immunologically ignorant' or 'immunological responsive' based on pre-existing antitumour immune response. Chimeric antigen receptor T (CAR-T) cells, now adopted in blood cancers, are finding their way in to the treatment of epithelial tumours and melanomas. CAR-T cells can identify tumour-associated antigens expressed on tumour cells and specifically eliminate these tumour cells (Kosti et al., 2018). Ongoing studies and early clinical trials are evaluating the role of CAR-T cells targeting EGFR, HER2, and mesothelin.

CASE STUDY 11.1

A 62-year-old male presented with rectal bleeding and anaemia, diagnosed with caecal adenocarcinoma following colonoscopy and imaging staging. There was evidence of vascular invasion but no lymph node involvement (Figure 11.9a). Eighteen months later, a single tumour mass was noted in the left lower lobe of the lung, confirmed as metastatic colorectal carcinoma on biopsy by ICC (CK20+, CDX2+, CK7-, and TTF1 negative; Figure 11.9b).

Somatic mutation analysis showed *KRAS* wild type colorectal cancer and patient started chemotherapy with FOLFOX (FOL: folinic acid, F: fluorouracil (5FU); OX: oxaliplatin) and cetuximab (anti-EGFR). Despite treatment, 12 months later the patient developed a second colorectal cancer (sigmoid) requiring completion

colectomy, and an inguinal mass (3 months after colectomy). The inguinal mass was confirmed as metastatic colorectal cancer (Figure 11.9c).

Somatic mutation analysis again showed *KRAS* wild type colorectal cancer and patient started chemotherapy with FOLFIRI (FOL: folinic acid, F: fluorouracil (5FU); IRI: irinotecan) and cetuximab (anti-EGFR), but progressed with new lung metastatic deposits. In view of the presence of metachronous colorectal cancer ICC for mismatch repair proteins was performed and the tumours were found to be MMR deficient (loss of MLH1 and PMS2 expression) (Figure 11.9d). The patient started immunotherapy with pembrolizumab (PD-1 inhibitor) achieving complete response of the lung disease and is currently free of disease.

(a)

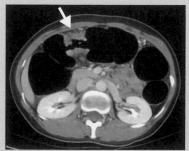
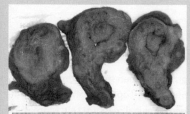
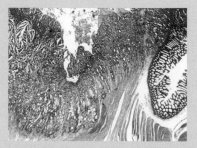

FIGURE 11.9 (*CONTINUED*)

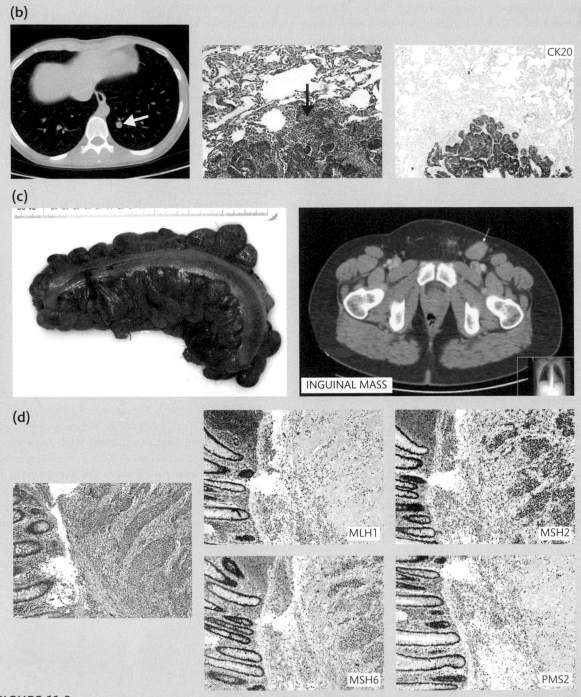

FIGURE 11.9

(a) Left: scan showing tumour (arrow). Centre: gross specimen showing tumour. Right: microscopic appearance of adenocarcinoma. (b) Left: scan showing mass in lung (arrow). Centre: microscopic appearance of tumour. Right: ICC showing positive staining for CK20. (c) Left: gross colectomy. Right: scan showing inguinal mass. (d) Left: Morphological appearance of colorectal cancer. Centre and right: ICC for MMR expression. Note normal positive staining for MSH2 and MSH6 in the tumour cells and absence of MHL1 and PMS2, indicating the loss of expression of these MMR proteins.

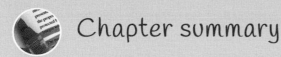

Chapter summary

Epithelial malignant tumours and melanoma are prevalent cancers in the general population involving a combination of complex molecular pathways and alterations. By identifying the molecular mechanisms of these diseases advances in appropriate therapeutic intervention are being made. In this context, key considerations for each cancer covered in this chapter are:

■ *Colorectal cancer*: The contribution of the Cancer Genomics Atlas and the International molecular consortium has identified different molecular subtypes in CRC associated with different prognosis and survival: molecular subtypes CMS1 (MSI-immune) and CMS4 (mesenchymal) have different prognosis and will benefit from different treatments. On the other hand, CMS2 (canonical) and CMS3 (metabolic) show similar disease-free survival and overall survival, and do not have specific treatments for them.

■ *Gastro-oesophageal cancer*: As in colorectal cancer, the international molecular consensus is replacing current classification of these type of tumours (intestinal versus diffuse type), which potentially could lead to improvement of patient selection for targeted therapy. EBV subtype appears to be associated with the best prognosis, while the gnomically stable subtype is probably associated with the worst prognosis.

■ *Lung cancer*: Despite being one of the leading cause of cancer death, the introduction of targeted therapies has led to an unparalleled improvement in overall survival. Predictive biomarkers (*EGFR* mutation status, *ALK* rearrangements, PD-L1 expression) are now routinely performed in referral/high-volume cancer centres. The application of new technologies, such as liquid biopsy, has the potential to identify early resistance to cancer therapy and monitor minimal residual disease after initial treatment.

■ *Melanoma*: Advances in understanding the molecular pathogenesis of malignant melanomas have helped to manage more effectively this aggressive neoplasm, particularly in the metastatic setting. The high prevalence of *BRAF*^{V600E} mutation in melanoma patients has proved to be a very efficient target for anti-melanoma therapy. In addition, novel immune checkpoint inhibitors, single, or in combination (anti-CTLA4 and anti-PD-1), are revolutionizing melanoma treatment, even becoming a first line of treatment, preferable to chemotherapy and radiotherapy with long-term response and less severe side effects.

■ The very recent introduction of PD-1 and PD-L1 immune checkpoint inhibitors is opening the way to the treatment of several cancers such as lung, head and neck, urothelial, and melanoma, where these molecules are expressed on tumour cells or in the tumour-associated lymphocytes. The introduction and approval of these new treatments in cancer is seeing an unprecedented durable response in a variety of cancers.

Further reading

● Dietel M, Jöhrens K, Laffert M, et al. Predictive molecular pathology and its role in targeted cancer therapy: a review focussing on clinical relevance. Cancer Gene Therapy 2013; 20: 211–221.

● Emens LA, Ascierto PA, Darcy PK, et al. Cancer immunotherapy: opportunities and challenges in the rapidly evolving clinical landscape. European Journal of Cancer 2017; 81: 116–129.

● Harris TJR, McCornick F. The molecular pathology of cancer. Nature Reviews: Clinical Oncology 2010; 7: 251–265.

- Kumar V, Abbas AK, Aster JC. (Eds). *Robbins Basic Pathology*, 10th edn. Oxford, Elsevier, 2017.

- McGranahan N, Swanton C. Clonal heterogeneity and tumour evolution; past, present and the future. Cell 2017; 168: 613–628.

 # References

- AJCC Colorectal Taskforce: Colon and rectum. In: Greene FL, Page AL, Fleming ID, et al. (Eds), *AJCC Cancer Staging Manual*, 6th edn. New York, NY, Springer, 2002, pp. 113–124.

- Amin MB, Edge SB, Greene FL, et al. (Eds). *AJCC Cancer Staging Manual*. 8th ed. New York: Springer; 2017.

- Ang PW, Loh M, Liem N, et al. Comprehensive profiling of DNA methylation in colorectal cancer reveals subgroups with distinct clinicopathological and molecular features. BMC Cancer 2010; 10: 227.

- Arnold M, Sierra MS, Laversanne M, et al. Global patterns and trends in colorectal cancer incidence and mortality. Gut 2017; 66: 683–691.

- Ascierto PA, Kirkwood JM, Grob J-J, et al. The role of BRAF V600 mutation in melanoma. Journal of Translational Medicine 2012; 10: 85.

- Bae JM, Kim JH, Cho N-Y, Kim T-Y, Kang GH. Prognostic implication of the CpG island methylator phenotype in colorectal cancers depends on tumour location. British Journal of Cancer 2013; 10: 1004–1012.

- Bae JM, Kim JH, Kang GH. Molecular subtypes of colorectal cancer and their clinicopathologic features, with an emphasis on the serrated neoplasia pathway. Archives of Pathology & Laboratory Medicine 2016; 140: 406–412.

- Bang YJ, Van Cutsem E, Feyereislova A, et al. Trastuzumab in combination with chemotherapy versus chemotherapy alone for treatment of HER2-positive advanced gastric or gastro-oesophageal junction cancer (ToGA): a phase 3, open-label, randomised controlled trial. Lancet 2010; 376: 687–697.

- Benson AB, Schrag D, Somerfield MR, et al. American Society of Clinical Oncology recommendations on adjuvant chemotherapy for stage II colon cancer. Journal of Clinical Oncology 2004; 22: 3408–3419.

- Bosman FT, Carneiro F, Hruban RH, Theise ND. *WHO Classification of Tumours of the Digestive System. IARC WHO Classification of Tumours*, 4th edn. Lyon, IARC Press, 2010.

- Bradish JR, Richey JD, Post KM, et al. Discordancy in BRAF mutations among primary and metastatic melanoma lesions: clinical implications for targeted therapy. Modern Pathology 2015; 28: 480–486.

- Bucheit AB, Chen G, Siroy A, et al. Complete loss of PTEN protein expression correlates with shorter time to brain metastasis and survival in stage IIIB/C melanoma patients with BRAFV600 mutations. Clinical Cancer Research 2014; 20: 5527–5536.

- Burt R, Jass JR. Hyperplastic Polyposis. In: Hamilton SR, Aaltonen LA (Eds), *Pathology and Genetics of Tumours of the Digestive System*. Lyon, ARC Press, 2000, pp. 20–65.

- Burton S, Normal AR, Brown AG, Abulafi AM, Swift RI. Predictive poor prognostic factors in colonic carcinoma. Surgical Oncology 2006; 15: 71–78.

- Cagir B, Nagy MW, Topham A, Rakinic J, Fry RD. Adenosquamous carcinoma of the colon, rectum, and anus: epidemiology, distribution, and survival characteristics. Diseases of the Colon and Rectum 1999; 42: 258–263.

- Cao GD, Chen K, Xiong MM, Chen B. HER3, but Not HER4, plays an essential role in the clinicopathology and prognosis of gastric cancer: a meta-analysis. PLoS One 2016; 11: e0161219.

- Carr NJ, Mahajan H, Tan KL, Hawkins NJ, Ward RL. Serrated and non-serrated polyps of the colorectum: their prevalence in an unselected case series and correlation of BRAF mutation analysis with the diagnosis of sessile serrated adenoma. Journal of Clinical Pathology 2009; 62: 516–518.

- Choi HJ, Park KJ, Shin JS, et al. Tumor budding as a prognostic marker in stage-III rectal carcinoma. International Journal of Colorectal Diseases 2007; 22: 863–868.

- Colling RT. The diagnostic molecular pathology of colorectal carcinoma using automated PCR. Thesis submitted for the degree of Doctor of Medicine, University College London 2017. Available at: http://discovery.ucl.ac.uk/view/theses/UCL=5FThesis/2017.html#group (accessed 3 October 2018).

- Cremolini C, Di Bartolomeo M, Amatu A, et al. BRAF codons 594 and 596 mutations identify a new molecular subtype of metastatic colorectal cancer at favorable prognosis. Annals of Oncology 2015; 26: 2092–2097.

- Curtin JA, Busam K, Pinkel D, Bastian BC. Somatic activation of KIT in distinct subtypes of melanoma. Journal of Clinical Oncology 2006; 24: 4340–4346.

- Curtin K, Slattery ML, Samowitz WS. CpG island methylation in colorectal cancer: past, present and future. Pathology Research International 2011; 2011: 902674.

- De Roock W, Claes B, Bernasconi D, et al. Effects of KRAS, BRAF, NRAS, and PIK3CA mutations on the efficacy of cetuximab plus chemotherapy in chemotherapy-refractory metastatic colorectal cancer: a retrospective consortium analysis. Lancet Oncology 2010; 11: 753–762.

- de Vries E, Bray F, Coebergh JW et al. Melanocytic Tumours. In: LeBoit PE, Burg G, Weedon D, Sarasin A (Eds), *Pathology and genetics of skin tumours*. World Health Organization Classification of Tumours. Lyon, IARC Press, 2006, pp. 49–65.

- Dong Y, Wang J, Ma H, Zhou H, Lu G, Zhou X. Primary adenosquamous carcinoma of the colon: report of five cases. Surgery Today 2009; 39: 619–623.

- East JE, Atkin WS, Bateman AC, et al. British Society of Gastroenterology position statement on serrated polyps in the colon and rectum. Gut 2017; 66: 1181–1196.

- Edge SE, Byrd DR, Carducci MA, et al. (eds.) *AJCC Cancer Staging Manual*, 7th edn. New York, NY: Springer; 2009.

- Ferlay J, Soerjomataram I, Dikshit R, et al. Cancer incidence and mortality worldwide: sources, methods and major patterns in GLOBOCAN 2012. International Journal of Cancer 2015; 136: E359–E386.

- Ferlay J, Steliarova-Foucher E, Lortet-Tieulent J, et al. Cancer incidence and mortality patterns in Europe: estimates for 40 countries in 2012. European Journal of Cancer 2013; 49(6): 1374–1403.

- Fisher R, Pusztai L, Swanton C. Cancer heterogeneity: implications for targeted therapeutics. British Journal of Cancer 2013; 108: 479–485.

- Frizelle FA, Hobday KS, Batts KP, Nelson H. Adenosquamous and squamous carcinoma of the colon and upper rectum: a clinical and histopathologic study. Diseases of the Colon and Rectum 2001; 44: 341–346.

- Garcia-Aguilar J, Chen Z, Smith DD, Li W, et al. Identification of a biomarker profile associated with resistance to neoadjuvant chemoradiation therapy in rectal cancer. Annals of Surgery 2011; 254, 486–492.

- Guinney J, Dienstmann R, Wang X, et al. The consensus molecular subtypes of colorectal cancer. Nature: Medicine 2015; 21: 1350–1356.

- Hamilton SR, Rubio CA. Carcinoma of the colon and rectum. In: Hamilton SR, Aaltonen LA (Eds), *WHO Classification of Tumours. Tumours of the digestive system*. Lyon, IARC Press, 2010, pp. 101–119.

- Hase K, Shatney C, Johnson D, et al. Prognostic value of tumour 'budding' in patients with colorectal cancer. Diseases of the Colon and Rectum 1993; 36: 627–635.

- Huang JQ, Zheng GF, Sumanac K, Irvine EJ, Hunt RH. Meta-analysis of the relationship between cagA seropositivity and gastric cancer. Gastroenterology 2003; 125: 1633–1644.

- Hutchins G, Southward K, Handley K, et al. Value of mismatch repair, KRAS, and BRAF mutations in predicting recurrence and benefits from chemotherapy in colorectal cancer. Journal of Clinical Oncology 2011; 29: 1261–1270.

- Issa JP. CpG-island methylation in aging and cancer. Current Topics in Microbiology and Immunology 2000; 249: 101–118.

- Jakob JA, Bassett RL Jr, Ng CS, et al. NRAS mutation status is an independent prognostic factor in metastatic melanoma. Cancer 2012; 118: 4014–4023.

- Janku F, Lee JJ, Tsimberidou AM, et al. PIK3CA mutations frequently coexist with RAS and BRAF mutations in patients with advanced cancers. PLoS ONE 2011; 6(7): e2276.

- Jones JC, Renfro LA, Al-Shamsi HO, et al. Non-V600 BRAF mutations define a clinically distinct molecular subtype of metastatic colorectal cancer. Journal of Clinical Oncology 2017; 35; 2624–230.

- Jover R, Nguyen TP, Pérez-Carbonell L, et al. 5-Fluorouracil adjuvant chemotherapy does not increase survival in patients with CpG island methylator phenotype colorectal cancer. Gastroenterology 2011; 130: 1174–1181.

- Katoh M, Katoh M. Notch signalling in gastrointestinal tract. International Journal of Oncology 2007; 30: 247–251.

- Kawabata Y, Tomita N, Monden T, et al. Molecular characteristics of poorly differentiated adenocarcinoma and signet-ring cell carcinoma of colorectum. International Journal of Cancer 1999; 84: 33–38.

- Kim JH, Moon WS, Kang MJ, Park MJ, Lee DG. Sarcomatoid carcinoma of the colon: a case report. Journal of Korean Medical Sciences 2001; 16: 657–660.

- Klein O, Clements A, Menzies AM, et al. BRAF inhibitor activity in V600R metastatic melanoma. European Journal of Cancer 2013; 49: 1073–1079.

- Kong CS, Welton ML, Longacre TA. Role of human papillomavirus in squamous cell metaplasia-dysplasia-carcinoma of the rectum. American Journal of Surgical Pathology 2007; 31: 919–925.

● Kosti P, Maher J, Arnold JN. Perspectives on chimeric antigen receptor T-cell immuno-therapy for solid tumours. Frontiers in Immunology 2018; 9: 1104.

● Lash RH, Genta RM, Schuler CM. Sessile serrated adenomas: prevalence of dysplasia and carcinoma in 2139 patients. Journal of Clinical Pathology 2010; 63: 681–686.

● Lauren P. The two histological main types of gastric carcinoma: diffuse and so-called intestinal-type carcinoma. An attempt at a histo-clinical classification. Acta Pathologica et Microbiologica Scandinavica 1965; 64: 31–49.

● Le DT, Uram JN, Wang H, et al. PD-1 blockade in tumors with mismatch-repair defi-ciency. New England Journal of Medicine 2015; 372: 2509–2520.

● Leedham SJ, Graham TA, Oukrif D, et al. Clonality, founder mutations, and field canceri-zation in human ulcerative colitis-associated neoplasia. Gastroenterology 2009; 136: 542–550.

● Leggett B, Whitehall V. Role of the serrated pathway in colorectal cancer pathogenesis. Gastroenterology 2010; 138: 2088–2100.

● Littleford SE, Baird A, Rotimi O, Verbeke CS, Scott N. Interobserver variation in the reporting of local peritoneal involvement and extramural venous invasion in colonic cancer. Histopathology 2009; 55: 407–413.

● Lordick F, Kang YK, Chung HC, et al. Capecitabine and cisplatin with or without cetuxi-mab for patients with previously untreated advanced gastric cancer (EXPAND): a ran-domised, open-label phase 3 trial. Lancet Oncology 2013,14: 490–499.

● Loughrey M, Quirke P, Shepherd N. Standards and datasets for reporting cancers. Dataset for Colorectal cancer histopathology reports 2014. London, Royal College of Pathologists. Available at: https://www.rcpath.org/asset/E94CE4A2-D722-44A7-84B9D68294134CFC/ (accessed 1 October 2018).

● Mei ZB, Duan CY, Li CB, Cui L, Ogino S. Prognostic role of tumor PIK3CA mutation in colorectal cancer: a systematic review and meta-analysis. Annals of Oncology 2016; 27: 1836–1848.

● Merker JD, Oxnard GR, Compton C, et al. Circulating tumour DNA analysis in patients with cancer: American Society of Clinical Oncology and College of American Pathologists joint review. Journal of Clinical Oncology 2018; 36: 1631–1641.

● Milagre C, Dhomen N, Geyer FC, et al. A mouse model of melanoma driven by onco-genic KRAS. Cancer Research 2010; 70: 5549–5557.

● Minoo P, Jass JR. Senescence and serration: a new twist to an old tale. Journal of Pathology 2006; 210: 137–140.

● NICE diagnostic guidance. Molecular testing strategies for Lynch syndrome in people with colorectal cancer. National Institute for Health and Care Excellence, February 2017: Available at: https://www.nice.org.uk/guidance/dg27 (accessed 1 October 2018).

● NICE. Technology appraisal guidance (TA416). Osimertinib for treating locally advanced or metastatic EGFR T790M mutation-positive non-small-cell lung cancer. October 2016. https://www.nice.org.uk/guidance/ta416 (accessed 1 October 2018).

● Nosho K, Baba Y, Tanaka N, et al. Tumour-infiltrating T-cell subsets, molecular changes in colorectal cancer, and prognosis: cohort study and literature review. Journal of Pathology 2010; 222: 350–366.

● Ogino S, Nosho K, Irahara N, et al. Lymphocytic reaction to colorectal cancer is associated with longer survival, independent of lymph node count, microsatellite instability, and CpG island methylator phenotype. Clinical Cancer Research 2009a; 15: 6412–6420.

● Ogino S, Nosho K, Kirkner GJ, et al. CpG island methylator phenotype, microsatellite instability, BRAF mutation and clinical outcome in colon cancer. Gut 2009b; 58: 90–96.

● Park SJ, Rashid A, Lee JH, et al. Frequent CpG island methylation in serrated adenomas of the colorectum. American Journal of Pathology 2003; 162: 815–822.

● Pectasides E, Bass AJ. ERBB2 emerges as a target for colorectal cancer. Cancer Discoveries 2015; 5: 799–901.

● Poeschl EM, Pollheimer MJ, Kornprat P, et al. Perineural invasion: correlation with aggressive phenotype and independent prognostic variable in both colon and rectum cancer. Journal of Clinical Oncology 2010; 28: e358–e360.

● Pyo JS, Kang G, Cho H. Clinicopathological significance and diagnostic accuracy of c-MET expression by immunohistochemistry in gastric cancer: a meta-analysis. Journal of Gastric Cancer 2016; 16: 141–151.

● Qizilbash AH. Pathologic studies in colorectal cancer. A guide to the surgical pathology examination of colorectal specimens and review of features of prognostic significance. Pathology Annual 1982; 17: 1–46.

● Razis E, Pentheroudakis G, Rigakos G, et al. EGFR gene gain and PTEN protein expression are favorable prognostic factors in patients with KRAS wild-type metastatic colorectal cancer treated with cetuximab. Journal of Cancer Research, Clinical Oncology 2014; 140: 737–748.

● Richtig G, Aigelsreiter A, Kashofer K, et al. Two case reports of rare BRAF mutations in exon 11 and exon 15 with discussion of potential treatment options. Case Reports on Oncology 2016; 9: 543–546.

● Rosty C, Hewett DG, Brown IS, Leggett BA, Whitehall VL. Serrated polyps of the large intestine: current understanding of diagnosis, pathogenesis, and clinical management. Journal of Gastroenterology 2013; 48: 287–302.

● Rowland A, Dias MM, Wiese MD, et al. Meta-analysis comparing the efficacy of anti-EGFR monoclonal antibody therapy between KRASG13D and other KRAS mutant metastatic colorectal cancer tumours. European Journal of Cancer 2016; 55: 122–130.

● Rubio CA, Stemme S, Jaramillo E, Lindblom A. Hyperplastic polyposis coli syndrome and colorectal carcinoma. Endoscopy 2006; 38: 266–270.

● Sartore-Bianchi A, Trusolino L, Martino C, et al. Dual-targeted therapy with trastuzumab and lapatinib in treatment-refractory, KRAS codon 12/13 wild-type, HER2-positive metastatic colorectal cancer (HERACLES): a proof-of-concept, multicentre, open-label, phase 2 trial. Lancet Oncology 2016; 17: 738–746.

● Schatton T, Scolyer RA, Thompson JF, Mihm MC Jr Tumor-infiltrating lymphocytes and their significance in melanoma prognosis. Methods in Molecular Biology 2014; 1102: 287–324.

● Sclafani F, Gullo G, Sheahan K, Crown J. BRAF mutations in melanoma and colorectal cancer: A single oncogenic mutation with different tumour phenotypes and clinical implications. Critical Reviews in Oncology and Hematology 2013, 87: 55–68.

- Secco GB, Fardelli R, Campora E, et al. Primary mucinous adenocarcinoma and signet -ring cell carcinomas of colon and rectum. Oncology 1994, 51: 30-34.

- Shackelford RE, Vora M, Mayhall K, Cotelingam J. ALK-rearrangements and testing methods in non-small cell lung cancer: a review. Genes Cancer 2014; 5: 1-14.

- Shah MA, Cho JY, Tan IB, et al. A randomized phase II study of FOLFOX with or without the MET inhibitor onartuzumab in advanced adenocarcinoma of the stomach and gastroesophageal junction. Oncologist 2016; 21: 1085-1090.

- Sharma SV, Bell DW, Settleman J, Haber DA. Epidermal growth factor receptor mutations in lung cancer. Nature Reviews: Cancer 2007; 3:169-181.

- Sohn BH, Hwang JE, Jang HJ, et al. Clinical significance of four molecular subtypes of gastric cancer identified by the Cancer Genome Atlas Project. Clinical Cancer Research 2017; 23(15): 4441-4449.

- Sorich MJ, Wiese MD, Rowland A, et al. Extended RAS mutations and anti-EGFR monoclonal antibody survival benefit in metastatic colorectal cancer: a meta-analysis of randomized, controlled trials. Annals of Oncology 2015; 26: 13-21.

- Spier, Holzapfel S, Altmüller J, et al. Frequency and phenotypic spectrum of germline mutations in POLE and seven other polymerase genes in 266 patients with colorectal adenomas and carcinomas. International Journal of Cancer 2015; 137: 320-331.

- Sternberg A, Amar M, Alfici R, Groisman G. Conclusions from a study of venous invasion in stage IV colorectal adenocarcinoma. Journal of Clinical Pathology 2002; 55: 17-21.

- Stewart EL, Tan SZ, Liu G, Tsao MS. Known and putative mechanisms of resistance to EGFR targeted therapies in NSCLC patients with EGFR mutations—a review. Translations in Lung Cancer Research 2015; 1: 67-81.

- Su X, Zhan P, Gavine PR, Morgan S et al. FGFR2 amplification has prognostic significance in gastric cancer: results from a large international multicentre study. British Journal of Cancer 2014; 110: 967-975.

- Suzuki H, Igarashi S, Nojima M, et al. IGFBP7 is a p53-responsive gene specifically silenced in colorectal cancer with CpG island methylator phenotype. Carcinogenesis 2010; 31: 342-349.

- Sy J, Fung CL, Dent OF, et al. Tumor budding and survival after potentially curative resection of node-positive colon cancer. Diseases of the Colon and Rectum 2010; 53: 301-730.

- Taipale J, Beachy PA. The hedgehog and Wnt signalling pathways in cancer. Nature 2001; 411: 349-354.

- Takahama T, Sakai K, Takeda M, Azuma K. Detection of the T790M mutation of EGFR in plasma of advanced non-small cell lung cancer patients with acquired resistance to tyrosine kinase inhibitors (West Japan oncology group 8014LTR study). Oncotarget 2016; 7(36): 58492-58499.

- The Cancer Genome Atlas Network. Comprehensive molecular characterization of human colon and rectal cancer. Nature 2012; 487: 330-337.

- The Cancer Genome Atlas Research Network. Comprehensive molecular characterization of gastric adenocarcinoma. Nature 2014; 513: 202–209.

- United States Department of Health & Human Services. Reducing the Health Consequences of Smoking: 25 Years of Progress. A Report of the Surgeon General. 2012. Available at: http://profiles.nlm.nih.gov/ps/retrieve/ResourceMetadata/NNBBXS (accessed 1 October 2018).

- Ugurel S, Rohmel J, Ascierto PA, et al. Survival of patients with advanced metastatic melanoma. The impact of novel therapies. European Journal of Cancer 2016; 53: 125–134.

- Vogelstein B, Fearon ER, Hamilton SR, Kern SE, et al. Genetic alterations during colorectal-tumour development. New England Journal of Medicine 1988, 319: 525–532.

- Wang Q, Liu F, Liu L. Prognostic significance of PD-L1 in solid tumor: An updated meta-analysis. Medicine (Baltimore) 2017; 96(18): e6369.

- Wong NA, Hunt LP, Novelli MR, Shepherd NA, Warren BF. Observer agreement in the diagnosis of serrated polyps of the large bowel. Histopathology 2009; 55: 63–66.

- Wu J, Rosenbaum E, Begum S, Westra WH. Distribution of BRAF T1799A(V600E) mutations across various types of benign nevi: implications for melanocytic tumorigenesis. American Journal of Dermatopathology 2007; 29: 534–537.

- Yano T, Sano Y, Iwasaki J, et al. Distribution and prevalence of colorectal hyperplastic polyps using magnifying pan-mucosal chromoendoscopy and its relationship with synchronous colorectal cancer: prospective study. Journal of Gastroenterology and Hepatology 2005; 20: 1572–1577.

- Yi ES, Boland JM, Maleszewski JJ, et al. Correlation of IHC and FISH for ALK gene rearrangement in non-small cell lung carcinoma: IHC score algorithm for FISH. Journal of Thoracic Oncology 2011; 6: 459–465.

 ## Useful websites

- *American Cancer Society*: https://www.cancer.org/cancer/melanoma-skin-cancer/about/key-statistics.html.

- *Cancer Research UK Cancer Statistics*: http://www.cancerresearchuk.org/health-professional/cancer-statistics/statistics-by-cancer-type.

- *National Cancer Institute, Cancer Statistics*: https://seer.cancer.gov/statfacts/.

- *Surveillance, Epidemiology and End Results*: https://seer.cancer.gov/statfacts/html/lungb.html

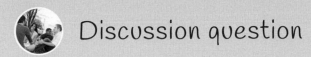 # Discussion question

11.1 Microsatellite instability (MSI) and/or mismatch repair proteins (MMR) deficiency is used for screening of Lynch syndrome, but also for prognostic purposes as well as a biomarker for response to immunotherapy. Which techniques could you use to assess MSI or MMR? What are the advantages or disadvantages?

Answers to the self-check questions and tips for responding to the discussion question are provided on the book's accompanying website:

Visit www.oup.com/uk/warford

12

Mesenchymal Tumours

Fernanda Amary

Learning objectives

After studying this chapter you should be able to:

- Describe most of the genetic abnormalities found in mesenchymal tumours.
- Describe applications of molecular diagnostic assays in clinical scenarios.
- Identify the most common abnormalities found in the main sarcoma subtypes.

12.1 Introduction

Mesenchymal tumours are relatively common neoplasms derived from cells of the mesenchyma: bone, muscle, fat, and connective tissue, including that present within the organs. The real estimate of benign mesenchymal tumours is difficult to ascertain, as many cases are silent and present as a hidden growth within the body, frequently unaccompanied by symptoms (Fletcher, 2013). A haemangioma, a benign tumour of the blood vessels, is an example of a common benign mesenchymal tumour. The so-called uterine fibroids, which are benign tumours of smooth muscle cells (leiomyomas), are also fairly frequent benign mesenchymal tumours.

Benign tumours usually are not life-threatening and can be cured with simple excision if symptomatic.

The same is not true for malignant mesenchymal tumours. These are called sarcomas and are defined as rare malignant neoplasms, comprising approximately 1% of the cancers (Borden and Goldblum, 2002; Fletcher, 2013). The word 'sarcoma' is likely originated from the Greek *sarkoun* 'to produce flesh'.

The rarity and variety of histological subtypes makes this group of tumours one of the most challenging in the diagnostic fields with a high rate of misdiagnoses. Fortunately, many of these tumours have specific genetic events that allow the development of molecular tests to help achieve an accurate diagnosis (Flanagan et al., 2010a,b). Currently, it is recommended that molecular tests are used, particularly when the clinical and/or histopathological features are not typical (Chang and Shidham, 2003; Bridge, 2014).

The genetic alterations in mesenchymal tumours can, in the majority of cases, be included in one of four categories:

- chromosomal translocations that generate a chimeric fusion transcript;
- copy number change (gene amplifications);
- point mutations;
- complex genomic events.

Over the years, molecular techniques have improved and most of these abnormalities can be detected in **FFPE**, used in routine diagnostic histopathology (Cheah and Billings, 2012; Cerrone et al., 2014).

FISH- and PCR-based techniques are currently available in a large number of histopathology diagnostic laboratories, allowing those tests to be performed routinely to be used as ancillary diagnostic methods.

It is always crucial that the molecular results are interpreted in the light of the appropriate clinical, radiological, and morphological information.

> ## Key Point
>
> Sarcomas are a group of malignant mesenchymal tumours that are rare and display a large variety of histological subtypes. Molecular tests to detect specific abnormalities related to each subtype are helpful to aid the diagnostic pathway.

12.1 SELF-CHECK QUESTION

What are the main types of genetic alterations that are found in mesenchymal tumours?

Formalin-fixed paraffin embedding (FFPE)

For diagnostic histopathology, formalin fixation is followed by embedding in paraffin wax. The former irreversibly halts degradation of the sample; the latter provides an internal and external support to the tissue so that sections, typically between 2- and 4-μm thick, can be cut for staining and microscopic examination.

Cross reference

See Chapter 3 for discussion of *in situ* hybridization and Chapter 4 for discussion of PCR-based techniques.

12.2 Mesenchymal tumours associated with chromosomal translocations

Chromosomal translocations are the genetic hallmark of numerous malignant mesenchymal tumours (sarcomas). Less frequently, benign tumours may also be associated with specific chromosomal translocations. These cytogenetic events usually lead to the formation of novel 'chimeric' genes that are transcribed and translated into chimeric proteins. Some of these chimeric proteins may act as transcription factors.

The chromosomal translocations are reciprocal (balanced) in the majority of the cases: there is reciprocal exchange of genetic material between two chromosomes without any loss as shown in Figure 12.1. This type of reciprocal gene rearrangement may be easily detected by FISH with the use of break-apart probes. The probes are generally labelled in spectrum red/orange and spectrum green, and placed on each side of the gene (telomeric and centromeric side). When there is no rearrangement or when normal tissue is targeted, each allele is visible as one

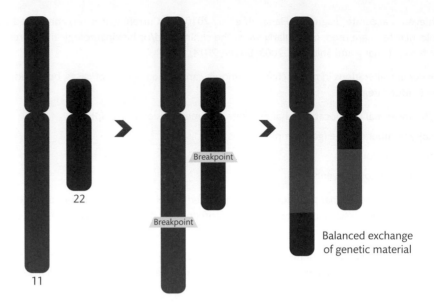

FIGURE 12.1

An illustration of reciprocal exchange of genetic material between two chromosomes: 11 and 22. No loss of genetic material is seen.

green signal close to one red/orange signal as demonstrated in Figure 12.2. Depending on the proximity of the probes, the combination of the spectrum green and orange/red may appear as a yellow signal. When gene rearrangement is present, a combined signal (yellow, or red and green very close together) is seen in one allele and a break-apart signal (red/orange distant from green) is seen representing the rearrangement of the gene (second allele; see Figure 12.3).

In a few sarcoma subtypes, however, the translocation is unbalanced with loss of chromosomal material (deletions). In the unbalanced translocations, the use of break-apart probes is not

FIGURE 12.2

Interphase FISH image showing normal diploid cells with no rearrangement: spectrum green close to spectrum orange/red flanking the target region.

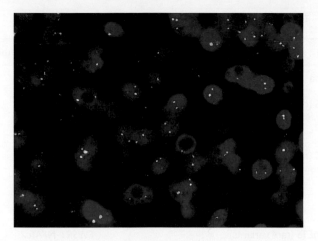

FIGURE 12.3

Interphase FISH image showing gene rearrangement using break-apart probes: a combined signal (yellow, or red and green very close together) is seen in one allele and a break-apart signal (red/orange distant from green) is seen in the second allele representing the rearrangement of the target gene.

ideal, as one of the probes may be lost, resulting in inconclusive molecular diagnosis (Klebe et al., 2006; Cerrone et al., 2014). In these cases, a fusion probe assay may be used.

RT-PCR-based assays are frequently used to detect the chimeric fusion transcript. These assays are more specific, as they target both gene partners in a rearrangement through the use of specific oligonucleotide sequences (primers) designed to align to both parts of the fusion gene. The amplification is performed using cDNA (complementary DNA generated from single-stranded RNA via reverse transcriptase). Nevertheless, due to multiple possible breakpoints and transcript variations, the use of multiple primers sets is sometimes required, making it time-consuming.

In most scenarios, clinical, histological, and immunohistochemical features, associated with a gene rearrangement detected by FISH, are sufficient to give the pathologist the tools for a specific diagnosis.

12.2.1 Ewing's sarcoma

Ewing's sarcoma was probably the first sarcoma to be associated with a chromosomal translocation. The t(11;22)(q24;q12) was characterized through karyotyping. Later, a specific fusion transcript was detected involving the *EWSR1* gene in chromosome 22 and the *FLI1* gene in chromosome 11. *EWSR1-FLI1* fusion is detected in approximately 85% of the cases. Other translocations generating different fusion transcript were subsequently described.

Ewing's sarcoma is the second most common bone sarcoma in children and young adults, after osteosarcoma. It occurs in long bones, flat bones, the spine, or in soft tissues. The classical histology shows sheets of monomorphic small round cells with scant cytoplasm. The cells consistently express CD99 on immunohistochemistry (membranous pattern). Molecular genetics tests are currently requested in all cases of Ewing's sarcoma. The specific fusion transcript is detected through RT-PCR. FISH for *EWSR1* gene rearrangement is not specific, but when associated with the clinical, histological, and immunohistochemical features, supports the diagnosis of Ewing's sarcoma.

TABLE 12.1 Tumour types with associated *EWSR1* gene rearrangement

Tumour type	EWSR1 and partner genes
Ewing sarcoma	
EWSR1-ETS (common variants)	*EWSR1-FLI1*
	EWSR1-ERG
EWSR1-ETS (rare variants)	*EWSR1-ETV1*
	EWSR1-ETV4
	EWSR1-FEV
EWSR1-nonETS (rare variants)	*EWSR1-NFATc2*
	EWSR1–POU5F1
	EWSR1–SMARCA5
	EWSR1–PATZ
	EWSR1–SP3
Other tumour types where EWSR1 *may be rearranged*	
Angiomatoid fibrous histiocytoma	*EWSR1-ATF1*
	EWSR1-CREB1
Clear cell sarcoma	*EWSR1-CREB1*
Desmoplastic small round cell tumour	*EWSR1-WT1*
Extraskeletal myxoid chondrosarcoma	*EWSR1-NR4A3 (TAF15-NR4A3)*
Mixed tumour/myoepithelioma	*EWSR1-POU5F1*
	EWSR1-PBX1
Myxoid liposarcoma	*EWSR1-DDIT3* (up to 10% of the cases—90% are *FUS-DDIT3*)
Sclerosing epithelioid fibrosarcoma	*EWSR1-CREB3L1(FUS-CREB3L2)*
Odontogenic clear cell carcinoma and hyalinizing clear cell carcinoma of salivary origin	*EWSR1-ATF1*

EWSR1 gene is rearranged in different sarcoma subtypes with completely different morphological and immunohistochemical patterns (Chang and Shidham, 2003; Bridge, 2014) as illustrated in Table 12.1. As this is a common gene, rearranged in mesenchymal tumours, it is a powerful diagnostic tool to have in a molecular pathology laboratory. Usually associating histological, immunohistochemical, and clinical features with the presence of *EWSR1* gene rearrangement detected using break-apart probes on a FISH assay, a definitive histological subtype can be given. A large numbers of laboratories have *EWSR1* break-apart probes as part of their diagnostic arsenal (see Case study 12.1).

CASE STUDY 12.1

A 13-year-old boy presented with a gradually enlarging swelling on his left shoulder. Plain radiographs of his shoulder showed a sclerotic and lucent destructive bone lesion with a large extra-osseous component (see Figure CS1.1). MRI of the left humerus showed a heterogeneous intra-osseous tumour extending into soft tissue (Figure CS1.2). The needle core biopsy performed revealed a small round cell tumour (Figure CS1.3) that diffusely expressed CD99, membranous pattern, on immunohistochemistry (Figure CS1.4). FISH analysis for *EWSR1* rearrangement using break-apart probes has been requested to confirm the diagnosis of Ewing's sarcoma (Figure CS1.5).

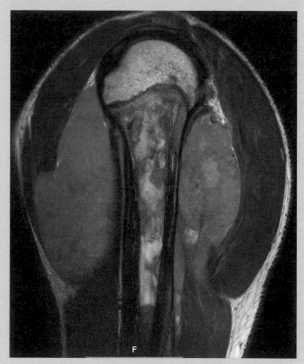

FIGURE CS1.2
MRI of left humerus, Sagittal T2.

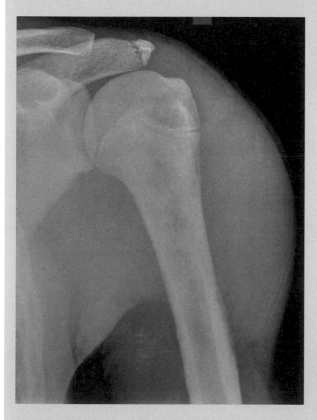

FIGURE CS1.1
Plain radiograph of the left shoulder.

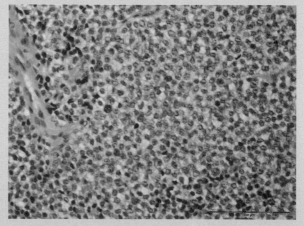

FIGURE CS1.3
Haematoxylin & eosin (H&E) stain of the needle biopsy showing sheets of relatively monomorphic small round cells with scant cytoplasm.

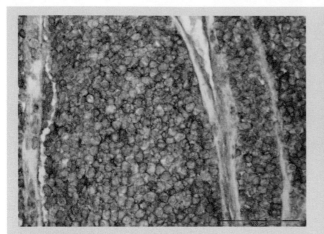

FIGURE CS1.4
CD99 expression, membranous pattern
(immunohistochemistry).

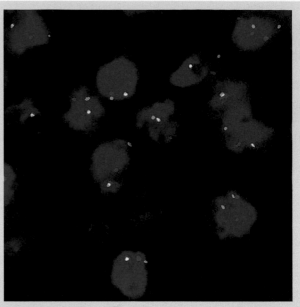

FIGURE CS1.5
FISH, *EWSR1* rearrangement using break-apart probes.
One normal signal (fused green and red = yellow) and a
break-apart (red separated from green), indicating presence
of gene rearrangement.

12.2.2 Other sarcomas with *EWSR1* gene rearrangement

The EWSR1 gene is involved with multiple partners in different rearrangements, associated with a variety of sarcoma subtypes that are morphologically dissimilar and, most importantly, have different prognosis and treatment options. Reaching the correct diagnosis of a specific subtype is paramount in providing the appropriate treatment.

12.2.2.1 Angiomatoid fibrous histiocytoma

This affects children and young adults being rare in patients over 40 years of age, and is present as haemorrhagic subcutaneous tumours in the extremities with blood-filled spaces. Morphologically, they show a haemorrhagic centre lined by sheets of histiocytic-like cells that express desmin and/or epithelial membrane antigen (EMA) on immunohistochemistry. The outer surface of the tumour usually shows a heavy lymphocytic and plasma cell inflammatory component, with occasional germinal centre formation. This inflammatory component may be mistaken for a lymph node. The fusion transcripts associated with angiomatoid fibrous histiocytoma are: *ESWR1-CREB1* (most commonly) and *EWSR1-ATF1*(Thway and Fisher, 2015). Although these fusion types are also observed in clear cell sarcoma of soft tissue (see later), those are completely different tumours with distinct histological features and biological behaviour. Angiomatoid fibrous histiocytoma is considered a tumour of intermediate biological potential (rarely metastasizing). This is a category created in the *WHO Classification of Tumours of Bone and Soft Tissue* (Fletcher, 2013) to define tumours that are 'borderline' between malignant and benign in their behaviour.

12.2.2.2 Clear cell sarcoma of soft tissues

This typically affects adults in the third or fourth decades of life. They are frequently slow-growing masses in the extremities, the majority in the foot and ankle regions. The prognosis is poor with metastases to lymph nodes and lungs. Morphologically, they show a nested growth pattern, the cells are plump with vesicular nucleolated nuclei. The tumour cells express S100 protein and melanocytic markers on immunohistochemistry (Fletcher, 2013). *EWSR1* rearrangement, found in over 90% of cases, is useful to differentiate this tumour from melanoma. The fusion transcripts associated with clear cell sarcoma are *EWSR1-ATF1* and *ESWR1-CREB1*.

12.2.2.3 Desmoplastic small round cell tumour

This affects young males predominantly and, typically, presents as a large tumour in the abdomen or pelvis. Morphologically, they are composed of sheets or nests of monomorphic small round cells within an abundant desmoplastic (fibrous) stroma. The tumour cells express cytokeratins and EMA. Desmin expression as dot-like cytoplasmic (perinuclear) labelling is also a frequent and helpful finding. The fusion transcript associated with this tumour type is *ESWR1-WT1* (Sandberg and Bridge, 2002).

12.2.2.4 Extraskeletal myxoid chondrosarcoma

This can affect any age, but is more common in male adults, in the deep soft tissues of the extremities. Initially considered cartilaginous in origin, these tumours are now known to be unrelated to the chondrosarcoma of bone. There is no unequivocal cartilaginous differentiation and they are classified under the umbrella of 'uncertain differentiation'. Morphologically, these are multinodular tumours composed of cords, ribbons, or nests of small round to spindle cells embedded in a myxoid stroma (Fletcher, 2013). The cells tend to connect to each other. Mitotic activity is usually low. Immunohistochemistry is usually unhelpful in reaching the diagnosis—S100 is expressed in less than one-third of cases. EMA and CD99 are frequently expressed. The fusion transcripts associated to this tumour type are *EWSR1-NR4A3* or *TAF15-NR4A3*. If a FISH test is employed, the use of *NR4A3* break-apart probes will give a more sensitive and specific result for this tumour type when compared to *EWSR1* gene rearrangement analysis.

12.2.2.5 Myxoid liposarcoma

EWSR1 gene rearrangement (*EWSR1-DDIT3*) is involved in ~5% of myxoid liposarcomas. The remaining have a *FUS-DDIT3* gene rearrangement (see specifications later in this chapter).

Key Point

EWSR1 gene rearrangement is associated with different tumour types in sarcoma. Each tumour type has its morphological and immunohistochemical features. The biological behaviour is also distinct amongst the tumour types. A specific integrated diagnosis (morphology, immunohistochemistry, and molecular genetics) is paramount.

12.2.3 Sarcomas with *FUS* gene rearrangement

FUS gene is fused with different partners in diverse sarcoma subtypes. This event is similar to what has been described previously with the tumours associated with *EWSR1* gene rearrangement.

12.2.3.1 Low grade fibromyxoid sarcoma (LGFMS)

The majority of cases are in young adults, but any age may be affected. Tumours are typically sited in the proximal extremities or trunk. Morphologically, these are generally bland-looking tumours with no striking cytological atypia. Short fascicles of monomorphic slender fibroblastic cells are set in a collagenized stroma with myxoid areas (Doyle et al., 2012). Mitotic activity is generally low. MUC4 is a very helpful immunohistochemical marker used to confirm the diagnosis of LGFMS. It is highly sensitive and specific in the context of mesenchymal tumour. The same marker is also expressed by sclerosing epithelioid sarcoma, a tumour that shares morphological and molecular similarities with LGFMS. The fusion transcripts associated with this tumour type are *FUS-CREB3L2* and *FUS-CREB3L1*.

12.2.3.2 Myxoid liposarcoma

The second most common type of liposarcoma (after well-differentiated liposarcoma/atypical lipomatous tumour). The majority are sited in the extremities, predominantly thigh, with a peak incidence in the fifth decade. Morphologically, these are multinodular tumours composed of small round to stellated cells dispersed in an abundant myxoid stroma containing a delicate network of capillary-type blood vessels. Numerous signet-ring lipoblasts are usually present. The presence of solid round cell cellular areas indicates a transition to a higher grade tumour and a poorer prognosis (Fletcher, 2013). The fusion transcripts associated to this tumour type are *FUS-DDIT3* and, in the minority of cases, *EWSR1-DDIT3*.

12.2.4 Mesenchymal tumours with other gene rearrangements

The list of mesenchymal tumour associated with a chromosomal translocation is consistently growing. This section discusses the main aspects of some of these tumours. For a comprehensive list see Tables 12.1 and 12.2.

12.2.4.1 Alveolar rhabdomyosarcoma (ARMS)

This primarily affects the extremities of adolescents and young adults, representing 20% of all paediatrics rhabdomyosarcomas. These are high-grade tumours with a more aggressive clinical course when compared with the most common paediatric rhabdomyosarcoma subtype—embryonal rhabdomyosarcoma (ERMS). Morphologically, ARMS are round cell tumours in which an 'alveolar' pattern (discohesive cell nests separated by fibrous septa) is classically observed, although solid variants may occur (Fletcher, 2013). The tumour cells express desmin and myogenin on immunohistochemistry. The fusion transcripts associated with this tumour type are *PAX3-FOXO1* or *PAX7-FOXO1*.

12.2.4.2 Dermatofibrosarcoma protuberans (DFSP)

These are superficial tumours that usually grow within the dermis and subcutaneous tissue, presenting as a nodular and/or indurate plaque-like cutaneous tumour mass. They are locally

TABLE 12.2 Mesenchymal tumour types, with respect to the most common gene rearrangements and FISH probes that may be used as ancillary diagnostic methods (for rearrangements involving the *EWSR1* gene, refer to Table 12.1)

Tumour type	Fusion transcript	FISH probe
Alveolar rhabdomyosarcoma	*PAX3-FOXO1* *PAX7-FOXO1*	*FOXO1** *PAX3-FOXO1*† *PAX3-FOXO1*†
Alveolar soft part sarcoma	*ASPSCR1-TEF3*	*TFE3**
Aneurysmal bone cyst	*CDH11-USP6* *COL1A1-USP6* *THRAP3-USP6* *CNBP-USP6* *OMD-USP6*	*USP6**
Dermatofibrosarcoma protuberans	*COL1A1-PDGFB*	*COL1A1-PDGFB*†
Endometrial stromal sarcoma, low grade	*JAZF1-SUZ12*	*JAZF1**
Endometrial stromal sarcoma, high grade	*YWHAE-FAM2*	*YWHAE**
Epithelioid haemangioendothelioma	*WWTR1-CAMTA1*	
Infantile fibrosarcoma	*ETV6-NTRK3*	*ETV6**
Inflammatory myofibroblastic tumour	*ALK* with various partners	*ALK**
Low grade fibromyxoid sarcoma	*FUS-CREB3L2* *FUS-CREB3L1*	*FUS**
Mesenchymal chondrosarcoma	*HEY1-NCOA2*	
Myxoid liposarcoma	*FUS-DDIT3*	*FUS** *DDIT3**
Myxo-inflammatory fibroblastic sarcoma/haemosiderotic fibrolipomatous tumour	*TGFBR3-MGEA5*	
Nodular fasciitis	*MYH9-USP6*	*USP6**
Solitary fibrous tumour	*NAB2–STAT6*	
Synovial sarcoma	*SS18-SSX1* *SS18-SSX2*	*SS18**

*Break-apart probes.
†Fusion probes.

aggressive tumours with a high incidence of local recurrence. High-grade transformation occurs in up to 15% of the cases, and in these, the tumour has metastatic potential. Any age may be affected, but it is more common in young to middle-aged patients. Morphologically, they are monomorphic spindle cell tumours with a striking storiform growth pattern. The tumour is usually infiltrative at the edges, permeating the adipose tissue of the subcutis in a honeycomb pattern (Fletcher, 2013). Mitotic activity is generally low, unless associated with high grade transformation. CD34 is diffusely expressed by the tumour cells. The fusion transcript associated with this tumour type is *COL1A1-PDGBF*, originating from an unbalanced translocation t(12;22). For this reason, FISH break-apart probes are not ideal as a diagnostic tool. In DFSP, and other tumours associated with unbalanced translocation, the use of fusion-type FISH probe is indicated.

12.2.4.3 Synovial sarcoma

This affects predominantly young adults in the deep soft tissues in the vicinity of large joints. However, almost any anatomical site and patients in any age group may be affected. Morphologically, they are divided in biphasic or monophasic subtypes related to the presence or absence of well-defined epithelial component, usually separated from the spindle cell component by basal membranes. The monophasic/spindle cell component features closely packed short fascicles of monomorphic spindle cells with oval nuclei and scant cytoplasm. These are frequently high grade tumours (Amary et al., 2007a). The fusion transcripts associated with this tumour type are *SS18-SSX1* and *SS18-SSX2*. *SS18* gene rearrangement detected by FISH break-apart probes is considered specific to this tumour type (see Case study 12.2).

CASE STUDY 12.2

A 48-year-old man presented with a 10-month history of a tender palpable mass on the posterior of the right proximal arm. MRI showed a deep heterogeneous tumour mass. Needle core biopsy revealed a biphasic tumour composed of atypical epithelial structure embedded in a spindle cell component, featuring short fascicles of monomorphic spindle cells. The epithelial component and scattered cells in the spindle cell component express cytokeratin (MNF116). FISH for *SS18* gene rearrangement has been requested to confirm the diagnosis of synovial sarcoma, biphasic type.

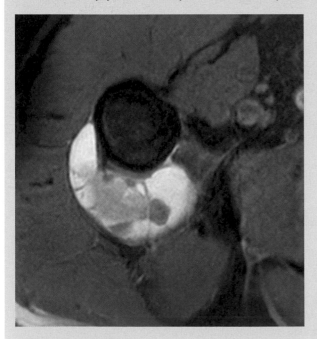

FIGURE CS2.1
Axial MRI of the right arm showing a deep tumour with hyperintense, hypointense, and intermediate signal.

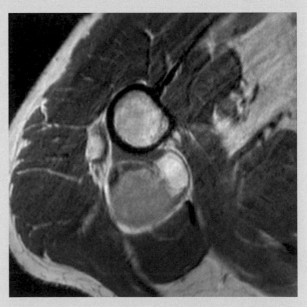

FIGURE CS2.2
Axial MRI inferior to the imaging seen in Figure CS2.1.

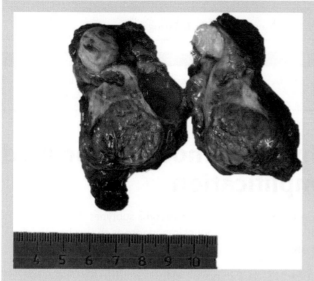

FIGURE CS2.3

Macroscopy of the resection specimen showing a heterogeneous tumour with fleshy areas, covered by skeletal muscle and fat.

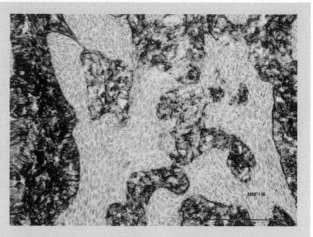

FIGURE CS2.5

Cytokeratin immunohistochemistry (MNF116) highlighting the epithelial component and showing expression on scattered spindle cells.

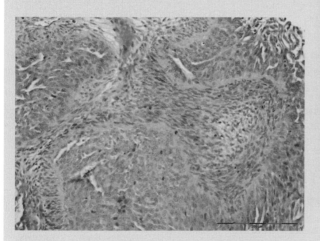

FIGURE CS2.4

Microphotography showing a malignant biphasic tumour composed of spindle cells and epithelial elements.

FIGURE CS2.6

FISH for *SS18* gene rearrangement using break-apart probes. One normal signal (fused green and red = yellow) and a break-apart (red separated from green), indicating presence of gene rearrangement.

Key Point

Numerous mesenchymal tumours are associated with gene rearrangements. The detection of specific rearrangement associated with morphological and immunohistochemical features usually allows the precise subtyping of these tumours.

12.2 SELF-CHECK QUESTION

Is the presence of *EWSR1* gene rearrangement specific to Ewing's sarcoma?

12.3 SELF-CHECK QUESTION

What type of FISH probe should be used to detect gene rearrangements in unbalanced translocations?

12.3 **Mesenchymal tumours associated with gene amplification**

Gene amplification is the cytogenetic hallmark of some sarcoma subtypes. The detection of gene amplification using specific probes is of paramount importance as an ancillary diagnostic tool, particularly in small biopsies.

The main clinical scenarios in which gene amplification analysis can help resolve a diagnostic dilemma are the following:

- A mature adipocytic tumour when the differential diagnosis lies between lipoma (benign) and atypical lipomatous tumour/well-differentiated liposarcoma.
- A high-grade sarcoma in the retroperitoneum when the differential diagnosis includes a dedifferentiated liposarcoma.
- A surface bone lesion when the differential diagnosis includes a parosteal osteosarcoma.
- A vascular lesion in the skin of the breast, developed after radiotherapy for breast cancer, when the differential diagnosis includes a radiotherapy-related non-cancerous vascular lesion and a radiation-induced angiosarcoma.

These are examples of clinical scenarios that could alter significantly the management of those patients.

12.3.1 **Tumour types**

12.3.1.1 *Atypical lipomatous tumours (ALTs)/well-differentiated liposarcomas (WDLs)*

These are tumours associated with *MDM2* gene amplification. Clinically, these tumours may be misdiagnosed and are considered to be benign fatty tumours—lipomas. ALTs, similarly to lipomas, are composed of lobules of well-differentiated adipocytes. Scattered large cells with hyperchromatic nuclei are usually seen in these tumour types, triggering the correct diagnosis. In small biopsies, or in sites in which there is a high risk of dedifferentiation (see below), detection of *MDM2* gene amplification is helpful and may lead to the correct treatment choice (Kalimuthu et al., 2015).

12.3.1.2 *Dedifferentiated liposarcomas*

The concept of 'dedifferentiation' was first described as a late complication of well-differentiated chondrosarcomas. In liposarcomas, dedifferentiation is classically defined as a biphasic tumour, comprising an area of well-differentiated adipocytic tumour juxtaposed with areas of high grade non-adipocytic sarcoma. In rare cases, the well-differentiated component is effaced by the high grade tumour and the diagnosis can only be reached with the help of *MDM2* gene amplification analysis. In fact, the majority of high grade sarcomas in the retroperitoneal region are dedifferentiated liposarcomas.

12.3.1.3 Parosteal osteosarcomas

This is a low grade osteosarcoma arising from the outer periosteum (in the surface of the bone). The majority occur adjacent to the metaphyseal region of the posterior surface of distal femur, followed by the proximal humerus and proximal tibia. These are densely mineralized lesions and, in some cases, the differential diagnosis includes myositis ossificans, a benign ossifying lesion. On histology, long bone trabeculae are admixed with a moderately cellular stroma featuring spindle cells with minimal cytological atypia. On biopsies, the diagnosis can be difficult, and the detection of *MDM2* gene amplification is the only diagnostic feature to support a non-atypical histology in reaching a diagnosis of malignancy (Salinas-Souza et al. 2015).

FISH analysis for *MDM2* gene amplification is quite straightforward in the tumour types mentioned previously. Due to the formation of one or more supernumerary ring chromosomes that contain amplified material from the 12q 13–15 region, the amplification is usually detected at high level (uncountable) copy numbers in the form of clusters (Figure 12.4 and Case study 12.3).

Key Point

Some sarcomas, malignant mesenchymal tumours, are associated with gene amplification. The detection of specific gene amplification in a specific clinical and histological context helps in the diagnosis of the tumour subtype.

12.4 SELF-CHECK QUESTION

How can you safely differentiate a large lipoma from an atypical lipomatous tumour?

FIGURE 12.4

FISH image showing *MDM2* gene amplification: multiple copies of the *MDM2* locus (spectrum green) are seen in clusters.

CASE STUDY 12.3

A 52-year-old woman presented with enlargement of her left thigh. MRI shows a homogeneous large fatty tumour crossed by fine septa.

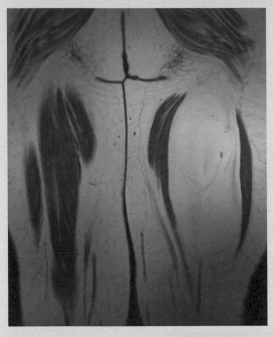

FIGURE CS3.1
MRI, coronal T1, posterior thigh, showing a large fatty mass.

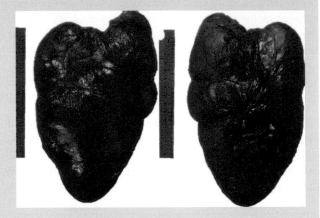

FIGURE CS3.3
Resection specimen. Large fatty tumour, completely covered by a thin membrane and strands of skeletal muscle.

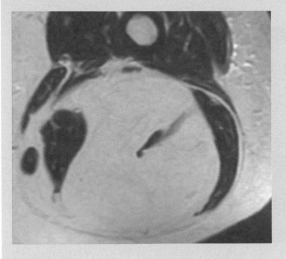

FIGURE CS3.2
MRI, axial T1, left thigh mass, encasing semitendinosus muscle.

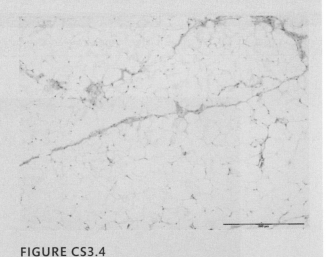

FIGURE CS3.4
Microphotography showing lobules of well-differentiated adipocytes, crossed by long thin fibrous septa with scattered hyperchromatic cells.

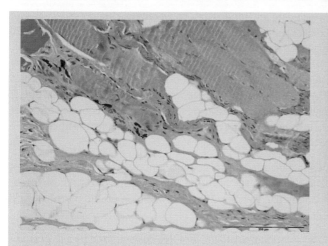

FIGURE CS3.5
Entrapped skeletal muscle with atrophy.

FIGURE CS3.6
FISH assay for *MDM2* gene amplification. *MDM2* gene targeted by red probes showing multiple copies. CEP12 in green labelling 2 centromeres per nucleus.

12.4 Mesenchymal tumours associated with point mutations

Point mutations, a single-base substitution, may result in change of an amino acid (missense mutation) and, consequently, in translation to an altered protein with change of function.

Detection of specific point mutations is used, in a subset of mesenchymal tumours, as an axillary diagnostic tool.

Detection of mutations in the genome is also being used to focus on targeted therapy. Several types of mutations are currently considered 'actionable', meaning that the effect of the mutation can be targeted by drugs. This is the basis of personalized/precision medicine, which is the promising future of oncological therapy.

The chief example of an actionable mutation in mesenchymal tumours is the KIT (receptor tyrosine kinase) mutation detected in 85% of *gastro-intestinal stromal tumours (GISTs)*. KIT mutation occurs in different sites within the gene, and the specific site may be relevant to predicting response to therapy and to prognosis. The majority are found in exon 11. A subset of GISTs may carry a mutation in PDGFR alpha (platelet-derived growth factor receptor alpha; Torres-Mora et al., 2014; Pogorzelski et al., 2016). The drug that targets mutation in these genes is called imatinib and it is used as a primary treatment option in non-resectable and or metastatic GISTs. The use is also indicated as adjuvant option in high-risk patients.

Gastro-intestinal stromal tumours are considered the most frequent malignant mesenchymal tumours of the gastro-intestinal tract, but they are very rare when compared with epithelial tumours: carcinomas.

Central cartilaginous tumours, including enchondromas, conventional chondrosarcomas and dedifferentiated chondrosarcomas were found to harbour an isocitrate dehydrogenase (*IDH1*) type 1 and *IDH2* missense mutation in the R132 and R172 position, respectively (Amary et al., 2011a,b). The mutant IDH1/IDH2 enzyme fails to convert isocitrate to alpha-ketoglutarate and gains a new function that leads to the accumulation of d-2-hydroxyglutarate (2HG), which acts as an oncometabolite. These mutations occur in approximately 60% of cases. Although the detection of such mutation is usually not necessary for the diagnosis of chondral tumours, they may have a therapeutic hole through the use of drug inhibitors that alter the ability of the mutant IDH1 to produce 2HG.

Osteochondroma, a type of peripheral/surface cartilaginous tumour, is associated with point mutation in *EXT1* or *EXT2*. These findings are not used for diagnostic purposes, as osteo-chondromas are fairly common tumours (Fletcher, 2013). The radiological and morphological aspects are diagnostic in the vast majority of cases.

Giant cell tumours of bone (GCTs) are subarticular lytic tumours characterized by a large number of osteoclasts. Recently a whole genomic analysis using massively parallel DNA sequencing revealed a recurrent mutation in *H3F3A* gene [p.Gly34Trp (p.G34W)] present in over 95% of GCTs. In the same studies, more than 95% of *chondroblastomas* were shown to harbour a mutation in either the *H3F3A* or the *H3F3B* genes involving p.Lys36Met (p.K36M; Behjati et al., 2013; Presneau et al., 2015). Interestingly, chondroblastomas also occur in the subarticular location of long bones, but in a slightly younger age group. The detection of these point mutations in those tumours has a hole in diagnosis in atypical locations and small biopsy samples. K36M substitution may be detected at protein level through immunohistochemistry (Amary et al., 2016).

After KIT and PGDFR mutation in GISTs, maybe the most clinically relevant molecular assay to detect a point mutation in mesenchymal tumour for diagnostic purposes, is the detection of *CTNNB1* gene (beta catenin) mutation in *desmoid-type fibromatosis*. These are fibrous tumours that occur in the abdominal wall, intra-abdominal, or deep soft tissue sites. They are considered non-cancerous growths, as there is no associated ability to metastasize. Nevertheless, they can be quite aggressive locally with a high rate of recurrence when surgically excised. Currently, a non-surgical treatment is usually a preferable reason why the precision in the diagnosis is paramount and, in small biopsies, the detection of *CTNNB1* gene mutation supports the diagnosis of fibromatosis and the clinical treatment. The three most common substitutions (p.T41A, p.S35P, and p.S45F) occur in exon 3 in ~86% of cases of sporadic desmoid-type fibromatosis (Amary et al., 2007b). The mutations are detected by direct sequencing (capillary Sanger sequencing), next-generation sequencing, or PCR followed by restriction enzyme digestion (MSRED).

Key Point

Missense mutation in specific genes may be used to determine specific tumour types in mesenchymal tumours.

12.5 SELF-CHECK QUESTION

Gastro-intestinal stromal tumour has specific mutations that can be therapeutically targeted. Check that you know the main genes involved.

12.5 Sarcomas associated with complex genomic events

High-grade central osteosarcoma, the most common primary bone tumour and yet accounting for less than 1% of all cancers, is a good example of a sarcoma associated with complex genomic abnormalities (Szuhai et al., 2012). Many soft tissue sarcomas also have complex genomic changes, which are difficult to assess, and to use the information for diagnostic or therapeutic purposes (Henderson-Jackson and Bui, 2015). One frequent example is the undifferentiated pleomorphic sarcoma (previously named malignant fibrous histiocytoma). This tumour type is a high-grade sarcoma with pleomorphic phenotype in which the line of differentiation cannot be determined.

While osteosarcomas are more frequent in bone of children and adolescents, undifferentiated pleomorphic sarcomas usually occur in patients over 50 years of age in the deep soft tissues of the extremities.

The genetic abnormalities in this tumour group include numerous mutations, rearrangements and copy number alterations throughout the genome. Large areas can be found to be amplified and other areas may be absent (deleted). The deleted areas may include important tumour suppressors such as *CDKN2A*, *PTEN*, *RB1*, *NF1*, and *TP53*. Non-diploid karyotypes are usually seen (Quesada and Amato, 2012).

Chapter summary

- Sarcomas are rare malignant mesenchymal tumours with complex and diverse histological subtypes.

- Genetic alterations in mesenchymal tumours are usually within these categories: chromosomal translocations that generate a chimeric fusion transcript; copy number change (gene amplifications); point mutations or complex genomic events.

- FISH- and PCR-based techniques are frequently used as ancillary diagnostic tools in mesenchymal tumours.

Further reading

- App for mobile devices (iOS and Google play): BoSTT—Bone and soft tissue tumour case studies. Royal National Orthopaedic Hospital.

- Borden EC, Goldblum JR. Molecular characterization of soft tissue tumors. Current Oncology Reports 2002; 4(6): 497–498.

- Bridge JA. The role of cytogenetics and molecular diagnostics in the diagnosis of soft-tissue tumors. Modern Pathology 2014; 27(Suppl 1): S80–S97.

- Cerrone M, Cantile M, Collina F, et al. Molecular strategies for detecting chromosomal translocations in soft tissue tumors (review). International Journal of Molecular Medicine 2014; 33(6): 1379–1391.

- Chang CC, Shidham VB. Molecular genetics of pediatric soft tissue tumors: clinical application. Journal of Molecular Diagnosis 2003; 5(3): 143–154.

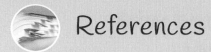

References

● Amary MF, Bacsi K, Maggiani F, et al. IDH1 and IDH2 mutations are frequent events in central chondrosarcoma and central and periosteal chondromas but not in other mesenchymal tumours. Journal of Pathology 2011a; 224(3): 334–343.

● Amary MF., Berisha F, Bernardi Fdel C, et al. Detection of SS18-SSX fusion transcripts in formalin-fixed paraffin-embedded neoplasms: analysis of conventional RT-PCR, qRT-PCR and dual color FISH as diagnostic tools for synovial sarcoma. Modern Pathology 2007a; 20(4): 482–496.

● Amary MF, Berisha F, Mozela R, et al. The H3F3 K36M mutant antibody is a sensitive and specific marker for the diagnosis of chondroblastoma. Histopathology 2016; 69(1): 121–127.

● Amary MF, Damato S, Halai D, et al. Ollier disease and Maffucci syndrome are caused by somatic mosaic mutations of IDH1 and IDH2. Nature: Genetics 2011b; 43(12): 1262–1265.

● Amary MF, Pauwels P, Meulemans E, et al. Detection of beta-catenin mutations in paraffin-embedded sporadic desmoid-type fibromatosis by mutation-specific restriction enzyme digestion (MSRED): an ancillary diagnostic tool. American Journal of Surgical Pathology 2007b; 31(9): 1299–1309.

● Amary F, Pillay N, Flanagan AM. Molecular testing of sarcomas. Diagnostic Histopathology 2017; 23(10): 431–441.

● Behjati S, Tarpey PS, Presneau N, et al. Distinct H3F3A and H3F3B driver mutations define chondroblastoma and giant cell tumor of bone. Nature: Genetics 2013; 45(12): 1479–1482.

● Borden EC, Goldblum JR. Molecular characterization of soft tissue tumors. Current Oncology Reports 2002; 4(6): 497–498.

● Bridge JA. The role of cytogenetics and molecular diagnostics in the diagnosis of soft-tissue tumors. Modern Pathology 2014; 27(Suppl 1): S80–S97.

● Cerrone M, Cantile M, Collina F, et al. Molecular strategies for detecting chromosomal translocations in soft tissue tumors (review). International Journal of Molecular Medicine 2014; 33(6): 1379–1391.

● Chang CC, Shidham VB. Molecular genetics of pediatric soft tissue tumors: clinical application. Journal of Molecular Diagnosis 2003; 5(3): 143–154.

● Cheah AL, Billings SD. The role of molecular testing in the diagnosis of cutaneous soft tissue tumors. Seminars in Cutaneous Medical Surgery 2012; 31(4): 221–233.

● Doyle LA, Wang WL, Dal Cin P, et al. MUC4 is a sensitive and extremely useful marker for sclerosing epithelioid fibrosarcoma: association with FUS gene rearrangement. American Journal of Surgical Pathology 2012; 36(10): 1444–1451.

● Flanagan AM, Delaney D, O'Donnell P. The benefits of molecular pathology in the diagnosis of musculoskeletal disease: part I of a two-part review: soft tissue tumors. Skeletal Radiology 2010a; 39(2): 105–115.

● Flanagan AM, Delaney D, O'Donnell P. Benefits of molecular pathology in the diagnosis of musculoskeletal disease: Part II of a two-part review: bone tumors and metabolic disorders. Skeletal Radiology 2010b; 39(3): 213–224.

- Fletcher CDM. *WHO Classification of Tumours of Soft Tissue and Bone.* Lyon, IARC Press, 2013.

- Henderson-Jackson EB, Bui MM. Molecular pathology of soft-tissue neoplasms and its role in clinical practice. Cancer Control 2015; 22(2): 186–192.

- Kalimuthu SN, Tilley C, Forbes G, et al. Clinical outcome in patients with peripherally-sited atypical lipomatous tumours and dedifferentiated liposarcoma. Journal of Pathology & Clinical Research 2015; 1(2): 106–112.

- Klebe S, Mead K, Brennan J, Skinner J, Sykes P, Allen P. Molecular testing for soft tissue tumours with known translocations. Pathology 2006; 38(4): 382–383.

- Pogorzelski M, Falkenhorst J, Bauer S. Molecular subtypes of gastrointestinal stromal tumor requiring specific treatments. Current Opinions in Oncology 2016; 28(4): 331–337.

- Presneau N, Baumhoer D, Behjati S, et al. Diagnostic value of H3F3A mutations in giant cell tumour of bone compared to osteoclast-rich mimics. Journal of Pathology & Clinical Research 2015; 1(2): 113–123.

- Quesada J, Amato R. The molecular biology of soft-tissue sarcomas and current trends in therapy. Sarcoma 2012; 2012: 849456.

- Salinas-Souza C, De Andrea C, Bihl M, et al. GNAS mutations are not detected in parosteal and low-grade central osteosarcomas. Modern Pathology 2015; 28(10): 1336–1342.

- Sandberg AA, Bridge JA. Updates on the cytogenetics and molecular genetics of bone and soft tissue tumors: desmoplastic small round-cell tumors. Cancer Genetics & Cytogenetics 2002; 138(1): 1–10.

- Szuhai K, Cleton-Jansen AM, Hogendoorn PC, Bovee JV. Molecular pathology and its diagnostic use in bone tumors. Cancer Genetics 2012; 205(5): 193–204.

- Thway K, Fisher C. Angiomatoid fibrous histiocytoma: the current status of pathology and genetics. Archives of Pathology & Laboratory Medicine 2015; 139(5): 674–682.

- Torres-Mora J, Fritchie KJ, Robinson SI, Oliveira AM. Molecular genetics of soft-tissue sarcomas: a brief overview for clinical oncologists. Cancer Journal 2014; 20(1): 73–79.

 Discussion questions

12.1 Numerous sarcomas are associated with chromosomal translocation. Which diagnostic tools could you use to aid the precise diagnosis in these cases?

12.2 Which techniques could you use to diagnose the main molecular alterations in sarcomas?

Answers to the self-check questions and tips for responding to the discussion questions are provided on the book's accompanying website:

Visit: www.oup.com/uk/warford

13

Prenatal and Neonatal Testing

Claire V. S. Brasted-Pike

Learning objectives

After reading this chapter you should be able to:

- Understand established and emerging techniques for taking pre- and neo-natal patient samples, and the associated risks.
- Describe the key biochemical, protein-based, and DNA-based methods for analysing pre- and neo-natal patient samples.
- Provide examples of genetic diseases and congenital conditions that can be diagnosed pre-natally and/or neo-natally.
- Discuss the ethical considerations that must accompany the realm of pre- and neo-natal testing, respecting individual rights and beliefs.

13.1 Introduction

Suppose a certain genetic disease runs in a family. A woman discovers she is pregnant and wants to find out whether her child will suffer from the disease. If the disease is a serious one, she may consider an abortion. Alternatively, she may seek support for raising a child with a life-limiting condition.

Already, one can see that the realm of pre- and neo-natal testing presents some of the key applications for molecular diagnostics and, arguably, some of the most controversial.

This chapter will discuss the various methods—both long-established and recently developed—that can be used to obtain samples from a foetus or newborn, with their advantages and risks. The diverse ways in which such samples can be diagnostically analysed will be considered, making links with several earlier chapters. Together with this, examples of diseases for which pre- and/or neo-natal molecular testing are relevant will be discussed, and a perspective upon some of the ethical considerations that unavoidably accompany this area of biomedical science provided.

Key Point
Pre- and neo-natal testing is a varied and complex field—both scientifically and ethically.

13.2 Invasive testing

Traditional methods for gaining information about a developing foetus rely upon sampling foetal or placental material directly. Such methods are termed *invasive* because they involve an instrument entering the uterus. Although it is now quite small, all invasive methods carry a risk of foetal loss (Corton et al., 2014).

13.2.1 Amniocentesis

Amniocentesis is a technique in which a sample of amniotic fluid—the protective fluid within the amniotic sac that cushions the developing foetus—is withdrawn from a pregnant woman. Sloughed-off foetal cells are present in the amniotic fluid and it is these that are recovered from the sample by centrifugation, for further analysis (Corton et al., 2014). Biochemical analysis of the amniotic fluid can also be used to diagnose a number of congenital conditions (see Section 13.3.1).

Cross reference
See Chapter 3 for further information on intact sample analysis.

Withdrawal of amniotic fluid during pregnancy was first reported in Germany in 1877, but this was in order to remove excess fluid that was applying undue pressure to the foetus, not for a diagnostic purpose. The procedure developed, however, such that, by the late 1950s, Dr Fritz Fuchs—an American obstetrician—and his colleague Polv Riis, first isolated and analysed foetal cells from amniotic fluid samples to aid in the prediction of genetic disease (Fuchs and Riis, 1956). They published a description of their method and findings in the journal *Nature*, and paved the way to amniocentesis being a widely-used diagnostic technique around the world.

In that first reported case of diagnostic use, the extracted foetal cells were identified as female or male, owing to the respective presence or absence of the Barr body (hypercondensed copy of the X-chromosome, which is present only in female cells) in their nuclei (Fuchs and Riis, 1956).

In 1960, the test was applied to a patient family that contained carriers and sufferers of haemophilia A, a bleeding disorder, whereby blood fails to clot when a vessel is broken and, which had a contemporary average life expectancy of about 11 years. Haemophilia A is caused by a recessive gene on the X-chromosome; most sufferers are males, as males receive only one copy of the haemophilia locus and will express the recessive gene if it is inherited. The results of the test therefore indicated the risk of the baby being a haemophiliac—a male foetus from a carrier mother and a normal father had a 50% chance of suffering from the disease, whereas a female foetus was expected not have haemophilia (see Section 13.5.4; Kliegman et al., 2015). Here is the first historical example of decisions about abortion being made owing to the results of a prenatal diagnostic test (see Section 13.7.2).

Modern amniocentesis is usually performed between the 15th and 20th weeks of pregnancy, although it may exceptionally be performed as early as the 11th week (Corton et al., 2014). Procedures prior to week 15 are known as 'early amniocentesis' and carry a greater risk of harm occurring to the foetus (Seeds, 2004). However, this enhanced risk must be weighed against the ethical concerns surrounding and trauma caused by a later-stage abortion, should the results from the foetal sample lead to a decision to discontinue the pregnancy.

The technique is carried out under local anaesthetic. A needle is inserted through the mother's abdominal wall, through the wall of the uterus, and through the amniotic sac, to withdraw approximately 20 mL of fluid. Since the mid-1970s, ultrasonic imaging had been used to guide the insertion, and thus decrease the chance that the foetus will be physically damaged by the needle (Kelley, 2013). Figure 13.1 shows the transabdominal sampling technique.

Despite modern imaging methods, however, the procedure still carries risks—estimates of the rate of foetal loss from complications of amniocentesis vary from 0.06% to 0.11% of women undergoing the technique (Akolekar et al., 2015). Such complications include infection carried by the needle, induction of pre-term labour, and alloimmunization of the mother (i.e. raising an immune reaction to the foetal cells in the mother's bloodstream), in cases where the mother and foetus differ at the *RhD* allele or certain other alleles, potentially leading to haemolytic disease (Corton et al., 2014).

Key Point

Amniotic fluid contains foetal cells and compounds produced by the foetus, which can be analysed to diagnose a range of genetic diseases and congenital conditions.

13.1 SELF-CHECK QUESTION

When was amniocentesis invented? How is it a less risky procedure now than when it was first used?

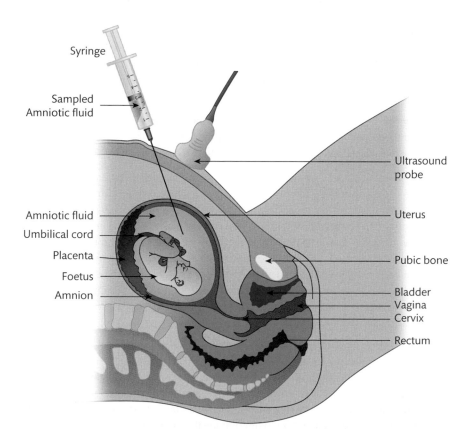

FIGURE 13.1
Transabdominal sampling of amniotic fluid.

13.2.2 Chorionic villus sampling

Like amniocentesis, chorionic villus sampling (CVS) is an invasive sampling technique involved in prenatal diagnostic testing. Instead of sampling the amniotic fluid, however, biopsies are taken of a part of the placental tissue—the chorionic villi—and this material is used for genetic tests (Gibbs et al., 2008). Placental material is suitable for diagnostic testing because it almost always has exactly the same genome as the foetus, as the fertilized zygote gives rise to them both (Forgács and Newman, 2005).

The procedure is carried out between the 10th and 12th gestational week, although its use as early as the 8th gestational week has been described (Bona et al., 1989; Gibbs et al., 2008). It is, therefore, commonly used in cases where family history suggests a relatively high risk of genetic disease in the foetus, allowing decisions about the management of the pregnancy to be made in the first trimester—earlier than amniocentesis affords.

Interest in a diagnostic technique that is effective in the first trimester of pregnancy has been present since amniocentesis became widely adopted. Indeed, chorionic villous material was successfully sampled by Mohr in 1968, but the methods used at this time (transcervical sampling, guided by a 5-mm straight endoscope) had an unacceptably high level of complications in terms of infection and bleeding. Figure 13.2 shows the transcervical sampling technique.

Technical advancements in the 1970s regarding the fineness and accuracy of instruments, together with the advent of ultrasound, provided opportunities for an improved complication rate in CVS and a better recovery rate of suitable placental material for analysis, leading to initial reports of prenatal sex determination (Tietung Hospital, 1975). However, a leap forward was made in 1984, with the introduction of transabdominal fine-needle aspiration biopsy of

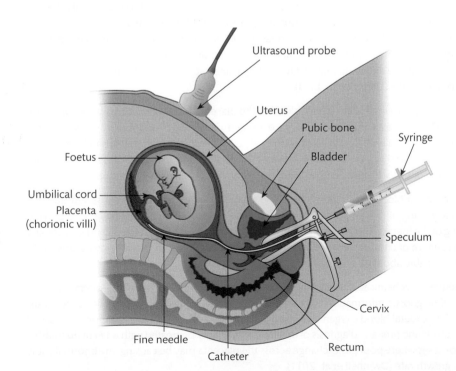

FIGURE 13.2
Transcervical chorionic villus sampling.

chorionic villi under ultrasonic guidance, by Smidt-Jensen and colleagues (Smidt-Jensen et al., 1985). Transabdominal CVS has a lower uterine infection rate than transcervical sampling, making the technique suitable for widespread clinical use (Silverman et al., 1994).

Today, the risk of foetal loss associated with complications from CVS is estimated at approximately 0.22% (Akolekar et al., 2015); complications are usually related to loss of amniotic fluid, the triggering of premature birth or accidental damage of the foetus by the needle (Gibbs et al., 2008). For this reason, the procedure is generally offered only to women whose foetuses are at higher risk of genetic abnormality. In addition to those with known family history of disease, such patients may include women aged over 35, whose pregnancies are at greater risk of chromosome abnormality (Coppedè, 2016), and patients whose early routine ultrasound scans show some irregularities.

Key Point

The chorionic villi (parts of the placenta) have the same genetic make-up as the foetus. Therefore, biopsies of chorionic villi can be taken for use in genetic tests that provide information about the unborn child.

13.2 SELF-CHECK QUESTION

Describe the similarities and differences between transabdominal and transcervical chorionic villus sampling.

13.2.3 Percutaneous umbilical cord blood sampling

Like amniocentesis and CVS, percutaneous umbilical cord blood sampling (PUBS)—also known as cordocentesis—is a technique for obtaining a sample of genetic material from a foetus. Unlike the previously-mentioned techniques, however, it is a second-line sampling method, performed only when results from earlier genetic tests are inconclusive, or when it was not possible to retrieve adequate material for testing (Gibbs et al., 2008). This procedure was first developed in 1983 (Daffos et al., 1983).

The umbilical cord carries the foetus' blood to the placenta for oxygenation, via two umbilical arteries. Once gas exchange has taken place, oxygen-rich blood is returned from the placenta to the foetus via an umbilical vein (Gordon et al., 2007). Remember that only small molecules, such as oxygen and glucose, cross the placenta; the mother's and foetus' blood supplies do not come into contact with one another. Figure 13.3 shows the placental blood supply.

In PUBS, foetal blood is withdrawn from the umbilical vein. The needle is inserted transabdominally, using ultrasonic guidance. It may be precoated with an anti-coagulant, to prevent clotting of the foetal blood as it is withdrawn (Welch et al., 1995). Genetic tests and karyotyping via the culture of chromosomes (see Box 1.2) can be performed on the WBCs recovered, with results available about 48 hours after sampling.

In addition, biochemical tests can be performed on the foetal blood; PUBS offers a greater range of diagnostic information than does sampling techniques used solely for genetic tests. Using PUBS, foetal blood oxygen levels and blood pH may be assayed. It is, therefore, a useful technique in the prenatal diagnosis of anaemia, which may be treated with an *in utero* transfusion, or cases where poor gas exchange across the placenta may be causing an abnormally low foetal growth rate (Dwinnell et al., 2011).

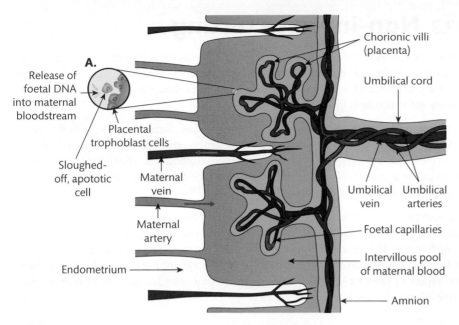

FIGURE 13.3

Placental blood supply. Maternal and foetal blood supplies approach closely within the placenta, but the vessels do not join. Deoxygenated blood flows from the foetus to the placenta via the umbilical arteries. In the chorionic villi, the blood in foetal capillaries absorbs oxygen from the maternal intervillous blood pools. Oxygenated blood returns to the foetus via the umbilical vein. (A) Cells of the placenta, which have the same genotype as the foetus, may become apoptotic—sloughing-off into the maternal bloodstream and releasing cell-free foetal DNA (cffDNA).

PUBS is the riskiest modern prenatal sampling technique in general use, carrying a 1–2% rate of procedure-associated miscarriage (Wilson et al., 2015). This fact underlines its place as a technique that is used only in particular circumstances or where other sampling techniques have failed. The most common complication is excessive bleeding at the site of venepuncture. The earlier the stage of pregnancy, the more severe bleeding is likely to be (Bigelow et al., 2016). Therefore, PUBS is rarely used at a gestational age younger than 18 weeks and may well be administered at a stage where the foetus has a chance of life outside of the mother, if delivered prematurely. For this reason, PUBS procedures are typically performed alongside a facility where an emergency Caesarean section can be carried out, should complications arise.

Key Point

Foetal blood can be sampled from the umbilical vein, for use in genetic and biochemical tests.

13.3 SELF-CHECK QUESTION

What are the main risks of PUBS? Given these, when is the technique used?

13.3 **Non-invasive testing**

Although much work has been done to minimize and mitigate against the risks involved with the sampling techniques described so far in this chapter, amniocentesis, CVS and PUBS remain invasive procedures. It is difficult to see how the risk of induced foetal loss could ever be entirely eliminated from these methods.

A great leap forward in the safety of prenatal testing was made, therefore, with the advent of tests that are performed on a maternal blood sample—thus gaining valuable information about the foetus without physically entering the uterus (Gibbs et al., 2008). The tests described below all use a small (1–20 mL) sample of maternal blood, collected using standard capillary puncture or venepuncture techniques.

It should be noted, however, that the maternal blood tests discussed here are all classed as *screening*, rather than *diagnostic tests*. A screening test provides a level of risk that the developing foetus will be abnormal, rather than a definitive diagnosis; such tests are likely to be performed routinely on the general population of pregnant patients. Only those patients who report as having a high level of risk will be recommended for further invasive, diagnostic testing.

13.3.1 **Factor levels in maternal blood**

Routinely offered to pregnant women in developed countries, the Triple Test and Quadruple Test screen maternal blood is for three or four specific factors, respectively (Wald et al., 2006; Dashe, 2016). These tests are typically performed in the 16th to 18th weeks of pregnancy. Together with biometric data about the foetus, such as its age and size, the tests have a 79% sensitivity and a 96% specificity for detecting chromosomal abnormalities and neural tube defects (see Section 13.5 in the case of the Triple Test, which rises to a 86% sensitivity and a 96% specificity with the addition of a further marker, in the Quadruple Test (Tu, et al., 2016). The specific markers screened are explained in Sections 13.3.1.1–13.3.1.4; the first three (alpha-fetoprotein, oestriol, and human chorionic gonadotropin (hCG) are used in both the Triple and Quadruple tests, and the fourth (inhibin A) is a part of the Quadruple test, only. Alpha-fetoprotein (AFP; see Section 13.3.1.1), oestriol (see Section 13.3.1.2) and inhibin A (see Section 13.3.1.4) levels are classically measured by ELISA (see Chapter 4, Section 4.3.1.2 and consider Section 13.4.4), but in the modern laboratory, mass-spectrometry-based methods are increasingly being developed (see Chapter 4, Section 4.3.2 and consider Section 13.4.5). hCG levels are measured by immunoassay.

Alternatively, or additionally, a Combined Test may be offered earlier in pregnancy, at 10-14 weeks gestational age. This test is targeted at detection of Down syndrome and other aneuploidies (see Section 13.5.2) and combines findings from ultrasound scans (typically measurements of nuchal translucency) with maternal serum tests for two hormones, hCG and pregnancy-associated plasma protein A (PAPP-A; see Section 13.3.1.5; Breathnach and Malone, 2007). For detection of Down syndrome, the Combined Test has an 85% sensitivity and 97% specificity (Huang et al., 2015). For normal and abnormal levels of all of the maternal serum markers described in Section 13.3.1, see Table 13.1.

TABLE 13.1 Maternal serum markers. Values are expressed in multiples of the median (MoM), where an 'ideal' normal result is 1 MoM. MoM tend to be used, rather than empirical units, as the normal reference ranges for each factor change throughout the gestational period

Maternal serum factor	Pathological associations with low levels	Pathological associations with high levels
Alpha-fetoprotein	< 0.4: chromosomal defects	> 2.5: neural tube defects; abdominal wall defects
Oestriol	< 0.5: chromosomal defects; pre-eclampsia; foetal anaemia	–
Human chorionic gonadotropin	< 0.4: increased risk of miscarriage; ectopic pregnancy	> 2.5: chromosomal defects
Inhibin A	–	> 2.5: chromosomal defects
Pregnancy-associated plasma protein A	< 0.4: chromosomal defects; pre-eclampsia; placental immaturity	–

13.3.1.1 Alpha-foetoprotein (AFP)

AFP is the major protein component of foetal blood plasma and is thought to perform an equivalent role to adult serum albumin in the foetus. That is, AFP plays a significant role in maintaining the osmolarity of foetal blood, such that an appropriate osmotic balance is maintained between the blood and other tissues (Mizejewski, 2001). Furthermore, this glycoprotein acts as a carrier protein for various hydrophobic and ionic molecules that must be transported in the foetal circulation, reversibly binding to fatty acids, haemin (an iron-containing porphyrin molecule that is a key component of haemoglobin), bilirubin (a waste product that results from broken-down haemoglobin), various metal ions, and lipophilic hormones (Mizejewski, 2001). AFP is produced by the foetal liver and the yolk sac. Levels rise throughout gestation until approximately the 32nd week of pregnancy and then begin to decline. The decline accelerates rapidly after birth; negligible levels of AFP are detectable in healthy adults (Ball et al., 1992).

In a healthy foetus, a small amount of AFP from the foetus' blood crosses the placenta and is detectable in the mother's bloodstream; the amount detected is directly proportional to the AFP foetal blood level. Expected levels of AFP in maternal blood can, therefore, be calculated from the gestational age of the foetus. AFP levels are empirically measured in ng/mL units, but given the fact that expected levels change considerably according to gestational age, test results are typically reported as multiples of the median (MoM), in comparison with other pregnancies of similar temporal advancement (David and Jauniaux, 2016).

Deviations from expected levels for healthy pregnancies can be indicative of foetal abnormalities. For example, higher than usual levels of AFP in maternal serum or amniotic fluid correlate with malformations of the foetus, such as neural tube defects (e.g. spina bifida and anencephaly) and abdominal wall defects (see Section 13.5.5). This is thought to be because, in the case of such malformations, the foetus' skin is not intact, so foetal blood escapes into the amniotic fluid, bathing the placenta and resulting in higher levels of AFP in maternal blood (Krantz et al., 2016).

Conversely, maternal serum AFP levels below the reference ranges indicate an increased risk of aneuploidies, most commonly Down syndrome (see Section 13.5.2). This is because aneuploidy foetuses are typically small for their gestational age, so are producing less AFP than is typical at the relevant developmental stage (Alldred et al., 2015).

13.3.1.2 Oestriol

Oestriol is a type of oestrogen steroid hormone that is produced primarily by the placenta; maternal serum levels of oestriol are typically 1000 times greater during pregnancy than in non-pregnant women. Similar to the other oestrogens, oestrone and oestradiol, oestriol binds to the intracellular oestrogen receptor, and stimulates the expression of a wide range of genes across many tissues (Pasqualini and Chetrite, 2016). These hormones are essential to normal development and tissue maintenance in both males and females, throughout human life.

The precursor molecule for oestriol is 16-hydroxydehydroepiandrosterone sulfate (16-OH DHEAS), and this is produced by the foetal liver and adrenal glands. Where a foetus is developmentally impaired, therefore, 16-OH DHEAS levels are limiting, and maternal serum oestriol levels are lower than expected. Low oestriol levels are thus associated with chromosomal abnormalities, such as Down syndrome and Edwards syndrome (see Sections 13.5.2.1 and 13.5.2.2), in addition to other conditions involving foetal developmental delay (Ren et al., 2016).

Other conditions that do not involve gross development defects in the foetus, such as pre-eclampsia and foetal anaemia, may also give rise to low maternal oestriol levels (Settiyanan et al., 2016). Thus, this reading alone should not be taken as strongly indicative of chromosomal abnormality.

13.3.1.3 Human chorionic gonadotropin

hCG is a hormone that is specifically produced by the embryo upon implantation in the uterine wall, from the syncytiotrophoblast cells (Cole, 2009); it is, therefore, commonly used as the target molecule in over-the-counter pregnancy tests. Maternal serum hCG levels continue to increase throughout pregnancy and return to minimal levels following childbirth (Ren et al., 2016).

Essential for placental health, hCG interacts with the ovarian LHCGR, stimulating it to produce progesterone. In turn, progesterone promotes angiogenesis in the uterine lining, thus allowing proper placental development (Hill, 2016b).

The hormone molecule is composed of a protein dimer; the α subunit is common with the α subunits of both FSH and LH, while the β subunit is unique to hCG. It is for this reason that analysis of hCG levels is performed using an immunoassay (typically, a sandwich ELISA; see Figure 4.7) with an antibody against the β subunit—to avoid accidental detection of FSH and LH from artificially inflating the reported levels. hCG levels are typically reported in thousandth international units per millilitre (mIU/mL), where one international unit is equal to approximately 6×10^{-8} g (Liu et al., 2016).

Down syndrome pregnancies typically have a maternal serum hCG levels double that of normal pregnancies; elevated hCG is, therefore, considered a risk factor for chromosomal abnormalities (see Section 13.5.2; Tørring, 2016). Such results are not fully indicative, however; there is a wide range of natural variation in hCG within healthy pregnancies (with particular variations between different ethnic groups), and elevated hCG levels are also found in multiple pregnancies and molar pregnancies (an embryo that is non-viable at a very early stage of development and may be present alongside a normally-developing foetus (Kirk et al., 2013)). Such test results should, therefore, not be interpreted alone.

hCG testing forms part of the first trimester Combined Test for aneuploidies, together with testing for PAPP-A by ultrasound.

13.3.1.4 Inhibin A

Inhibin is a glycoprotein hormone produced by ovaries and, during pregnancy, the placenta. It represses the production of FSH and, indeed, is part of a negative feedback loop; inhibin production is stimulated by the presence of FSH. Inhibin is present in two isoforms—A and B.

Inhibin is an antagonist of activin, a stimulatory hormone that acts to upregulate transcription of a wide variety of genes in various cell and tissue types. Inhibin is thought to function by a mechanism of competitive inhibition of the binding of activin to the transmembrane activin receptors (Namwanje and Brown, 2016).

Although the exact details of inhibin's functional mechanism are yet to be elucidated, there is an observed correlation between higher than average maternal serum inhibin A levels and foetal Down syndrome. Indeed, inhibin A is thought to be a more accurate marker for Down syndrome risk and placental health than hCG, as it has a shorter half-life in the maternal blood-stream, and thus observed levels are more closely tied to production levels (Muttukrishna, 2004).

Taken together with AFP, oestriol and hCG levels, inhibin A levels form part of the predictive Quadruple blood test, that is offered to pregnant women in several developed countries.

13.3.1.5 Pregnancy-associated plasma protein A

PAPP-A levels in maternal serum are screened as part of the first trimester Combined Test, together with hCG levels, and ultrasound findings. PAPP-A is a metalloproteinase that plays a role in the insulin-like growth factor (IGF) pathway. Specifically, it cleaves insulin-like growth factor binding proteins (IGFBPs), and thus contributes to a wide range of physiological and developmental functions, such as bone morphogenesis and wound healing (Amiri and Christians, 2015).

In pregnancy, PAPP-A is produced by the syncytiotrophoblast. Lower than usual PAPP-A in maternal serum has been found to correlate with increased risk of aneuplodies, presumably because the foetus is under-developed for its gestational age. Like other serum tests, however, PAPP-A results should not be interpreted in isolation as an indicator of foetal aneuploidy; low PAPP-A levels are also linked with pre-eclampsia and placental immaturity in genetically normal pregnancies (Duckworth et al., 2016).

Key Point

Tests for maternal serum levels of key factors—AFP, oestriol, hCG, inhibin A, and PAPP-A—can, in combination, indicate the risk that a foetus will suffer from a congenital malformation or a disorder of aneuploidy. Such tests are not 100% sensitive or specific, however, and it is recommended that a result that indicates a high risk of abnormality is followed-up by an invasive diagnostic test.

What are the normal physiological functions of AFP, oestriol, hCG, inhibin A, and PAPP-A? In an abnormal pregnancy, how would levels of these proteins in maternal serum probably differ from the normal ranges?

13.3.2 Foetal cells in maternal blood

The physical separation of foetal and maternal blood across the placenta is not perfect—a fact that may come as a surprise, given that you have learned about the importance of maintaining distinct maternal and foetal circulatory systems, which do not mix. However, the numbers of foetal cells found in maternal blood, and maternal cells found in foetal blood, are extremely small, estimated at between 1 in 10,000 and 1 in 1 million, in both cases (Choolani et al., 2012).

Foetal cells were first detected in the maternal circulation in 1969 (Walknowska et al., 1969) and have since been found to be predominantly erythrocytes, which can be distinguished with some confidence from maternal red blood cells by their possession of a nucleus and high levels of foetal haemoglobin (Hill, 2016a). These nucleated cells—together with foetal lymphocytes and granulocytes, which are also present, but in lower frequencies—can be used in traditional *in-situ* cytogenetic analysis techniques, such as karyotyping and hybridization (Choolani et al., 2012). Such use of foetal cells from maternal blood was, therefore, a tantalizing prospect for the diagnosis of prenatal disease without invasion into the uterus, when the circulating cells were first discovered.

Although some proof-of-principle results were obtained, the scarcity of foetal cells in maternal circulation proved to be a very significant challenge; on average, fewer than 20 foetal cells are recovered from 20 mL maternal blood, and isolation and enrichment techniques have not yet developed sufficiently for this approach to enter standard clinical practice, supplanting invasive sampling methods (Norwitz and Levy, 2013). The simultaneous advancement of molecular biological techniques using cell-free DNA has seen the latter supplant circulating foetal cell analysis as the foremost non-invasive type of prenatal test (see Section 13.3.3 and Guetta et al., 2004).

However, an interesting observation has recently been made, relating to circulating foetal cells; the number of foetal cells present in maternal blood has been found to correlate with foetal sex, whereby more foetal cells were found in the bloodstream of women pregnant with male foetuses than women pregnant with female foetuses (Schlütter et al., 2014). Circulating foetal cells may, therefore, be an area ripe for further study, particularly when enrichment techniques have developed further.

Key Point

Intact foetal cells can be recovered from maternal blood and cytogenetically analysed for prenatal diagnosis. The very low circulating foetal cell numbers limit the practical applications of this idea, however.

Which tests could be performed on foetal cells that have been recovered from maternal blood?

13.3.3 Foetal cell-free DNA

As mentioned in Section 13.2.2 the placenta has the same genetic make-up as the foetus; they arise from the same fertilization event. During pregnancy, placental cells—trophoblasts—undergo apoptosis as they age and are replaced in the maintenance of the normal placenta. Such cellular remnants are shed into the maternal bloodstream (together with the remnants of rare circulating foetal cells), and it is principally from this source that the foetal genome may be detected in the mother's blood (see Figure 13.3; Alberry et al., 2007); cffDNA was first identified in maternal circulation in 1997 (Lo et al., 1997).

Indeed, foetal DNA can be identified in maternal blood from 7 weeks of gestational age and the amount of foetal DNA present continues to increase throughout pregnancy, as the maturing placenta continues to shed apoptotic cells. cffDNA has a high latency, however—disappearing just 2 hours after delivery of the child (Lo et al., 1998).

There is, of course, a significant amount of *maternal* cfDNA circulating in the mother's bloodstream, shed from a wide range of apoptotic cells in tissues other than the placenta. How, then, can the foetal cfDNA be specifically distinguished?

The answer lies in the size of DNA fragments present: cffDNA fragments tend to be around 200 bp in length, whereas cfDNA of maternal origin is typically present in longer fragments; over 200 bp and up to 23 kbp in size (Li et al., 2004; Arbabi et al., 2016). Several analysis techniques exploit this difference in fragment length or the presence of certain genetic markers that are present in the foetal genome, but not the maternal genome; for example, if the foetus is male, the presence of genes from the Y-chromosome can be used to reconfirm the sampling of cffDNA in maternal blood (Boon et al., 2007).

Once the cffDNA has been extracted, downstream molecular analysis techniques may include: PCR-based approaches, such as digital PCR and Quantitative/RT-PCR DNA sequencing of the whole foetal genome; and comparative genomic hybridization (CGH). Naturally, these techniques are suited to different testing applications, such as detection of single-gene disorders, detection of chromosome breakage or rearrangement, or detection of aneuploidies (see Section 13.5.2).

Please bear in mind that all such testing approaches generally rely upon identification of key characteristics of foetal DNA during the analysis phase, rather than the complete physical purification of cffDNA from maternal cfDNA in the sample preparation phase. The latter remains challenging, owing to the relative scarcity of foetally derived molecules—approximately 1 μg of cffDNA is present per 20 mL of maternal blood (Sekizawa et al., 2000)—whereas many of the analysis methods mentioned previously contain integral DNA amplification phases, resulting in sufficient foetally-derived material for nuanced analysis.

Physical methods to enrich cffDNA do exist, however. Once total cfDNA has been purified from maternal serum, size-selection of DNA molecules is possible using gel electrophoresis; after extracting only DNA molecules of smaller size from the relevant gel, the sample will be around 70% of foetal origin (Li et al., 2004; see Figure 13.4). Additionally, there are some reports that percentage yield of cffDNA may be increased by treating maternal serum samples with formaldehyde prior to DNA extraction. Given that fragmenting cells shed DNA, formaldehyde fixation may minimize maternal cfDNA in the sample by stabilizing any maternal cells present in the serum, thus making them resistant to DNA extraction. The technique can give rise to cffDNA yields of around 20% of DNA extracted (Dhallan et al., 2004). These two approaches

Cross reference

See Chapter 7, Section 7.3 for more details on dPCR.

Cross reference

See Chapter 4, Section 4.2.1.2 for more details on real time PCR.

Cross reference

See Chapter 4, Section 4.2.1.4 for more details on CGH.

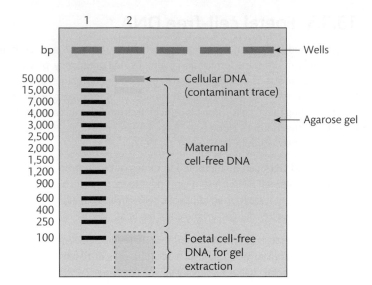

FIGURE 13.4

Size selection of cffDNA from maternal serum, following agarose gel electrophoresis. *Lane 1*: DNA ladder of size standards. *Lane 2*: cfDNA extracted from a maternal blood sample. cfDNA fragments of 200 bp in size or smaller are likely to be of foetal origin.

may be used together to further increase the percentage cffDNA yield and thus to enhance the efficiency of prenatal screening tests.

In comparison with the more traditional protein-based Triple Test and Quadruple Test (see Section 13.3.1), cffDNA screening offers the potential of a great leap forward in terms of sensitivity and specificity: both criteria are found to be over 99% for the majority of commonly-screened chromosomal defects, such as Down syndrome (see Section 13.5.2.1; Palomaki et al., 2012). Next-generation sequencing is typically used to analyse cffDNA samples for evidence of trisomies; numbers of reads originating from each foetal chromosome are found, and ratios between the different chromosomes of origin are calculated. Where a foetal chromosome is over-represented by a ratio of around 3:2, a high risk of trisomy is reported. cffDNA screening also has almost perfect recognition for dominant single-gene disorders—so much so that further confirmatory invasive tests are deemed unnecessary in these cases (NHS RAPID, 2016).

At the time of writing, the UK NHS offers cffDNA tests for the dominant single-gene disorders achondroplasia, thanatophoric dysplasia, and Apert syndrome, which may arise *de novo* or be paternally inherited. Furthermore, such tests are in development, together with tests for recessive single-gene disorders, such as cystic fibrosis, but only where the recessive mutant allele carried by the father is different from that carried by the mother. Note that it is not currently possible to test for recessive genes in the foetus, which are also carried by the mother, as it would not be possible to distinguish whether detection of the allele in question arose from maternal cfDNA or cffDNA.

Tests for the common aneuploidies (Down syndrome, Edwards syndrome, Patau syndrome—see Section 13.5.2) have been available to the public from commercial suppliers since 2013. Products include The Harmony Test (Ariosa Diagnostics, San Jose, CA), *Verifi Prenatal Test* (Illumina, Saffron Walden) and *The Iona Test* (Premaitha, Manchester). The NHS is currently (2016) undertaking an evaluation study to determine whether to offer such tests to all pregnant women in the UK.

Key Point

Placental cells shed foetal DNA into the maternal bloodstream. This can be purified and analysed, allowing non-invasive prenatal screening of pregnancies for genetic abnormalities.

13.6 SELF-CHECK QUESTION

How can cffDNA be distinguished from cfDNA?

13.4 Neonatal testing

In addition to a physical examination and hearing tests, newborn children in developed countries are routinely screened for a range of rare diseases using molecular techniques. Such tests aim to detect diseases that can lead to serious illness and impairment if left untreated, but which can be successfully managed if detected early. National programmes of newborn screening began in the 1960s, with the development of a screening test for phenylketonuria by Robert Guthrie (see Section 13.6.1; Mitchell et al., 2011), and have now expanded to screen for up to 50 disorders in some jurisdictions, although 5–20 disorders is more common (Lindner et al., 2011).

Bear in mind that these newborn screening tests are indicative, not diagnostic. If a positive result is obtained from such a test, further, usually genetic, tests need to be carried out in order to confirm a clear diagnosis.

13.4.1 Neonatal blood sample collection and analysis

A spot of blood is collected from the newborn child using capillary puncture—this is commonly referred to as a 'heel prick'—and the blood is collected on a piece of filter paper known as a Guthrie card—named after the practitioner who first developed the system (Mitchell et al., 2011). Sample collection is usually carried out when the child is between 1 and 7 days of age. The collected blood is then analysed using a range of techniques suited to the panel of diseases for which the system will screen. Infant urine may also be sampled for molecular testing.

13.4.2 Bacterial growth inhibition assay

The classical neonatal test for inborn metabolic conditions as developed by Guthrie, designed to detect conditions that result in excess amounts of phenylalanine (or the related compounds phenylpyruvate and phenyllactate) in the infant's blood (Mitchell et al., 2011).

Small circles of blood-soaked filter paper are placed onto agar plates that have been seeded with the bacterium *Bacillus subtilis*. The agar also contains the compound B-2-thienylalanine, which will inhibit growth of the bacteria, unless threshold quantities of phenylalanine (or a related compound) are present. Bacterial growth in the vicinity of a blood sample thereby indicates that high levels of a relevant phenyl-containing compound have diffused from the neonatal blood sample to the agar and are thus indicative of potential metabolic disease in the newborn child (Partington and Sinnott, 1964).

Increasingly, the classical bacterial growth inhibition assays are being supplanted by mass-spectrometry-based methods (see Section 13.4.5), which can detect a wider range of metabolites in the infant blood sample and thus screen for a greater number of disorders (Bruno et al., 2016).

13.4.3 Isoelectric focusing

Cross reference

See Chapter 4, Section 4.3.1.3 for more information about two dimensional gel electrophoresis.

Isoelectric focusing is an extension of the principle of protein electrophoresis, whereby a complex mixture can be separated into different molecular species owing to differences in molecular properties. Unlike standard protein electrophoresis, however, isoelectric focusing separates protein species not by molecular weight, but by differences in charge.

The technique uses acrylamide gels that have been polymerized to contain a stable (i.e. immobilized) pH gradient along their length. Protein samples are loaded into the gel, then an electric current is passed through it, such that one side of the gel is positively charged, and one side is negatively charged. Owing to the various positively and negatively charged residues that constitute the protein, it will move along the pH gradient until its isoelectric point is reached.

This method can produce finely resolved bands, even distinguishing between proteins that differ by only one unit of charge. In the context of neonatal testing, isoelectric focusing is principally used to detect haemoglobinopathies (see Section 13.6.2; Basset et al., 1978; Bender and Douthitt Seibel, 2014).

13.4.4 Immunoassays

Cross reference

See Chapter 4 for more discussion of immunoassays.

Immunoassays are typically used to screen infant blood for appropriate levels of several hormones in the first week of life; the range and number of hormones screened varies between jurisdictions. The conditions, congenital hypothyroidism (see Section 13.6.3.1) and congenital adrenal hyperplasia (see Section 13.6.3.2) are the most common targets of these screening programmes, for which levels of the thyroid hormones, and the precursor steroid hormone 17α-hydroxyprogesterone are measured, respectively (Pass and Neto, 2009).

Immunoassays are also used to test for levels of immunoreactive trypsinogen (IRT), which are indicative of the risk that the newborn has cystic fibrosis (see Section 13.5.1.3; Farrell and Sommerburg, 2016).

These tests originally used radioimmunoassay methods, which were supplanted by the chromogenic readouts typical in ELISA. In more modern laboratories, electrochemiluminescent immunoassay (ECLIA) systems may be used. ECLIA tests typically operate using the 'sandwich' principle also used in ELISA, but with the difference that a positive result will be indicated by the emission of light. Mass spectrometry may be used alongside or, in some cases, instead of immunoassay protocols.

13.4.5 Mass spectrometry

Mass spectrometry is increasingly used for the detection of levels of certain biochemicals in neonatal blood samples. In particular, metabolic disorders may be detected using this technique; it has the potential for fine discrimination between different amino acids, fatty acids, and protein derivatives, and can produce quantitative results in a high-throughput manner (Bruno et al., 2016).

Key Point

Public health programmes in developed countries routinely screen newborns for serious diseases that can be successfully managed if detected early. Such tests use a range of laboratory techniques, including bacterial growth inhibition assays, isoelectric focusing, immunoassays, and mass spectrometry.

13.7 SELF-CHECK QUESTION

Outline the principles of the bacterial growth inhibition assay, isoelectric focusing, and immunoassays.

13.5 Applications for prenatal testing

13.5.1 Detection of disease-causing alleles

Prenatally sampled foetal genetic material may be used to detect specific single-gene mutations—on a general screening basis, or perhaps because a particular mutation is known to be carried within the family in question. Testing is typically carried out using a PCR-based method or Sanger sequencing of the locus in question. Diseases caused by one mutant gene are typically 'rare diseases', often occurring *de novo* in a pregnancy; this is in contrast to the 'common diseases' found in developed countries—such as diabetes, asthma, cancer, and cardiovascular disease, which are caused by complex interactions of many different genes and the environment. In addition to being individually rare, named single-gene diseases are very numerous. Only three examples of such conditions are described here; the examples are chosen for their relatively large number of sufferers.

Cross reference

See Chapter 5 for further details about sequencing technologies used in testing.

13.5.1.1 Apert syndrome

Apert syndrome is caused by a mutation of the *fibroblast growth factor receptor 2* (*FGFR2*) gene, leading to oversensitivity of the receptor to its ligand, and cross-reaction with other, structurally similar ligands. In the developing embryo, the FGF signalling pathway acts to downregulate apoptosis and promote growth and differentiation of osteoblasts—cells which form bone (Fernandes et al., 2016). Over-activity of such functions can be linked to the symptoms of Apert syndrome: infants are generally born with syndactyly—that is, fingers and toes that are joined together, as the webbing that linked them in the early embryo was not properly broken down in development—and with small, pointed heads and dysmorphic facial features, owing to the bony plates of the skull joining together too early. Sadly, such cranial malformations usually lead to mental impairment in the child, as intracranial pressure is too high to support proper development of the foetal brain (Breik et al., 2016).

Interestingly, *de novo* mutations in the *FGFR2* gene seem to be almost entirely paternal in origin, and their rate of incidence correlates with increasing paternal age. Therefore, Apert syndrome is a good candidate for prenatal population screening using cffDNA isolated from maternal blood (see Section 13.3.3).

13.5.1.2 Achondroplasia

Achondroplasia is similar to Apert syndrome, in that it is caused by a mutation that gives rise to over-activity of an FGFR: in this case, FGFR3. In contrast to FGFR2, however, FGFR3 promotes ossification of the long bones of the skeleton, rather than the cranial bones. Such premature ossification of the foetal limbs, which are normally mostly made from cartilage as they continue to grow and lengthen, will give rise to infants born with abnormally short arms and legs (Teven et al., 2014). Therefore, from the symptoms, the common name for achondroplasia can be surmised—the condition is the most common cause of disproportional dwarfism.

Both Apert syndrome and achondroplasia are autosomal dominant genetic disorders, meaning that an affected parent has a 50% chance of passing the condition to their child (Teven et al., 2014). Where an affected parent and a normal parent wishes to have a normal child, pre-natal genetic testing, possibly combined with the practice of elective abortion (see Section 13.7.1), may be effective.

13.5.1.3 Cystic fibrosis

In contrast to Apert syndrome and achondroplasia, cystic fibrosis is an autosomal recessive disorder. The causative allele may, therefore, be carried by healthy individuals who do not realize that they are heterozygous, perhaps until such time as two carrier parents have a child who suffers from the disease.

The gene mutated in cystic fibrosis encodes the cystic fibrosis transmembrane conductance regulator (CFTR) protein. The normal function of this protein is to pump chloride ions out of epithelial cells, into luminal spaces that typically contain lubricating mucus. The presence of chloride in such spaces causes the osmotic movement of water into the mucus, thus making the mucus a suitable 'thin' texture for its function (Saint-Criq and Gray, 2017).

In the case of cystic fibrosis, however, the CFTR transporter is non-functional. Therefore, chloride—and hence water—does not move out of the epithelial cells, and the mucus remains abnormally thick and sticky. This causes blockages in the respiratory, gastrointestinal, and urogenital systems, and a fertile breeding ground for infectious bacteria (Castellani and Assael, 2017).

There is no cure for cystic fibrosis, although symptoms can be managed by a range of chemical and physical therapies. Gene therapies for cystic fibrosis are an area of active current research (Alton et al., 2016). Childhood mortality rates are high, but the average life expectancy of a sufferer who survives the early years is over 40 years (Elborn, 2016). In cases where a couple already has one child with cystic fibrosis, they may choose prenatal genetic testing to help ensure that any future pregnancies are either free from the mutant allele, or are heterozygous, and thus will not suffer from the disease (Drury et al., 2016).

In addition to the emerging use of prenatal testing to detect cystic fibrosis, the disease is commonly included as part of the panel for which neonatal screening is employed, by measurement of immunoreactive trypsinogen (IRT) levels (Kharrazi et al., 2015). IRT is a precursor molecule to a pancreatic enzyme. In cystic fibrosis, the pancreatic ducts may be blocked, leading to excess amounts of IRT accumulating in the infant's blood. If neonatal screening is suggestive of cystic fibrosis, the infant will be recommended for a sweat test, which measures chloride levels in perspiration; this test is definitively diagnostic for cystic fibrosis (Salvatore et al., 2016).

Key Point

Prenatal testing may be used to detect mutations in specific single genes that cause a rare disorder. Such disorders may be inherited in a dominant or a recessive pattern.

13.8 SELF-CHECK QUESTION

How can mutations in the *FGFR* family of genes cause a genetic disorder?

13.5.2 Detection of aneuploidy

An *aneuploidy* is a genetic condition in which a cell has an abnormal number of chromosomes. Healthy human cells have 23 pairs of chromosomes and are thus referred to as *disomic*; i.e. each chromosome is present in two matched copies (or two unmatched copies in the case of the XY sex chromosomes in males). Aneuploidies typically fall into the categories of *monosomies*—where one of the chromosomes is present in only one copy—and *trisomies*—where one of the chromosomes is present in three copies, rather than two. Aneuploidies typically arise from meiotic non-disjunction during the production of gametes in parents who are genetically normal, which is demonstrated in Figure 13.5; the risk of this error in division occurring increases with advancing maternal age (Handyside et al., 2012).

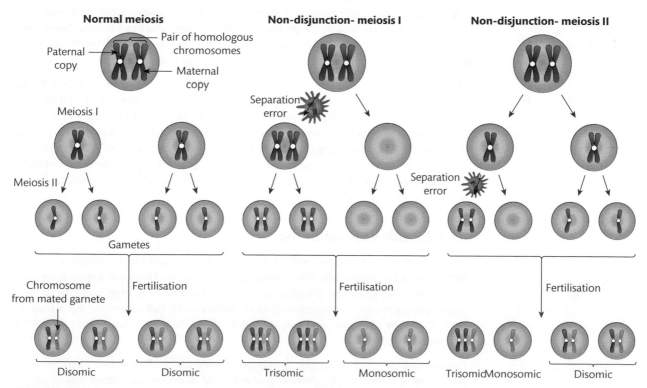

FIGURE 13.5

Aneuploidies may arise from non-disjunction during gametogenesis.

The vast majority of aneuploidies in human conceptions will lead to spontaneous abortion, usually at a very early stage of development, and often before the woman has even realized that she is pregnant. This is because a trisomy or monosomy of most of the chromosome types is incompatible with life (Jia et al., 2015).

The aneuploidies that can manifest as severe developmental disorders at birth therefore, are—perhaps counter-intuitively—the mildest versions of aneuploidies in humans. Here, three autosomal aneuploid disorders that are the focus of much of the effort of prenatal testing are reviewed—in terms of both diagnostic tests and screening tests, and encompassing many of the techniques discussed here, such as: invasive sampling (see Section 13.2) coupled with karyotyping (see Box 1.2); and maternal serum tests for cffDNA (see Section 13.3.3) coupled with PCR-based methods and next-generation DNA sequencing. Aneuploidies of the sex chromosomes will also be considered.

13.5.2.1 Down syndrome

Down syndrome is caused by trisomy of chromosome 21, and is the most well-known of the aneuploidy disorders. The reasons for this are two-fold—it is the most common aneuploidy in humans at birth and in prenatal detection, and Down syndrome individuals can commonly live into middle age with modern medical and societal support (Kliegman et al., 2015). The disorder is named after John Langdon Down, who was the first to thoroughly study the condition, in 1866. The extra material from chromosome 21 may be present either as a stand-alone additional chromosome or in the form of a Robertsonian translocated chromosome, inherited from a balanced-carrier parent.

Down syndrome is characterized by characteristic facial features, mild to moderate intellectual disability and growth retardation. Sufferers also have a predisposition to a wide array of other physical and behavioural problems, including congenital heart defects, hearing and vision impairments, a weakened immune system, gastrointestinal disorders, leukaemia, dental problems, autism, mood disorders, and dementia (Arumugam et al., 2016).

The broad range of symptoms associated with Down syndrome can be explained by the genetic complexity of the disorder; there is an overdose of all 310 genes on the 21st chromosome, of which approximately 50% are thought to be overexpressed (Lana-Elola et al., 2011). Current research is focused upon identifying the genes that are specifically responsible for the phenotypes observed. The pro-proliferative transcription factor C-ets-2, the antioxidant protein superoxide dismutase, and the amyloid protein are all thought to be likely candidates for pathology-causing overexpression (Lott et al., 2006), along with several chromosome 21 microRNAs (Lin et al., 2016).

13.5.2.2 Edwards syndrome

Caused by trisomy of chromosome 18, Edwards syndrome was first described in 1960, by British medical geneticist John Hilton Edwards (Edwards et al., 1960). It is a more severe disorder than Down syndrome, probably because chromosome 18 is larger and thus contains more overexpressed genetic material than chromosome 21; half of affected individuals die in the first week after birth; and only around 7.5% of sufferers live longer than 12 months (Roberts et al., 2016).

Edwards syndrome is characterized by widespread body-plan malformations, including severe heart defects, microcephaly, aberrant gastrointestinal development, kidney malformations, and characteristic facial features. Breathing and feeding difficulties are also usually present, together with severe intellectual disabilities (Roberts et al., 2016).

13.5.2.3 Patau syndrome

Patau syndrome is caused by trisomy of chromosome 13 and the clinical features associated with it, although almost always severe, can be quite varied. It is named after the American geneticist Klaus Patau who established the karyotype for the condition in 1960 (Patau et al., 1960).

Complex neuropathological features are common in Patau syndrome and these can include holoprosencephaly, the failure of the forebrain to divide into two hemispheres. Holoprosencephaly can, in turn, give rise to abnormal facial patterning such as cyclopaedia, the presence of one central eye structure, and a proboscis instead of normal eyes and nose. Severe heart defects, and malformations of the gastrointestinal, urogenital, and renal systems are also common (Caba et al., 2013); thus, Patau and Edwards (see Section 13.5.2.2) syndromes are sometimes difficult to tell apart on initial clinical examination. One distinguishing characteristic is the frequent presence of supernumerary digits (extra fingers and toes) in Patau syndrome, however. Prognosis for Patau syndrome is poor, with only 10% of children living longer than 12 months (Peroos et al., 2012).

13.5.2.4 Sex chromosome aneuploidies

In comparison to the somatic chromosomes, aneuploidies of the X and Y chromosomes give milder phenotypes, usually compatible with survival beyond the infant stage.

Turner syndrome is a monosomy of the X chromosome, with no Y chromosome present. Affected individuals will develop as female, but will not develop secondary sexual characteristics, remaining infertile. The condition is characterized by a short stature, 'webbing' of the neck (extra skin joining the neck and shoulders), and a predisposition to heart problems and diabetes. Turner syndrome patients are usually intellectually normal or have slight intellectual disability (Granger et al., 2016).

Kleinfelter syndrome may be described as the presence of an extra X chromosome in a male, giving the karyotype 47, XXY. Affected individuals are typically intellectually normal, but may have some difficulties with speech and reading. Symptoms are often not detected until puberty, when Kleinfelter boys may begin to show feminized physical characteristics, such as breast growth (Bird and Hurren, 2016). The clearest symptom of the disorder in males is infertility; Kleinfelter sufferers may, however, be able to father children using modern assisted reproduction techniques (Fullerton et al., 2010).

The karyotypes 47 XXX and 47 XYY both tend to produce individuals who are largely phenotypically normal, but are likely to be taller than average and may suffer from psychosociological problems and difficulty conceiving (Otter et al., 2010). Indeed, these karyotypes often come to light only when the affected individual visits a fertility specialist with the latter concern.

Key Point

Aneuploidies of the somatic chromosomes are usually fatal at a very early stage of development, with the exception of trisomies of the chromosomes 13, 18, and 21, which give rise to the severe disorders Patau, Edwards, and Down syndromes, respectively. Aneuploidies of the sex chromosomes produce milder phenotypes.

13.9 SELF-CHECK QUESTION

Why does Down syndrome have a wide range of associated phenotypes, across many physiological systems?

13.5.3 Detection of chromosomal structural change

Cross reference

See Chapter 3 for discussion of ISH and Chapter 4 for discussion of PCR.

In contrast to aneuploid disorders there is a large class of genetic disorders that arise owing to damage to the structure of one or more chromosomes, not owing to an abnormal chromosome number, per se. Such defects may be detectable in karyotype tests—depending upon the physical size of the chromosomal anomaly—or may require more sensitive methods of detection, such as ISH or PCR-based methods.

Damage to chromosomes may be classed as deletion, amplification, inversion, or translocation (see Figure 13.6), and may lead to inappropriate dosage or expression patterns of particular genes.

Given that such chromosomal damage can occur at random, essentially anywhere in the genome, the number of possible permutations is vast; a significant number of patients suffering from diseases of chromosomal breakage and rearrangement have syndromes that are so rare that they are essentially unique (Bird and Tan, 2016). Having said that, there are also a large number of rare conditions that have been medically described, linking characteristic symptoms to certain chromosomal anomalies. Mechanistic understanding of how exactly the genetic aberrations in question result in the phenotypes observed is an active field of research. Some examples of disorders of chromosomal structural disturbance are described below.

13.5.3.1 Cri du chat syndrome

Cri du chat syndrome is so named because an affected infant characteristically suffers from malformations of the larynx, giving a cry that sounds similar to the meowing of a kitten. It is caused by a deletion in the p (long) arm of chromosome 5, and typically, results in microcephaly and

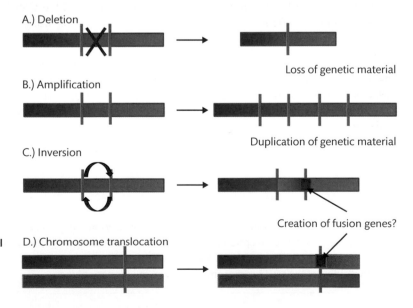

FIGURE 13.6

Schematic representations of (A) Chromosomal deletion. (B) Chromosomal amplification. (C) Inversion of one chromosomal section. (D) Chromosomal translocation.

severe intellectual disability, in addition to characteristic facial features, and potentially defects of the heart and kidney (Rodríguez-Caballero et al., 2010).

13.5.3.2 Smith–Magenis syndrome

Smith–Magenis syndrome is caused by a deletion in chromosome 17; typically, the 17p11.2 region. The deletion is usually accompanied by low-copy repeats of adjacent, non-coding material (Neira-Fresneda and Potocki, 2015). The disorder is characterized by characteristic facial features coupled with disturbed sleep patterns and widespread behavioural problems. Body-plan defects affecting the spine, ears, and eyes may also associated with the condition (Chen et al., 2015).

13.5.3.3 Charcot–Marie–Tooth disease

Charcot–Marie–Tooth disease is a genetic neuropathy caused by a duplication of a particular section of chromosome 17. Unlike the other disorders discussed here, it is not recognizable at birth, but symptoms—characteristically, progressive atrophy of the peripheral nervous system and degeneration of muscular tissue—begin any time from the second year of life to adulthood (Hoyle et al., 2015). Among other genes, the duplicated chromosomal section contains the gene *MFN2*, which encodes the protein mitofusin. An overexpression of mitofusin causes cellular mitochondria to fuse and aggregate excessively, leading to mitochondrial clots in the axons of nerve cells, and thus preventing neuronal synapses from functioning properly (Kawarai et al., 2016).

Key Point

Chromosomal deletions, duplications, inversions, and translocations can occur *de novo* in gametogenesis, and give rise to many rare genetic disorders.

13.10 SELF-CHECK QUESTION

What do *cri du chat* syndrome and Smith–Magenis syndrome have in common?

13.5.4 Sex determination and sex-linked disorders

Before the advent of molecular testing, prenatal sex determination was performed exclusively by ultrasound, at around 12–20 weeks' gestation. This is done by the ultrasound practitioner's observation of foetal external genitalia in the ultrasound image, but it can be somewhat unreliable, owing to differences in the foetus' position with, the uterus and varied interpretation of the scans (Youssef et al., 2011).

CffDNA testing (see Section 13.3.3) allows sex determination at a much earlier stage of pregnancy, however, and with considerably greater confidence. A test for the presence or absence of Y-chromosomal material in the maternal bloodstream can be performed at or after the 7th week, with up to 98% accuracy (Zimmermann et al., 2005). See Section 13.7.2 for a discussion of the ethics and legal status of prenatal sex testing.

Although knowing the sex of an unborn child can satisfy mere curiosity for many parents-to-be, or may inform some aesthetic purchase choices, there are cases in which it can have considerable medical relevance.

Sex-linked genetic disorders are typically caused by recessive mutant alleles that are carried on the X-chromosome. Females have two copies of the X chromosome, so if a girl inherits one disease-causing allele, she will probably still be healthy, as the normal allele on her other X chromosome will mask the effect of the mutant allele. If a boy inherits the mutant allele, he will probably suffer from the disease, however, as males are *haploid* for (i.e. have only one copy of) the X chromosome (see Figure 13.7 showing this inheritance pattern).

Haemophilia is a well-known X-linked disorder. It is characterized by excessive bleeding upon the puncture of a blood vessel, which without treatment may be fatal. Such bleeding may occur from the body to the outside environment, or may occur within a joint, soft tissue, or within the cranial cavity (Swystun and James, 2017).

The disease is caused by a recessive mutant allele of either the *F8* gene, which encodes clotting factor VIII (in the case of haemophilia A), or the *F9* gene, which encodes clotting factor IX (in the case of haemophilia B). Both of these genes reside on the X chromosome (Pavlova and Oldenburg, 2013). In normal individuals, clotting factors VIII and IX play essential roles in the formation of blood clots. They are part of a functional cascade that ends with thrombin cleaving the protein precursor fibrinogen to its active form, fibrin. Fibrin can then polymerize in such a way that platelets are attracted into a protein meshwork—together forming a haemostatic plug. This plug physically stops the passage of blood through the hole in the damaged vessel (Antovic et al., 2014).

Haemophiliac individuals lack this ability to form clots upon injury—a disorder that may be life-limiting when combined with even mild trauma. In developed countries, the condition is managed by injection of the absent clotting factor up to three times per week; the injected protein may be of isolated human or recombinant origin (Antovic et al., 2014).

Key Point

Prenatal sex determination may be useful in cases where a genetic disease is sex-linked, on the X-chromosome.

FIGURE 13.7

X-linked inheritance of a mutant allele in males and females.

$X^+ X^M$ × $X^+ Y$

Mother
- Carrier of recessive mutant allele

Father
- Normal genotype

♀ \ ♂	X^+	Y
X^+	X^+X^+ Daughter - Normal phenotype and genotype	X^+Y Son - Normal phenotype and genotype
X^M	X^+X^M Daughter - Normal phenotype carrier genotype	X^MY Son - Sufferer of the disorder

Why do males suffer from sex-linked diseases more commonly than females?

13.5.5 Detection of foetal malformations

In the case of single gene disorders (Section 13.5.1) and chromosomal disorders (Sections 13.5.2 and 13.5.3), a clear link can be made between the genome of the sufferer and the symptoms of the disorder. There is a further class of foetal malformation, however, that may be due to a complex interaction of genetic and environmental factors, which are only partially understood. Such malformations are relatively common, however, and have typically relied upon ultrasonographic methods and protein-based markers (see Section 13.3.1) for detection.

13.5.5.1 Neural tube defects

In the third week of embryonic life, a structure called the neural tube will form, following a stage of embryonic cell movement and rearrangement, known as gastrulation. The neural tube contains the cells that will form the foetal brain and spinal cord. In normal embryos, it will be completely sealed by an early developmental stage, with the skull and vertebrae later developing to completely enclose and protect the delicate neural matter (Cai and Shi, 2014).

Some pregnancies, however, suffer from a range of disorders, known as neural tube defects. These arise from improper sealing of the neural tube, such that in severe cases the foetus will have neural tissue protruding from between the vertebrae—spina bifida—or an improperly-formed brain—anencephaly (Arth et al., 2016).

Spina bifida can result in paralysis from the point of the defect downwards, as the projecting spinal cord is damaged to the point of non-functionality. It is also often associated with intellectual impairment, probably caused by associated subtle malformations of the brain. Although the implicated genes have not yet been fully described, there is evidence to suggest that spina bifida has a genetic component—if a couple has one child with spina bifida, they have a 3% chance of a second pregnancy also suffering from the condition (against a population average of four births in 10,000). Conversely, there is strong evidence that lack of folic acid in the mother's diet increases the incidence of spina bifida—hence, the widespread programme of folic acid supplementation in common foodstuffs from the late 1990s onwards, which is credited with reducing the incidence of spina bifida in current populations (Arth et al., 2016).

Spina bifida correlates with raised levels of AFP in maternal blood (see Section 13.3.1.1), so this prenatal test is routinely used to screen for the condition (Krantz et al., 2016). There is no treatment that can restore lost nervous function, but cases are typically treated with surgery to close the gap in the spine, either *in utero*, or shortly after birth.

Anencephaly is the most severe type of neural tube defect. It occurs when the part of the neural tube that forms the embryo's head does not close, leading to a foetus lacking the cerebral hemispheres—the major parts of the human brain, which are responsible for cognition. If they are born alive, infants with anencephaly typically survive for only hours or days, and have no sensory processing abilities (Dickman et al., 2016).

13.5.5.2 Abdominal wall defects

High levels of AFP in maternal blood can also be indicative of abdominal wall defects in the foetus; that is, cases where some or all of the abdominal organs protrude through an abnormal hole in the stomach wall of the foetus and are present in a flimsy sac. The most severe kind of abdominal wall defect is omphalocele, where the liver, intestines, and sometimes other abdominal organs are outside of the abdomen (Darouich et al., 2015).

The causes of abdominal wall defects are not well understood, but they are often part of the presentation of the trisomy disorders Edwards syndrome and Patau syndrome (see Section 13.5.2). Some environmental pollutants have also been implicated in raising population occurrence rates of these defects.

Key Point

Neural tube defects and abdominal wall defects have complex causes, both genetic and environmental. Maternal serum AFP levels are used to screen for these.

13.12 SELF-CHECK QUESTION

What is the neural tube? What may happen if it fails to close?

13.6 Applications for neonatal testing

13.6.1 Inherited metabolic disorders

Inherited metabolic diseases are typically caused by a recessive mutant allele that codes for a non-functional enzyme. In sufferers, therefore, a particular biochemical reaction fails to proceed, leading to a lack of the end-product of the reaction in question, and a potentially toxic build-up of one of the reactants. In the uterus, the foetus is usually protected from the effects of these disorders, because the mother's metabolic functions will supply the correct balance of biochemicals across the placenta (Sferruzzi-Perri and Camm, 2016). After birth, however, harmful effects from metabolic disorders can occur quickly and irreversibly, if not detected and treated at the earliest possible stage. Examples of metabolic diseases that appear frequently in newborn screening programmes are described below; they are typically inherited in an autosomal recessive pattern.

13.6.1.1 Disorders of amino acid metabolism

The metabolic disease phenylketonuria (PKU) was the first to be detected in general screening of newborn children. The test is described in Section 13.4.2.

In healthy individuals, the enzyme phenylalanine hydroxylase (PAH) converts the amino acid phenylalanine to the amino acid tyrosine and the pigment, melanin. Sufferers of PKU lack a functioning copy of the PAH enzyme, so phenylalanine levels increase in their bloodstream after birth (maternal PAH prevents high levels of phenylalanine from reaching the foetus *in utero*). Unfortunately, high levels of phenylalanine are toxic to brain cells, so undiagnosed PKU sufferers develop severe mental retardation and microcephaly. Other indicators of the disease are a lack of pigmentation—owing to the under-production of melanin—and a 'musty' odour, caused by the build-up of phenylalanine (Almannai et al., 2016).

When PKU is detected neonatally, sufferers are treated with a lifelong special diet that is low in phenylalanine; this leads to an excellent prognosis if medical instructions are followed carefully (Al Hafid and Christodoulou, 2015).

Maple syrup urine disease (MSUD) is similar to phenylketonuria in its pathophysiology; it results from recessive loss-of-function mutations in one or more genes that code for enzymes responsible for breaking down certain amino acids. The accumulation of these amino acids can be toxic, leading to brain damage and death. In the case of MSUD, the amino acids in question are those with branched chain R-groups: leucine, isoleucine, and valine. The four implicated genes all code for enzymes that form a part of the branched-chain alpha-keto acid dehydrogenase complex (Scaini et al., 2017).

Neonatal screening for the disorder is done using tandem mass spectrometry or gas chromatography/mass spectrometry, where excessive levels of branched amino acids may be detected in the infant blood or urine. Previously, diagnosis relied upon the characteristic colour and odour of sufferers' urine, from which the disorder gains its name. Once diagnosed, patients are managed with a carefully-prescribed diet, together with frequent analysis of blood amino acid level).

Isovaleric acidaemia (IVA), glutaric aciduria type 1 (GA1) and homocystinuria (pyridoxine unresponsive) (HCU) are also all rare conditions that result from a failure in amino acid metabolism (see Table 13.2 for further details).

13.6.1.2 Fatty-acid oxidation disorders

Fatty-acid oxidation disorders are a leading cause of sudden infant death when undetected, so are a highly-appropriate target for mass-spectrometry-based neonatal testing (Harpey et al., 1990). Medium-chain acyl-CoA dehydrogenase deficiency (MCADD) is a disorder in which sufferers have a genetic impairment in fatty-acid oxidation; they cannot cleave fatty acid chains that are between 6 and 12 carbon molecules in length. At times of sub-optimal glucose supply, such as fasting or illness, MCADD patients lack alternative means of generating acetyl-CoA, and thus cannot supply the citric acid cycle with this necessary reagent. The resulting lack of reduced hydrogen carriers leads to a deficiency in supply of ATP and thus cellular metabolism may fatally fail (Rice and Steiner, 2016).

MCADD individuals are treated with regular carbohydrate supplementation, particularly during periods of illness. Such vigilance and lifestyle adjustment is typically effective at managing the condition (Wilcken et al., 2009).

13.6.1.3 Lysosomal storage diseases

The lysosome is the 'recycling centre' of the cell; it is a membrane-bound organelle that contains a wide range of lytic enzymes, responsible for breaking down cellular components when they are no longer needed or inactivating infectious agents.

Where a genetic deficiency in one of these degradative enzymes occurs, a lysosomal storage disease may result. There are over 50 different lysosomal storage diseases, each of which results from a different genetic defect (Kuech et al., 2016). They share the feature that the

TABLE 13.2 Disorders of amino acid metabolism

Disorder	Amino acids whose breakdown is impaired	Mutated gene(s)	Symptoms when untreated	Treatment
Phenylketonuria (PKU)	Phenylalanine	Phenylalanine hydroxylase.	Brain damage, hypopigmentation	Low-phenylalanine diet
Maple syrup urine disease (MSUD)	Leucine, isoleucine, and valine	Branched chain keto acid dehydrogenase E1, alpha polypeptide (BCKDHA), Branched chain keto acid dehydrogenase E1, beta polypeptide (BCKDHB), Dihydrolipoamide branched chain transacylase E2 (DBT), Dihydrolipoamide dehydrogenase (DLD)	Brain damage, death	Diet low in branched amino acids
Isovaleric acidaemia (IVA)	Leucine	Isovaleryl-CoA Dehydrogenase (IVD)	Some sufferers are typically asymptomatic; some suffer severe seizures, brain damage and death	Low-leucine diet
Glutaric aciduria type 1 (GA1)	Tryptophan, lysine, hydroxylysine	Glutaryl-CoA dehydrogenase (GCDH)	Some sufferers are typically asymptomatic; some suffer macrocephaly and encephalopathic crisis	Carnitine supplementation, dietary restriction of tryptophan
Homocystinuria (pyridoxine unresponsive) (HCU)	Methionine	Cystathionine-beta-synthase (CBS), methylenetetrahydrofolate reductase (MTHFR), 5-methyltetrahydrofolate-homocysteine methyltransferase (MTR), 5-methyltetrahydrofolate-homocysteine methyltransferase reductase (MTRR), methylmalonic aciduria and homocystinuria, cblD type (MMADHC)	Complex pathologies of connective tissues, central nervous system and muscle	Low-methionine diet, vitamin B6 supplementation

lysosomes become progressively enlarged with undegraded material, which may inhibit cellular function to the point of cell death. The severity of lysosomal storage diseases can vary widely, from conditions that cause almost certain neonatal death, through those that cause progressive neuropathy in childhood, to cases that are largely asymptomatic throughout life (Kuech et al., 2016).

Lysosomal storage diseases are classified by the type of macromolecules that accumulate within cells—lipids, glycoproteins, mucopolysaccharides, and mucolipids. Tay–Sachs disease was the first such disease to be described, followed closely by Gaucher disease; they are both sphingolipid storage disorders (Ashida and Li, 2014).

Treatment for lysosomal storage largely addresses only the symptoms, although it may include bone marrow transplants in severe cases. Enzyme replacement, substrate reduction, and gene therapy are emerging treatments that offer hope to sufferers, however (Kelly et al., 2017).

Key Point

Inherited metabolic disorders typically result from recessive loss-of-function mutations in degradative enzymes. The consequent lack of energy release may cause cells
to cease functioning, or the excess substrate that accumulates can have severely toxic
consequences.

13.13 SELF-CHECK QUESTION

Which molecules accumulate in MSUD, Gaucher disease, and MCADD?

13.6.2 Haemoglobinopathies

A haemoglobinopathy is a mutation in one of the genes for globin—the key protein component of haemoglobin, which binds and releases oxygen in RBCs. The mutation results in a
change in the structure of the globin protein chains, impeding their function, and is inherited
in an autosomal recessive fashion (Moat et al., 2014).

Sickle cell anaemia is the haemoglobinopathy for which neonatal tests most commonly
screen. The condition is caused by a single missense point mutation in the gene for β-globin,
which results in the hydrophilic amino acid glutamic acid being substituted by the hydrophobic amino acid valine. The existence of a valine residue in an outward-facing position on the
molecule causes the globin chains to aggregate in long chains. The aggregates are poor at
carrying oxygen, and also distort the shape of the red blood cells, giving the typical 'sickle cell'
appearance (Oder et al., 2016).

Sickle cells have low elasticity and can become stuck in small capillaries, leading to haemolysis
and stroke. Sickling is particularly prominent in conditions of low oxygen tension and acidosis;
the most severe symptoms of the disorder tend to occur sporadically in 'sickle cell crises'.

The condition is managed using pain medication and antibiotics where appropriate, with
some patients offered regular blood transfusions and bone-marrow transplants. Hydroxyurea,
an anti-proliferative drug, can also produce some therapeutic benefit, by reactivating the gene
for γ-globin—a key component of foetal haemoglobin, which can, to a limited extent, compensate for sufferers' mutated β-globin chains. Hydroxyurea is often prescribed together with
erythropoietin, the hormone that stimulates red blood cell production (Little et al., 2006). Early
detection of sickle cell disease in newborns allows for appropriate management and monitoring strategies to be put in place, hopefully before a damaging sickling crisis has occurred.

Key Point

Sickle cell anaemia is the most common haemoglobinopathy and it is an inherited,
recessive condition.

13.14 SELF-CHECK QUESTION

What is the genetic and functional basis of sickle cell anaemia?

13.6.3 Hormonal disorders

Hormones are key cell–cell signalling molecules that act to co-ordinate the organ systems of the human body, both spatially and temporally. Over- or under-production of hormones at birth can, therefore, have systemic and potentially life-limiting consequences.

13.6.3.1 Congenital hypothyroidism

Congenital hypothyroidism (CH) was the second disorder for which widespread neonatal screening was introduced, after PKU, in the 1970s (Almannai et al., 2016). It is a condition in which insufficient levels of the thyroid hormone triiodothyronine and, by implication, its precursor molecule, thyroxine, are present. If left untreated, CH leads to permanent mental retardation and growth failure, but treatment is simple and effective—thyroxine can be given once a day by mouth to restore normal function (Zdraveska et al., 2016).

Iodine is an important trace element for human health, chiefly because it is a component of thyroxine. Cases of congenital hypothyroidism are frequently seen in areas where iodine is insufficient in the maternal diet. Although still a leading cause of the disorder in developing countries, iodine deficiency in developed countries is now rare; cases of CH detected through neonatal testing are more usually caused by a developmental defect of the infant thyroid gland (Fualal and Ehrenkranz, 2016).

13.6.3.2 Congenital adrenal hyperplasia

Congenital adrenal hyperplasia (CAH) is a term that encompasses a variety of different hormonal disorders. It classically presents with an overgrowth of the adrenal glands in the foetal stage, which is caused by a failure to produce adequate amounts of the hormone cortisol, owing to one of a range of autosomal recessive mutations in the genes that produce the required enzymes for hormone synthesis. The adrenal overgrowth leads to an overexpression of other hormones produced by the adrenal glands, however, including testosterone (Miller, 2017).

In XX foetuses, the overexpression of testosterone can lead to ambiguous genitalia forming; the baby may be designated 'intersex' at birth. In XY fetuses, no obvious physical symptoms are present at birth, but the condition may lead to a 'salt-losing crisis' within the first 1 or 2 weeks of life, which can be fatal. Salt loss is caused by insufficient production of the hormone aldosterone; aldosterone stimulates the kidneys to recover sodium ions from the urine, and without it, this essential electrolyte is lost from the body at a pathological rate (Bizzarri et al., 2016). It is this, otherwise undetectable, manifestation of the condition at which the neonatal screening programme is aimed.

Affected infants are treated with glucocorticoid and/or mineralocorticoid supplementation to remedy the underproduction of cortisol; female infants may also be treated with genital reconstructive surgery, although this practice is increasingly controversial, as it can be associated with gender dysphoria later in life (Silveira et al., 2016).

Key Point

Developed countries commonly screen at birth for the hormonal disorders CH and CAH, which can, respectively, lead to permanent intellectual impairment and sudden infant death, if left untreated.

13.15 SELF-CHECK QUESTION

Why do the symptoms of CAH differ between XX and XY newborns?

13.7 **Ethical considerations**

13.7.1 Elective abortion of affected foetuses

As the techniques for prenatally detecting aneuploidies and genetic mutations advance in sensitivity, specificity, the earliness of gestational age at which they are reliable, and availability to the general population, the ability of the pregnant woman to make an informed decision about whether to continue an affected pregnancy is constantly increasing.

In many cases, the decision made when it is reported that the unborn child will be adversely affected is to abort the pregnancy. Estimates suggest that around 90% of Down syndrome diagnoses made prenatally in Europe resulted in abortion (Mansfield et al., 1999), with a suggestion that the proportion is lower in the USA, at around 67% (Natoli et al., 2012). The latter is thought to result in an effective reduction of 30% in Down syndrome live births, accounting for the fact that a significant proportion of Down syndrome pregnancies naturally do not survive the full gestational period (de Graaf et al., 2015). Also consider the fact that the statistics previously reported outcomes for pregnant women who have chosen to have full diagnostic testing, including invasive procedures (see Section 13.2); when taking into account those women who elect not to be tested, the percentage of Down syndrome pregnancies ending in elective abortion is thought to be around 50%, in developed countries (Egan et al., 2011).

Elective abortion rates for other prenatally diagnosed conditions vary widely and are typically proportional to the severity of the disorder in question. For example, sex chromosome anomalies, such as Turner syndrome and Kleinfelter syndrome (see Section 13.5.4) have a lower rate of elective abortion among those receiving prenatal diagnosis than Down syndrome, at around 60% (Mansfield et al., 1999; Jeon et al., 2012).

There is a wide range within all of these average figures, however, with considerable variation in abortion frequency between nationalities, maternal age groups, gestational age at which diagnosis is made, and racial groups (Natoli et al., 2012). Indeed, abortion is considered unacceptable among many faith groups and is illegal in some countries. Some consider it an acceptable practice only when the mother's life would be at serious risk should the pregnancy continue, while others consider it unacceptable in all circumstances.

Whether or not religious factors come into play, the decision as to whether or not to abort a pregnancy where this is an option is a deeply personal one. Genetic counsellors can assist the decision-making process—not by persuading the pregnant woman to choose one path or the other, but by providing information about the risks to her and the unborn child, and the realities of raising a child with the disability in question—including the social, medical, and educational support that would be made available to the family, should she choose to continue the pregnancy (Neville, 2010).

Pregnant women and their partners facing such decisions may also find it useful to learn from the stories of others who have experienced a similar prenatal diagnosis, and are now living with their decision to abort the pregnancy or rear a disabled child. Many testimonials and support groups exist online for such purposes, in addition to in-person communities. Indeed, there is some suggestion that abortion rates for prenatally diagnosed foetuses have decreased in recent years, owing to improved support available for families with disabled children, and wider publication of positive stories of living with disability (Natoli et al., 2012).

Key Point

A significant percentage of foetuses prenatally diagnosed with a genetic or developmental disorder are electively aborted. The legality and social acceptability of abortion differs widely among different groups. Genetic counsellors can provide unbiased advice to pregnant women facing such a decision.

13.16 SELF-CHECK QUESTION

Why is the percentage of Down syndrome foetuses electively aborted after prenatal diagnosis likely to be different from the overall percentage of Down syndrome foetuses that do not result in a live birth?

13.7.2 Sex-selective abortion

As discussed in Section 13.5.4, inheritance of some sex-linked diseases will result in children of one sex (usually male) who will suffer from the condition in question, and children of the other sex (usually female) who will probably be healthy. Elective abortion of affected foetuses after sex testing is also subject to ethical considerations.

In other cases, sex-selective abortion of otherwise healthy foetuses may be practised for purely social reasons; this area is highly controversial. Parents may have a personal preference for a male or female child, but socio-economic issues may also loom large, here. In some cultures, the birth of a male child may be considered more prestigious and financially beneficial to the family than the birth of a female child. The combination of this attitude with access to prenatal sex determination technology has resulted in the widespread abortion of female embryos and, as a result, sex imbalance within some populations (Kant et al., 2015).

Some governments are keen to redress the imbalances caused. Indeed, sex-selective abortion has been illegal in China since 2005 and prenatal sex determination has been illegal in India since 1994. At the time of writing, both practices remain legal in the UK and USA, and may be practised for reasons of 'family balancing'.

With the advent of cffDNA analysis technology (see Section 13.3.3), which can detect foetal sex from 7 weeks of gestational age, it may be speculated that sex-selective abortions may become more widespread in future years—and, where the practice is illegal, even more difficult to police (Browne, 2017).

Key Point

Sex-selective abortion of otherwise healthy foetuses is a controversial practice that has led to a population gender imbalance in some countries.

13.17 SELF-CHECK QUESTION

Which technology offers the potential for pregnant women to opt for sex-selective abortion from the seventh week of pregnancy?

13.7.3 Disability rights activism

Hand-in-hand with the ongoing debate about the ethics of elective abortion in general, there are concerns that the normalization of aborting foetuses that are prenatally diagnosed with a disabling condition may lead to discrimination against disabled people who are living with the condition in question (Madeo et al., 2011).

Some people believe that such abortions send the message that affected lives are not worth living, and have a negative impact upon the roles and treatment of existing disabled people in society, and their families. Indeed, they argue further that society directly benefits from the existence of people with disabilities within communities, and that a disability-free world would be a societally impoverished one (Lobo and Genuis, 2014).

At the time of writing, this debate is particularly topical in relation to Down syndrome, as the UK NHS is undertaking an evaluation exercise to decide whether or not to offer cffDNA screening for chromosomal abnormalities to all pregnant women. If the test is made publicly available at no additional cost, it is predicted to increase the frequency of Down syndrome elective abortions, an outcome that opponents of the proposal feel is undesirable.

The contrary argument states, of course, that a pregnant woman should be offered as much accurate information as possible about the condition of her foetus, and that it is unethical for such information to be made inaccessible by economic factors. Furthermore, supporters of the cffDNA testing proposal argue that improved access to early and accurate prenatal testing need not increase the frequency of elective abortions; it may, instead, lead to more women feeling that the necessary support exists for them to choose to rear an affected child.

The solution seems to lie not with the test itself, per se, but with the organizational and supportive structure within which it is offered, and the ethical choices that we, as a society, continue to make.

Key Point

Some people believe that the increased availability of prenatal testing to detect disabling conditions will increase discrimination against existing disabled people. Others argue against the veracity of this claim and support the idea that all pregnant women should be offered prenatal testing for disabling conditions in the foetus.

13.18 SELF-CHECK QUESTION

Which technology may increase the rate of prenatal Down syndrome diagnosis, if it were to be made publicly available at no extra cost to the pregnant woman?

13.7.4 Availability of neonatal testing

As with all forms of medical care, citizens of different countries and different personal means have different levels of access to pre- and neo-natal testing (Ebener et al., 2015). The ethical arguments surrounding this fact are, perhaps, brought into even sharper relief in the case of diseases that can be successfully treated if detected in the newborn when no symptoms show, but cause lifelong disability if they go undetected until symptomatic. Indeed, a number of controversial cases have arisen where different local jurisdictions in the same country choose to fund or not fund screening for particular diseases at the neonatal stage, leading to vastly different outcomes for citizens of the same nation carrying the same genetic condition.

13.8 Future directions

As seen over recent years, the desire for accurate non-invasive prenatal testing is considerable, and this is likely to increase—as cffDNA testing becomes ever more sensitive, selective, and widely available, the use of invasive testing will continue its decline (Benn et al., 2004). Indeed, the field may reach a stage where invasive procedures for prenatal testing become redundant, such will be the accuracy of cffDNA-based techniques.

Most prenatal testing in the present day focuses on detecting the common aneuploidies, foetal malformations and perhaps specific conditions that are known to run in the family in question. As testing technology develops, and the understanding of the genetic bases for complex disorders increases, wider panels of disease, and disorder markers present in standard prenatal screening may be expected, thus increasing the information available about the health of a foetus before birth. Similarly, the trend is toward more and more thorough neonatal screening, with an increasing range of disorders detected in the first days of life.

Such an increase in diagnostic information does nothing to silence the ethical challenges that surround prenatal testing, however (Minear et al., 2015). Whatever the future may hold for technological innovation, we can be sure that the attendant thorny ethical issues will be as much a part of the future conversation as they are today.

CASE STUDY 13.1 Screening tests for chromosomal abnormalities

A woman presented to her GP as pregnant; based upon the date of her last menstrual period, the foetus already had a gestational age of 16 weeks. She was offered the Quadruple Test of maternal serum protein markers as part of routine screening, as appropriate for this gestational age. 10 mL of maternal blood were taken. Serum AFP, oestriol, and inhibin A levels were quantified by mass spectrometry, and hCG levels were quantified using a sandwich ELISA.

Test results revealed that AFP levels and oestriol levels were lower than reference ranges, and hCG levels and inhibin A levels were higher than expected. Taken together, these results suggested that there was a high risk that the unborn child had a chromosomal abnormality.

The patient was offered amniocentesis as a diagnostic test for chromosomal abnormalities, and the benefits and risks of the procedure were explained to her. Given that the amniocentesis procedure carries a risk of foetal loss, the patient declined the test.

However, the patient had read in the news about a relatively new kind of non-invasive test that is more accurate at detecting chromosomal abnormalities than the Quadruple Test and asked her healthcare provider about this. It was explained to her that tests for cffDNA in maternal blood are offered by private companies, but are not nationally-funded. Therefore, she decided to pay to have a cffDNA test.

The patient supplied a further blood sample, and cfDNA was purified from the sample. The cffDNA within the sample was analysed using next-generation sequencing. The test indicated a very high risk that the foetus had Down's Syndrome: it was a positive result from a test that has a 99% sensitivity and specificity. Although cffDNA testing is not actually diagnostic, the patient decided to refuse offers of further testing and accept the indicated result as true.

She was upset by the news, but decided to continue with the pregnancy, for religious reasons. She sought support from various groups to help her prepare for raising a disabled child.

CASE STUDY 13.2 Infant with failure to thrive

A male child was born at 40 weeks gestational age via a normal delivery. At 24 hours of age, a blood sample was taken from the infant by heel-prick, absorbed onto filter paper, and used for routine screening tests. The patient appeared healthy at birth, but over the first week of life presented as dehydrated and had difficulty feeding, indicating a failure to thrive.

The infant serum was analysed by mass spectrometry for levels of a panel of metabolites, to screen for inborn errors of metabolism. All results were within normal ranges, so the patient was deemed to have a low risk for disorders of amino acid metabolism, fatty acid oxidation disorders, and lysosomal storage diseases.

Part of the whole infant blood sample was lysed with a detergent and analysed using isoelectric focusing. The patient β-globin chains were observed to run in the same location as the normal control, so the infant tested negative for sickle cell anaemia.

Immunoassays were performed on the infant blood sample to quantify levels of thyroid hormones, 17α-hydroxyprogesterone and IRT. Readings for thyroxine and IRT were within the expected ranges, but 17α-hydroxyprogesterone were abnormally low. The child was, therefore, diagnosed with CAH, and treated with long-term glucocorticoid and mineralocorticoid replacement therapy, together with salt supplementation in the first 6 months of life.

With this treatment in place, the patient began to thrive, and was given an excellent prognosis, with normal life expectancy.

Chapter summary

- The principles of pre- and neo-natal testing are outlined, together with a description of the direct (invasive) methods that are used for obtaining samples from the foetus and newborn. These include amniocentesis, CVS, PUBS, and heel-prick blood sampling.

- Non-invasive sampling techniques are explained: where information can be gained about a foetus from maternal serum sampling, without increased risk to the pregnancy. The presence and implications of certain protein markers, foetal cells, and cffDNA in maternal serum are discussed.

- Links are made with several other chapters to highlight the molecular techniques that are used to analyse and process pre- and neo-natal samples in the laboratory.

- A wide range of applications for pre- and neo-natal testing are discussed, including single-gene disorders, aneuploidies, disorders arising from structural damage to chromosomes, structural defects of the foetus, and metabolic diseases.

- The ethical concerns that surround prenatal testing are also discussed.

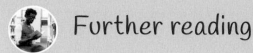

Further reading

The fields of pre- and neo-natal molecular testing have developed and evolved rapidly; specialist information should, therefore, be sought primarily from appropriate biomedical journals. However, other existing books can provide further useful grounding in the molecular techniques used to analyse patient samples, the medical genetics of disease applications of these techniques, and the ethical issues surrounding testing practice. The following can be recommended:

- Berliner J (Ed.). *Ethical Dilemmas in Genetics and Genetic Counseling: Principles Through Case Scenarios*. Oxford, Oxford University Press, 2014.

- Bradley Schaefer GB, Thompson JN Jr. *Medical Genetics—An Integrated Approach*. New York City, NY, McGraw-Hill Education, 2014.

- Hahn S, Jackson LG. (Eds). *Prenatal Diagnosis*. New York City, NY, Springer, 2008.

- Jorde LB, Carey JC, Bamshad MJ. *Medical Genetics*, 5th edn. Amsterdam, Elsevier, 2015.

- Patrinos GP, Ansorge W, Danielson PB (Eds). *Molecular Diagnostics*. New York City, NY, Associated Press, 2016.

- Tobias ES, Connor M, Ferguson-Smith M. *Essential Medical Genetics*, 6th edn. Hoboken, Wiley-Blackwell, 2011.

References

- Akolekar R, Beta J, Picciarelli G, Ogilvie C, D'Antonio F. Procedure-related risk of miscarriage following amniocentesis and chorionic villus sampling: a systematic review and meta-analysis. Ultrasound Obstetrics and Gynecology 2015; 45: 16–26.

- Al Hafid N, Christodoulou J. Phenylketonuria: a review of current and future treatments. Translations in Pediatrics 2015; 4: 304–317.

- Alberry M, Maddocks D, Jones M, et al. Free fetal DNA in maternal plasma in anembryonic pregnancies: confirmation that the origin is the trophoblast. Prenatal Diagnosis 2007; 27: 415–418.

- Alldred SK, Takwoingi Y, Guo B, et al. First trimester serum tests for Down's syndrome screening. Cochrane Database Systematic Reviews 2015; 30: CD011975.

- Almannai M, Marom R, Sutton VR. Newborn screening: a review of history, recent advancements, and future perspectives in the era of next generation sequencing. Current Opinions in Pediatrics 2016; 28: 694–699.

- Alton EW, Boyd AC, Davies JC, et al. Genetic medicines for CF: hype versus reality. Pediatric Pulmonology 2016; 51(S44): S5–S17.

- Amiri N, Christians JK. PAPP-A2 expression by osteoblasts is required for normal postnatal growth in mice. Growth Hormone IGF Research 2015; 25: 274–280.

- Antovic A, Mikovic D, Elezovic I, Zabczyk M, Hutenby K, Antovic JP. Improvement of fibrin clot structure after factor VIII injection in haemophilia A patients treated on demand. Thrombosis & Haemostasis 2014; 111: 656–661.

- Arbabi A, Rampášek L, Brudno M. Cell-free DNA fragment-size distribution analysis for non-invasive prenatal CNV prediction. Bioinformatics 2016; 32: 1662–1669.

- Arth A, Kancherla V, Pachón H, Zimmerman S, Johnson Q, Oakley GP Jr. A 2015 global update on folic acid-preventable spina bifida and anencephaly. Birth Defects Research A: Clinical & Molecular Teratology 2016; 106: 520–529.

- Arumugam A, Raja K, Venugopalan M, et al. Down syndrome—a narrative review with a focus on anatomical features. Clinical Anatomy 2016; 29: 568–577.

- Ashida H, Li YT. Glycosidases: inborn errors of glycosphingolipid catabolism. Advances in Neurobiology 2014; 9: 463–484.

- Ball D, Rose E, Alpert E. Alpha-fetoprotein levels in normal adults. American Journal of Medical Sciences 1992; 303: 157–159.

- Basset P, Beuzard Y, Garel MC, Rosa J. Isoelectric focusing of human hemoglobin: its application to screening, to the characterization of 70 variants, and to the study of modified fractions of normal hemoglobins. Blood 1978; 51: 971–982.

- Bender MA, Douthitt Seibel G. Sickle cell disease. *GeneReviews* 2014.

- Benn PA, Egan JF, Fang M, Smith-Bindman R. Changes in the utilization of prenatal diagnosis. Obstetrics and Gynecology 2004; 103: 1255–1260.

- Bigelow CA, Cinelli CM, Little SE, Benson CB, Frates MC, Wilkins-Haug LE. Percutaneous umbilical blood sampling: current trends and outcomes. European Journal of Obstetrics and Gynecology Reproductive Biology 2016; 200: 98–101.

- Bird RJ, Hurren BJ. Anatomical and clinical aspects of Klinefelter's syndrome. Clinical Anatomy 2016; 29: 606–619.

- Bird LM, Tan WH. Treatment of genetic disorders—a vision coming into focus. American Journal of Medical Genetics C: Seminars in Medical Genetics 2016; 172: 311–312.

- Bizzarri C, Olivini N, Pedicelli S, et al. Congenital primary adrenal insufficiency and selective aldosterone defects presenting as salt-wasting in infancy: a single center 10-year experience. Italian Journal of Pediatrics 2016; 42: 73.

- Bona G, Gallina MR, Dolfin G, Iavarone A, Perona A, Zaffaroni M. Prenatal diagnosis of heterozygosis in a pregnancy at risk for Wolman's disease at the 8th week of gestation. Panminerva Medicine 1989; 131: 180–182.

- Boon EM, Schlecht HB, Martin P, et al. Y chromosome detection by real time PCR and pyrophosphorolysis-activated polymerisation using free fetal DNA isolated from maternal plasma. Prenatal Diagnosis 2007; 27: 932–937.

- Breathnach FM, Malone FD. Screening for aneuploidy in first and second trimesters: is there an optimal paradigm? Current Opinions in Obstetrics and Gynecology 2007; 19: 176–182.

- Breik O, Mahindu A, Moore MH, Molloy CJ, Santoreneos S, David DJ. Central nervous system and cervical spine abnormalities in Apert syndrome. Childs Nervous System 2016; 32: 833–838.

- Browne TK. Why parents should not be told the sex of their fetus. Journal of Medical Ethics 2017; 43: 19–21.

- Bruno C, Dufour-Rainfray D, Patin F, et al. Validation of amino-acids measurement in dried blood spot by FIA-MS/MS for PKU management. Clinical Biochemistry 2016; 49: 1047–1050.

- Caba L, Rusu C, Butnariu L, et al. Phenotypic variability in Patau syndrome. Revista Medico-Chirurgicala a Societatii de Medici si Naturalisti din Iasi 2013; 117: 321–327.

- Cai C, Shi O. Genetic evidence in planar cell polarity signaling pathway in human neural tube defects. Frontiers in Medicine 2014; 8: 68–78.

- Castellani C, Assael BM. Cystic fibrosis: a clinical view. Cell and Molecular Life Sciences 2017; 74: 129–140.

- Chen L, Mullegama SV, Alaimo JT, Elsea SH. Smith–Magenis syndrome and its circadian influence on development, behavior, and obesity—own experience. Developmental Period Medicine 2015; 19: 149–156.

- Choolani M, Mahyuddin AP, Hahn S. The promise of fetal cells in maternal blood. Best Practice Research in Clinical Obstetrics and Gynaecology 2012; 26: 655–667.

- Cole LA. New discoveries on the biology and detection of human chorionic gonadotropin. Reproductive Biology and Endocrinology 2009; 7: 8.

- Coppedè F. Risk factors for Down syndrome. Archives of Toxicology 2016; 90: 2917–2929.

- Corton MM, Leveno K, Bloom S, Spong CY, Dashe JS, Hoffman, BL, Sheffield JS (Eds). *William's Obstetrics*, 24th edn. New York City, NY, McGraw-Hill Education, 2014.

- Daffos F, Capella-Pavlovsky M, Forestier F. A new procedure for fetal blood sampling in utero: preliminary results of fifty-three cases. American Journal of Obstetrics and Gynecolology 1983; 146: 985–987.

- Darouich AA, Liehr T, Weise A. et al. Alpha-fetoprotein and its value for predicting pregnancy outcomes—a re-evaluation. Journal of Prenatal Medicine 2015; 9: 18–23.

- Dashe JS. Aneuploidy screening in pregnancy. Obstetrics and Gynecolology 2016; 128: 181–194.

- David AL, Jauniaux E. Ultrasound and endocrinological markers of first trimester placentation and subsequent fetal size. Placenta 2016; 240: 29–33.

- de Graaf G, Buckley F, Skotko BG. Estimates of the live births, natural losses, and elective terminations with Down syndrome in the United States. American Journal of Medical Genetics A 2015; 167: 756–767.

- Dhallan R, Au W, Mattagajasingh S, et al. Methods to increase the percentage of free fetal DNA recovered from the maternal circulation. Journal of the American Medical Association 2004; 291: 1114–1119.

- Dickman H, Fletke K, Redfern RE. Prolonged unassisted survival in an infant with anencephaly. BMJ Case Reports 2016; doi: 10.1136/bcr-2016-215986.

- Drury S, Mason S, McKay F, et al. Implementing non-invasive prenatal diagnosis (NIPD) in a National Health Service Laboratory; from dominant to recessive disorders. Advances in Experimental Medical Biology 2016; 924: 71–75.

- Duckworth S, Griffin M, Seed PT, et al. Diagnostic biomarkers in women with suspected preeclampsia in a prospective multicenter study. Obstetrics and Gynecology 2016; 128: 245–252.

● Dwinnell SJ, Coad S, Butler B, et al. In utero diagnosis and management of a fetus with homozygous α-thalassemia in the second trimester: a case report and literature review. Journal of Pediatric Hematology and Oncology 2011; 33: e358–e360.

● Ebener S, Guerra-Arias M, Campbell J, et al. The geography of maternal and newborn health: the state of the art. International Journal of Health Geography 2015; 14: 19.

● Edwards JH, Harnden DG, Cameron AH, Crosse VM, Wolff OH. A new trisomic syndrome. Lancet 1960; 1: 787–790.

● Egan JFX, Campbell WA, Chapman A, et al. Distortions of sex ratios at birth in the United States; evidence for prenatal gender selection. Prenatal Diagnosis 2011; 31: 560–565.

● Elborn JS. Cystic fibrosis. Lancet 2016; 388(10059): 2519–2531.

● Farrell PM, Sommerburg O. Toward quality improvement in cystic fibrosis newborn screening: Progress and continuing challenges. Journal of Cystic Fibrosis 2016; 15: 267–269.

● Fernandes MB, Maximino LP, Perosa GB, Abramides DV, Passos-Bueno MR, Yacubian-Fernandes A. Apert and Crouzon syndromes—cognitive development, brain abnormalities, and molecular aspects. American Journal of Medical Genetics A 2016; 170: 1532–1537.

● Forgács G, Newman SA. *Biological Physics of the Developing Embryo*. Cambridge, Cambridge University Press, 2005.

● Fualal J, Ehrenkranz J. Access, availability, and infrastructure deficiency: the current management of thyroid disease in the developing world. Reviews of Endocrinology and Metabolic Disorders 2016; 17: 583–589.

● Fuchs F, Riis P. Antenatal sex determination. Nature 1956; 177: 330.

● Fullerton G, Hamilton M, Maheshwari A. Should non-mosaic Klinefelter syndrome men be labelled as infertile in 2009? Human Reproduction 2010; 25: 588–597.

● Gibbs RS, Karlan BY, Haney AF, Nygaard IE. *Danforth's Obstetrics and Gynecology*, 10th edn. Philadelphia, PA, Wolters Kluwer Lippincott Williams and Wilkins, 2008.

● Gordon Z, Elad D, Almog R, Hazan Y, Jaffa AJ, Eytan O. Anthropometry of fetal vasculature in the chorionic plate. Journal of Anatomy 2007; 211: 698–706.

● Granger A, Zurada A, Zurada-Zielińska A, Gielecki J, Loukas M. Anatomy of Turner syndrome. Clinical Anatomy 2016; 29: 638–642.

● Guetta E, Simchen MJ, Mammon-Daviko K, et al. Analysis of fetal blood cells in the maternal circulation: challenges, ongoing efforts, and potential solutions. Stem Cells Development 2004; 13: 93–99.

● Handyside AH, Montag M, Magli MC, et al. Multiple meiotic errors caused by pre-division of chromatids in women of advanced maternal age undergoing in vitro fertilisation. European Journal of Human Genetics 2012; 20: 742–747.

● Harpey JP, Charpentier C, Paturneau-Jouas M. Sudden infant death syndrome and inherited disorders of fatty acid beta-oxidation. Biology of the Neonate 1990; 58(Suppl 1): 70–80.

● Hill MA. 2016a. Fetal Cells in Maternal Blood. [online] Available at: https://embryology.med.unsw.edu.au/embryology/index.php/Fetal_Cells_in_Maternal_Blood (accessed 3 October 2018).

- Hill MA. 2016b. Human Chorionic Gonadotropin. [online] Available at: https://embryology.med.unsw.edu.au/embryology/index.php/Human_Chorionic_Gonadotropin (accessed 3 October 2018).

- Hoyle JC, Isfort MC, Roggenbuck J, Arnold WD. The genetics of Charcot–Marie–Tooth disease: current trends and future implications for diagnosis and management. Applied Clinical Genetics 2015; 8: 235–243.

- Huang T, Dennis A, Meschino WS, Rashid S, Mak-Tam, E, Cuckle H. First trimester screening for Down syndrome using nuchal translucency, maternal serum pregnancy-associated plasma protein A, free-β human chorionic gonadotrophin, placental growth factor, and α-fetoprotein. Prenatal Diagnosis 2015; 35: 709–716.

- Jeon KC, Chen L, Goodson P. Decision to abort after a prenatal diagnosis of sex chromosome abnormality: a systematic review of the literature. Genetic Medicine 2012; 14: 27–38.

- Jia CW, Wang L, Lan YL. Aneuploidy in early miscarriage and its related factors. Chinese Medical Journal (England) 2015; 128: 2772–2776.

- Kant S, Srivastava R, Rai SK, Misra P, Charlette L, Pandav CS. Induced abortion in villages of Ballabgarh HDSS: rates, trends, causes and determinants. Reproductive Health 2015; 12: 51.

- Kawarai T, Yamasaki K, Mori A, et al. MFN2 transcripts escaping from nonsense-mediated mRNA decay pathway cause Charcot–Marie–Tooth disease type 2A2. Journal of Neurology, Neurosurgery & Psychiatry 2016; 87: 1263–1265.

- Kelley K. 2013, Amniocentesis Prior to 1980. Embryo Project Encyclopedia. [online] Available at: https://embryo.asu.edu/pages/amniocentesis-prior-1980 (accessed 3 October 2018).

- Kelly JM, Bradbury A, Martin DR, Byrne ME. Emerging therapies for neuropathic lysosomal storage disorders. Progress in Neurobiology 2017; 152: 166–80.

- Kharrazi M, Yang J, Bishop T, et al. Newborn screening for cystic fibrosis in California. Pediatrics 2015; 136: 1062–1072.

- Kirk E, Bottomley C, Bourne T. Diagnosing ectopic pregnancy and current concepts in the management of pregnancy of unknown location. Human Reproduction Update 2013; 20: 250–261.

- Kliegman RM, Stanton BMD, St. Geme J, Schor NF. Nelson Textbook of Pediatrics, 20th edn. Oxford, Elsevier, 2015.

- Krantz DA, Hallahan TW, Carmichael JB. Screening for open neural tube defects. Clinical Laboratory Medicine 2016; 36: 401–406.

- Kuech EM, Brogden G, Naim HY. Alterations in membrane trafficking and pathophysiological implications in lysosomal storage disorders. Biochimie 2016; 130: 152–162.

- Lana-Elola E, Watson-Scales SD, Fisher EM, Tybulewicz VL. Down syndrome: searching for the genetic culprits. Disease Model Mechanisms 2011; 4: 586–595.

- Li Y, Zimmermann B, Rusterholz C, Kang A, Holzgreve W, Hahn S. Size separation of circulatory DNA in maternal plasma permits ready detection of fetal DNA polymorphisms. Clinical Chemistry 2004; 50: 1002–1011.

- Lin H, Sui W, Li W, et al. Integrated microRNA and protein expression analysis reveals novel microRNA regulation of targets in fetal Down syndrome. Molecular Medicine Reports 2016; 14: 4109–4118.

- Lindner M, Gramer G, Haege G, et al. Efficacy and outcome of expanded newborn screening for metabolic diseases—report of 10 years from South-West Germany. Orphanet Journal of Rare Diseases 2011; 6: 44.

- Little JA, McGowan VR, Kato GJ, et al. Combination erythropoietin-hydroxyurea therapy in sickle cell disease: experience from the National Institutes of Health and a literature review. Haematologica 2006; 91: 1076–1083.

- Liu Y, Liu Y, Li X, Jiao X, Zhang R, Zhang J. Predictive value of serum β-hCG for early pregnancy outcomes among women with recurrent spontaneous abortion. International Journal of Gynaecology and Obstetrics 2016; 135: 16–21.

- Lo YM, Corbetta N, Chamberlain PF, et al. Presence of fetal DNA in maternal plasma and serum. Lancet 1997; 350(9076): 485–487.

- Lo YM, Tein MS, Lau TK, et al. Quantitative analysis of fetal DNA in maternal plasma and serum: implications for noninvasive prenatal diagnosis. American Journal of Human Genetics 1998; 62: 768–775.

- Lobo R, Genuis G. Socially repugnant or the standard of care: Is there a distinction between sex-selective and ability-selective abortion? Canadian Family Physician 2014; 60: 212–216.

- Lott IT, Head E, Doran E, Busciglio J. Beta-amyloid, oxidative stress and Down syndrome. Current Alzheimer Research 2006; 3: 521–528.

- Madeo AC, Biesecker BB, Brasington C, Erby L, Peters KF. The relationship between the genetic counseling profession and the disability community: a commentary. American Journal of Medical Genetics A 2011; 155A: 1777–1785.

- Mansfield C, Hopffer S, Marteau TM. Termination rates after prenatal diagnosis of Down syndrome, spina bifida, anencephaly, and Turner and Klinefelter syndromes: a systematic literature review. European Concerted Action: DADA (Decision-making After the Diagnosis of a fetal Abnormality). Prenatal Diagnosis 1999; 19: 808–812.

- Miller WL. Disorders in the initial steps of steroid hormone synthesis. Journal of Steroid Biochemistry and Molecular Biology 2017; 165(Pt A): 18–37.

- Minear MA, Alessi S, Allyse M, Michie M, Chandrasekharan S. Noninvasive prenatal genetic testing: current and emerging ethical, legal, and social issues. Annual Review of Genomics and Human Genetics 2015; 16: 369–398.

- Mitchell JJ, Trakadis YJ, Scriver CR. Phenylalanine hydroxylase deficiency. Genetic Medicine 2011; 13: 697–707.

- Mizejewski GJ. Alpha-fetoprotein structure and function: relevance to isoforms, epitopes, and conformational variants. Experimental Biology and Medicine 2001; 226: 377–408.

- Moat SJ, Rees D, King L, et al. Newborn blood spot screening for sickle cell disease by using tandem mass spectrometry: implementation of a protocol to identify only the disease states of sickle cell disease. Clinical Chemistry 2014; 60: 373–80.

- Mohr J. Foetal genetic diagnosis: development of techniques for early sampling of foetal cells. Acta Pathologica Microbiologica Scandinavica 1968; 73: 73–77.

● Muttukrishna S. Role of inhibin in normal and high-risk pregnancy. Seminars in Reproductive Medicine 2004; 22: 227–234.

● Namwanje M, Brown CW. Activins and inhibins: roles in development, physiology, and disease. Cold Spring Harbor Perspectives in Biology 2016; 8: doi: 10.1101/cshperspect. a021881.

● Natoli JL, Ackerman DL, McDermott S, Edwards JG. Prenatal diagnosis of Down syndrome: a systematic review of termination rates (1995–2011). Prenatal diagnosis 2012; 32: 142–153.

● Neira-Fresneda J, Potocki L. neurodevelopmental disorders associated with abnormal gene dosage: Smith–Magenis and Potocki–Lupski Syndromes. Journal of Pediatric Genetics 2015; 4: 159–167.

● Neville A. 2010. Genetic Counselling in Pregnancy. [online] Available at: http://opendoors.com.au/education/pregnancy-loss/genetic-counselling-in-pregnancy/ (accessed 3 October 2018).

● Norwitz ER, Levy B. Noninvasive prenatal testing: the future is now. Reviews in Obstetrics and Gynecology 2013; 6: 48–62.

● Oder E, Safo MK, Abdulmalik O, Kato GJ. New developments in anti-sickling agents: can drugs directly prevent the polymerization of sickle haemoglobin *in vivo*? British Journal of Haematology 2016; 175: 24–30.

● Otter M, Schrander-Stumpel CT, Curfs LM. Triple X syndrome: a review of the literature. European Journal of Human Genetics 2010; 18: 265–271.

● Palomaki GE, Deciu C, Kloza EM, et al. DNA sequencing of maternal plasma reliably identifies trisomy 18 and trisomy 13 as well as Down syndrome: an international collaborative study. Genetics in Medicine 2012; 14: 296–305.

● Partington MW, Sinnott B. Case finding in phenylketonuria. II. The Guthrie test. Canadian Medical Association Journal 1964; 91: 105–114.

● Pasqualini JR, Chetrite GS. The formation and transformation of hormones in maternal, placental and fetal compartments: biological implications. Hormone and Molecular Biology and Clinical Investigations 2016; 27: 11–28.

● Pass KA, Neto EC. Update: newborn screening for endocrinopathies. Endocrinology and Metabolic Clinics of North America 2009; 38: 827–837.

● Patau K, Smith DW, Therman E, Inhorn SL, Wagner HP. Multiple congenital anomaly caused by an extra autosome. Lancet 1960; 1: 790.

● Pavlova A, Oldenburg J. Defining severity of hemophilia: more than factor levels. Seminars in Thrombosis and Hemostasis 2013; 39: 702–710.

● Peroos S, Forsythe E, Pugh JH, Arthur-Farraj P, Hodes D. Longevity and Patau syndrome: what determines survival? BMJ Case Reports 2012; 2012. pii: bcr0620114381.

● Ren F, Hu YU, Zhou H, et al. Second trimester maternal serum triple screening marker levels in normal twin and singleton pregnancies. Biomedical Reports 2016; 4: 475–478.

● Rice GM, Steiner RD. Inborn errors of metabolism (metabolic disorders). Pediatric Reviews 2016; 37: 3–15.

● Roberts W, Zurada A, Zurada-Zielińska A, Gielecki J, Loukas M. Anatomy of trisomy 18. Clinical Anatomy 2016; 29: 628–632.

- Rodríguez-Caballero A, Torres-Lagares D, Rodríguez-Pérez A, Serrera-Figallo MA, Hernández-Guisado JM, Machuca-Portillo G. *Cri du chat* syndrome: a critical review. Medicina Oral Patologia Oral y Cirugia Bucal 2010; 15: e473–478.

- Saint-Criq V, Gray, MA. Role of CFTR in epithelial physiology. Cell and Molecular Life Sciences 2017; 74: 93–115.

- Salvatore, M, Floridia G, Amato A, et al. The Italian pilot external quality assessment program for cystic fibrosis sweat test. Clinical Biochemistry 2016; 49: 601–605.

- Scaini G, Tonon T, de Souza CF, et al. Serum markers of neurodegeneration in maple syrup urine disease. Molecular Neurobiology 2017; 54: 5709–5719.

- Schlütter JM, Kirkegaard I, Petersen OB, et al. Fetal gender and several cytokines are associated with the number of fetal cells in maternal blood—an observational study. PLoS One 2014; 9: e106934.

- Seeds JW. Diagnostic mid trimester amniocentesis: How safe? American Journal of Obstetrics and Gynecology 2004; 191: 607–615.

- Sekizawa A, Samura O, Zhen DK, Falco V, Farina A, Bianchi DW. Apoptosis in fetal nucleated erythrocytes circulating in maternal blood. Prenatal Diagnosis 2000; 20: 886–889.

- Settiyanan T, Wanapirak C, Sirichotiyakul S, et al. Association between isolated abnormal levels of maternal serum unconjugated estriol in the second trimester and adverse pregnancy outcomes. Journal of Maternal and Fetal Neonatal Medicine 2016; 29: 2093–2097.

- Sferruzzi-Perri AN, Camm EJ. The programming power of the placenta. Frontiers in Physiology 2016; 7: 33.

- Silveira MT, Knobloch F, Silva Janovsky CC, Kater CE. Gender dysphoria in a 62-year-old genetic female with congenital adrenal hyperplasia. Archives of Sexual Behaviour 2016; 45: 1871–1875.

- Silverman NS, Sullivan MW, Jungkind DL, Weinblatt V, Beavis K, Wapner RJ. Incidence of bacteremia associated with chorionic villus sampling. Obstetrics and Gynecology 1994; 84: 1021–1024.

- Smidt-Jensen S, Hahnemann N, Jensen PKA, Therkelsen AJ. *First Trimester Fetal Diagnosis*. Berlin: Springer, 1985.

- Swystun LL, James PD. Genetic diagnosis in hemophilia and von Willebrand disease. Blood Review 2017; 31: 47–56.

- Teven CM, Farina EM, Rivas J, Reid RR. Fibroblast growth factor (FGF) signaling in development and skeletal diseases. Genes Disorders 2014; 1: 199–213.

- Tietung Hospital, Anshan, China. Fetal sex prediction by sex chromatin of chorionic villi cells during early pregnancy. Chinese Medical Journal 1975; 1: 117–126.

- Tørring N. First trimester combined screening—focus on early biochemistry. Scandinavian Journal of Clinical and Laboratory Investigations 2016; 76: 435–447.

- Wald NJ, Morris JK, Ibison J, Wu T, George LM. Screening in early pregnancy for preeclampsia using Down syndrome quadruple test markers. Prenatal Diagnosis 2006; 26: 559–564.

- Walknowska J, Conte FA, Grumbach MM. Practical and theoretical implications of fetal/maternal lymphocyte transfer. Lancet 1969; 293(7606): 1119–1122.

- Welch CR, Talbert DG, Warwick RM, Letsky EA, Rodeck CH. Needle modifications for invasive fetal procedures. Obstetrics and Gynecology 1995; 85: 113–117.

- Wilcken B, Haas M, Joy P, et al. Expanded newborn screening: outcome in screened and unscreened patients at age 6 years. Pediatrics 2009; 124: e241–248.

- Wilson RD, Gagnon A, Audibert F, et al. Prenatal Diagnosis Procedures and Techniques to Obtain a Diagnostic Fetal Specimen or Tissue: Maternal and Fetal Risks and Benefits. Journal of Obstetrics and Gynecology, Canada 2015; 37: 656–670.

- Youssef A, Arcangeli T, Radico D, et al. Accuracy of fetal gender determination in the first trimester using three-dimensional ultrasound. Ultrasound in Obstetrics and Gynecology 2011; 37: 557–561.

- Zdraveska N, Anastasovska V, Kocova M. Frequency of thyroid status monitoring in the first year of life and predictors for more frequent monitoring in infants with congenital hypothyroidism. Journal of Pediatric Endocrinology and Metabolism 2016; 29: 795–800.

- Zimmermann B, El-Sheikhah A, Nicolaides K, Holzgreve W, Hahn S. Optimized real-time quantitative PCR measurement of male fetal DNA in maternal plasma. Clinical Chemistry 2005; 51: 1598–1604.

Useful websites

- *Baby's First Test*: http://www.babysfirsttest.org/.

- *Down syndrome testing RAPID NHS*: http://www.rapid.nhs.uk/.

- *Genetic Counselling in Pregnancy*: http://opendoors.com.au/education/pregnancy-loss/genetic-counselling-in-pregnancy/.

Answers to the self-check questions and tips for responding to the discussion questions are provided on the book's accompanying website:

 Visit: www.oup.com/uk/warford

Discussion questions

13.1 How has the field of prenatal testing changed since such techniques were first developed? Consider the sampling and analysis methods available, and the advancing understanding of conditions that may be diagnosed prenatally.

13.2 Some conditions may be diagnosed either prenatally or neonatally. Give appropriate examples to illustrate this point, highlighting the testing methods involved. What are the advantages and disadvantages of these approaches?

Abbreviations

Ab	Antibody	**CLR**	Continuous long read
aCGH	Array-based comparative genomic hybridization	**CML**	Chronic myeloid leukaemia
AD	Alzheimer's disease	**CMV**	Cytomegalovirus
AFP	Alpha-foetoprotein	**CNV**	Copy number variation
Ag	Antigen	**COMET**	Co-differentially methylated regions
ALL	Acute lymphoblastic leukaemia	**CPD**	Continued professional development
ALT	Atypical lipomatous tumour	**CRC**	Colorectal cancer
AML	Acute myeloid leukaemia	**CSF**	Cerebrospinal fluid
APC	Allophycocyanin	**CSR**	Class switch recombination
APML	Acute promyelocytic leukaemia	**CTC**	Circulating tumour cell
APS	Adenosine 5´ phosphosulfate	**ctDNA**	Circulating tumour DNA
ARMS	Amplification Refractory Mutation System™	**CVS**	Chorionic villus sampling
ASIC	Application-specific integrated circuit	**2-DGE**	Two-dimensional gel electrophoresis
ATO	Arsenic trioxide	**DAPI**	4,6-diamidino-2-phenylindole
ATRA	All-trans retinoic acid	**dATP**	Deoxyadenosine triphosphate
BAC	Bacterial artificial chromosome	**dATPαS**	Deoxyadenosine alpha-thio triphosphate
BAM	Binary alignment map	**DCIS**	Ductal carcinoma *in situ*
BCR	Breakpoint cluster region	**ddNTP**	Dideoxynucleoside triphosphate
BMD	Becker muscular dystrophy	**DE**	Differentially expressed
BRCA	Breast cancer gene	**DEP**	Dielectrophorectic
BTK	Bruton tyrosine kinase	**DFSP**	Dermatofibrosarcoma protuberans
CAH	Congenital adrenal hyperplasia	**DGGE**	Denaturing gradient gel electrophoresis
CAR-T	Chimeric antigen receptor T	**2-DIGE**	Two-dimensional differential gel electrophoresis
CBF-AML	Core-binding factor leukaemia	**DLBCL**	Diffuse large B-cell lymphoma
CCD	Charge-coupled device	**DMD**	Duchenne muscular dystrophy
CD	Cluster of differentiation	**dMIQE**	Digital minimum information for publication of quantitative digital PCR experiments
cDNA	Complementary DNA		
CEA	Carcinoembryonic antigen	**DMP**	Differentially methylated point
CEBPA	CCAAT/enhancer binding protein α	**DMR**	Differentially methylated regions
CF	Cystic fibrosis	**dNTP**	Deoxynucleoside triphosphate
cfDNA	Cell-free DNA	**dPCR**	Digital PCR
cffDNA	Cell-free foetal DNA	**EBV**	Epstein–Barr virus
cfRNA	Cell-free RNA	**ECLIA**	Electrochemiluminescent immunoassay
CFTR	Cystic fibrosis transmembrane conductance regulator	**eDAR**	Ensemble-decision aliquot ranking
		EDTA	Ethylenediaminetetraacetic acid
CGH	Comparative genomic hybridization	**EGFR**	Epidermal growth factor receptor
CH	Congenital hypothyroidism	**ELISA**	Enzyme linked immunosorbent assay
CID	Collision-induced dissociation	**EMA**	Epithelial membrane antigen
CLL	Chronic lymphocytic leukaemia	**EMT**	Epithelial-to-mesenchymal transition

EMVI	Extramural venous invasion		**HL**	Hodgkin lymphoma
EQA	External quality assessment		**HPV**	Human papilloma virus
ER	Oestrogen receptor		**HSC**	Haematopoietic stem cells
ERMS	Embryonal rhabdomyosarcoma		**IBD**	Irritable bowel disease
ESI	Electrospray ionization		**ICC**	Immunocytochemistry
EV	Extracellular vesicles		**IDC**	Invasive ductal carcinoma
FACS	Fluorescence-activated cell sorting		**IDH1**	Isocitrate dehydrogenase 1
FAP	Familial adenomatosis polyposis		**IDH2**	Isocitrate dehydrogenase 2
FAST	Fibre-optic array scanning technology		**IEF**	Isoelectric focusing
FC	Flow cytometry		**Ig**	Immunoglobulin
FCS	Foetal calf serum		**IGF**	Insulin-like growth factor
FDA	US Food and Drug Administration		**IGFBP**	Insulin-like growth factor binding proteins
FFPE	Formalin fixed paraffin embedded		**IGHV**	Variable region of the IGH gene
FGFR	Fibroblast growth factor receptor		**IGK**	IGKappa
FISH	Fluorescent *In situ* hybridization		**IGL**	IGLambda
FISSEQ	Fluorescent *in situ* RNA sequencing		**IGV**	Integrated genome viewer
FITC	Fluorescein isothiocyanate		**ILC**	Invasive lobular carcinoma
FNA	Fine needle aspirate		**IMAC**	Immobilized metal affinity chromatography
FPKM	Fragments per kilobase of exon per million fragments mapped		**IMGT**	International Immunogenetics Information System
FS	Forward scatter		**IPC**	Internal positive control
FSH	Follicle-stimulating hormone		**IPI**	International Prognostic Index
FT	Fourier transform		**IRT**	Immunoreactive trypsinogen
GA1	Glutaric aciduria type 1		**IS**	International scale
GEM	Gene expression microarrays		**ISH**	*In situ* hybridization
GCT	Giant cell tumour (of bone)		**IT**	Information technology
gDNA	Genomic DNA		**ITD**	Internal tandem duplications
GIST	Gastro-intestinal stromal tumour		**IVA**	Isovaleric acidaemia
GLP	Good Laboratory Practice		**IVD**	*In vitro* diagnostic
GMC	Genetically modified crops		**LAMP**	Loop-mediated isothermal amplification
GMO	Genetically modified organism		**LB**	liquid biopsy
GPI	Glycosylphosphatidylinositol		**LC**	Liquid chromatography
GSEA	Gene set enrichment analysis		**LCIS**	Lobular carcinoma *in situ*
GWAS	Genome-wide association studies		**LGFMS**	Low grade fibromyxoid sarcoma
HCD	High energy collision dissociation		**LH**	Luteinizing hormone
hCG	Human chorionic gonadotropin		**LHCGR**	Luteinizing hormone/choriogonadotropin receptor
HCL	Hairy cell leukaemia		**LIMMA**	Linear models for microarray analysis
HCU	Homocystinuria		**LIMS**	Laboratory Information System
HD	High definition		**LN**	Lymph node
HDA	Heteroduplexes analysis		**LNA**	Locked nucleic acids
HER2	Human epidermal growth factor receptor 2		**LOH**	Loss of heterozygosity
HGF	Hepatocyte growth factor		**LPL**	Lymphoplasmacytic lymphoma
HHV-6	human herpesvirus 6		**MALDI**	Matrix-assisted laser desorption/ionization
HIER	Heat-induced epitope retrieval			

MALT	Mucosa associated lymphoid tissue
MAPK	Mitogen-activated protein kinase
MCADD	Medium-chain acyl-CoA dehydrogenase deficiency
MCL	Mantle cell lymphoma
MDS	Multi-dimensional scatter plots
MGS	Monoclonal gammopathy of uncertain significance
MM	Multiple myeloma
MMR	Mismatch repair
MNC	Mononuclear cell
MoM	Multiples of median
MOPS	3-(N-morpholino)propanesulfonic acid
MPN	Myeloproliferative neoplasms
MRD	Minimal residual disease
MS	Mass spectroscopy
MSI	Microsatellite instability
MSI-H/L	microsatellite instability high/low tumours
MSRED	Mutation specific restriction enzyme digestion
MSUD	Maple syrup urine disease
NBT/BCIP	Nitro blue tetrazolium/5-bromo-4-chloro-3-indolyl phosphate
NEQAS	National External Quality Assessment Service
NGS	Next-generation sequencing
NHL	Non-Hodgkin lymphoma
NICE	UK National Institute for Health and Care Excellence
NSCLC	Non-small cell lung cancer
NST	No special type
OAC	Oesophageal adenocarcinoma
ONT	Oxford Nanopore Technologies
OSNA	One-step nucleic acid amplification
PAH	Phenylalanine hydroxylase
PAPP-A	Pregnancy-associated plasma protein A
PARP	Poly (ADP-ribose) polymerase
PCR	Polymerase chain reaction
PDGFR	Platelet-derived growth factor receptor
PD-1	Programmed cell death receptor-1
PD-L1	Programmed death cell ligand-1
PE	Phycoerythrin
PERM	Paraffin-embedded RNA metric
PGM	Personal genome machine
PIER	Proteolytic-induced epitope retrieval
PKU	Phenylketonuria

PMF	Peptide mass fingerprinting
PML	Promyelocytic leukaemia
PNA	Peptide nucleic acids
POC	Point of care
PPi	Pyrophosphate
PR	Progesterone receptor
PSA	Prostate-specific antigen
PTD	Partial tandem duplication
PTM	Protein post-translational modification
PUBS	Percutaneous umbilical cord blood sampling
QD	Quantum dots
QMS	Quality management system
qPCR	Quantitative polymerase chain reaction
RARA	Retinoic acid receptor alpha
RB	Retinoblastoma
RBC	Red blood cells
RFLP	Restriction fragment length polymorphisms
RIN	Ribosome integrity number
RMA	Robust multi-array average
RPMI	Roswell Park Memorial Institute medium
RT	Reverse transcriptase
RT-LAMP	Reverse transcription loop-mediated isothermal amplification
RT-PCR	Reverse transcription polymerase chain reaction
SA	Serrated adenomas
SAM	Sequence alignment map
SBS	Sequencing by synthesis
SCC	Squamous cell carcinomas
SEC	Size-exclusion chromatography
SHM	Somatic hypermutation
SIMOA	Single-molecule arrays
SISH	Silver-enhanced ISH
SLL	Small lymphocytic lymphoma
SMA	Smooth muscle actin
SMM	Smouldering multiple myeloma
SMRT	Single-molecule real-time
SNP	Single nucleotide polymorphism
SOP	Standard operating procedure
SR	Signet ring
SS	Side scatter
SSP	Sessile serrated polyps
SSCP	Single-strand conformation polymorphism
TAE	Tris acetate EDTA
TAMRA	Tetramethylrhodamine

TCR	T-cell receptor		**TOF**	Time of flight
TDT	Terminal deoxynucleotide transferase		**TQ**	Triple quadrupole
TET	Tetrachlorofluorescein		**TTF**	Thyroid transcription factor
TIL	Tumour-infiltrating lymphocyte		**VCF**	Variant call format
TKI	Tyrosine kinase inhibitor		**WBC**	White blood cells
TMAP	Torrent Mapping Alignment Program		**WDL**	Well-differentiated liposarcoma
TMM	Trimmed mean of median		**WGBS**	Whole-genome bisulfite sequencing
TNBC	Triple negative breast cancers		**WM**	Waldenström's macroglobulinaemia
TNM	Tumour, node, metastasis		**ZMW**	Zero-mode waveguide
TOC	Tylosis with oesophageal cancer			

Subject Index

Notes, Tables, figures and boxes are indicated by an italic *t*, *f* or *b* following the page number.